Membrane Proteins: Function, Structure, and Dynamic

Membrane Proteins: Function, Structure, and Dynamic

Editors

Yosuke Senju
Shiro Suetsugu

MDPI

Basel • Beijing • Wuhan • Barcelona • Belgrade • Novi Sad • Cluj • Manchester

Editors

Yosuke Senju
Research Institute for Interdisciplinary Science (RIIS)
Okayama University
Okayama
Japan

Shiro Suetsugu
Division of Biological Science, Graduate School of Science and Technology
Nara Institute of Science and Technology
Nara
Japan

Editorial Office
MDPI
St. Alban-Anlage 66
4052 Basel, Switzerland

This is a reprint of articles from the Special Issue published online in the open access journal *Membranes* (ISSN 2077-0375) (available at: https://www.mdpi.com/journal/membranes/special_issues/Membrane_Proteins_Function_Structure_Dynamic).

For citation purposes, cite each article independently as indicated on the article page online and as indicated below:

Lastname, A.A.; Lastname, B.B. Article Title. *Journal Name* **Year**, *Volume Number*, Page Range.

ISBN 978-3-7258-0028-5 (Hbk)
ISBN 978-3-7258-0027-8 (PDF)
doi.org/10.3390/books978-3-7258-0027-8

Contents

Preface

This Special Issue of *Membranes*, entitled "Membrane Proteins: Function, Structure, and Dynamics", discusses the recent progress in the studies of membrane proteins. The technical approaches presented here will help us understand membrane proteins' spatiotemporal dynamics and physical properties. These studies provide new insights into the fundamental principles underlying membrane proteins' structure and physiological functions and could lead to the development of new drug designs. This Special Issue contains seven articles, five reviews, and one editorial. We express our gratitude to all authors and reviewers for their contributions.

Yosuke Senju and Shiro Suetsugu
Editors

membranes

Editorial

Membrane Proteins: Function, Structure, and Dynamics

Yosuke Senju [1,*] and Shiro Suetsugu [2,3,4,*]

1 Research Institute for Interdisciplinary Science (RIIS), Okayama University, Okayama 700-8530, Japan
2 Division of Biological Science, Graduate School of Science and Technology, Nara Institute of Science and Technology, Nara 630-0192, Japan
3 Data Science Center, Nara Institute of Science and Technology, Nara 630-0192, Japan
4 Center for Digital Green-innovation, Nara Institute of Science and Technology, Nara 630-0192, Japan
* Correspondence: yosuke.senju@okayama-u.ac.jp (Y.S.); suetsugu@bs.naist.jp (S.S.)

Plasma and intracellular membranes are characterized by different lipid compositions that enable proteins to localize to distinct subcellular compartments [1]. Many proteins interact with cellular and subcellular membranes. These include transmembrane proteins (TMEMs) (e.g., ion channels, transmembrane receptors, and transporters), which integrate into lipid bilayers to transport molecules and ions across membranes, and peripheral membrane proteins, which associate with membranes via electrostatic and/or hydrophobic interactions by penetrating into lipid bilayers. These lipid–protein interactions determine protein conformations and protein–protein interactions, which in turn precisely regulate the localization and activation of molecular complexes at the respective membranes. These signaling pathways are crucial for various cellular processes, such as membrane trafficking and signal transduction. Membrane proteins have also been implicated in many diseases, such as cancer and Alzheimer's disease, and can be targeted for drug design; however, given the complexity of the abundant lipid–protein/protein–protein interactions at membranes, the exact molecular and cellular mechanisms underlying the structure and function of membrane proteins remain unclear.

This Special Issue of *Membranes*, entitled "Membrane Proteins: Function, Structure, and Dynamics", discusses the recent progress in the studies of membrane proteins from various perspectives, including cell biology, biochemistry, biophysics, structural biology, and molecular dynamics (MD) simulations. These studies elucidate the structural and physiological functions of membrane proteins, providing new insights into their fundamental principles. A summary of these studies is as follows.

Kim et al. [2] studied the lipid-scrambling and ion-transport activities of human TMEM16C isoforms. They found that, among the isoforms 1–3, isoforms 1 and 3 transported phosphatidylserine (PS) to the outer leaflet of the membrane, whereas isoform 2 did not. They also found that these results were due to the differences in the surface expression levels of each isoform. The surface expression of isoform 2, which did not exhibit scrambling activity, was approximately five times lower than that of the other two isoforms. Unlike other TMEM16 proteins, TMEM16C isoforms did not show ion channel activity in flux assays or electrophysiological recordings. Thus, they concluded that the N-terminus of TMEM16C isoforms 1 and 3 determines whether they can translocate to the plasma membrane and facilitate the scrambling activity of transporting PS to the outer leaflet.

Li et al. [3] focused on the dimerization of TMEMs in cancer immunotherapy. They described the structures and functions of several TMEMs related to tumor immunity and analyzed the binding properties and functions of these immune checkpoint proteins and their receptors. They also discussed the regulation of TMEM dimerization and its potential as a target for cancer immunotherapy. They believe that understanding the mechanisms underlying TMEM dimerization will provide valuable insights into the development of novel antitumor drugs.

Citation: Senju, Y.; Suetsugu, S. Membrane Proteins: Function, Structure, and Dynamics. *Membranes* **2023**, *13*, 904. https://doi.org/10.3390/membranes13120904

Received: 7 November 2023
Accepted: 29 November 2023
Published: 9 December 2023

Anosov et al. [4] studied the influence of pH and cytochrome c (cyt c) on the electrical properties of asolectin bilayer lipid membranes. They found that acid phospholipids in asolectin membranes can bind to cyt c. They measured differences in cyt c-induced surface potential and found that as the pH decreased, the adsorption of cyt c molecules on the surface of asolectin membranes increased although no increase in membrane conductance. In contrast, the membrane conductance increased as the pH increased, indicating that these two variables were directly proportional.

Rodrigues et al. [5] aimed to provide an in-depth characterization of the proteins in the outer membrane of uropathogenic *Escherichia coli* (UPEC), with a focus on outer membrane proteins (OMPs) and their role in antimicrobial resistance. They discussed several OMPs were related to antimicrobial resistance. Fluoroquinolones and β-lactams were the antibiotics most affected by OMP-conferred antimicrobial resistance. The authors emphasized that the implementation of new strategies for administering antimicrobial agents, the development of improved antimicrobials, protective vaccines, and specific inhibitors of virulence and resistance factors are crucial in veterinary and human medicine to address UPEC resistance to antimicrobial agents.

Dengler et al. [6] investigated the molecular basis for the cholecystokinin (CCK) action of type 1 CCK receptor (CCK1R) positive allosteric modulator (PAM) ligands. Their findings expanded our understanding of the structure–activity relationships of molecules with tetracyclic scaffolds. They explored structural modifications present in 65 commercially available analogs and conducted an analog-by-catalog expansion of the scaffold. They eliminated the off-target effect observed in this scaffold while maintaining its activity as a PAM of CCK1R in normal and high-cholesterol membrane environments.

Isu et al. [7] reviewed the binding sites of cholesterol within the structures of class C G-protein-coupled receptors (GPCRs) and found that cholesterol is typically bound between the transmembrane dimers of GPCRs and within the surrounding grooves of their transmembrane helices.

Bykhovskaia [8] discussed MD simulations to investigate the pre-fusion protein–lipid complex. The author discussed the dynamics of the SNARE complex between lipid bilayers, interactions of synaptotagmin-1 with lipids and SNARE proteins, and the regulation of complexin in the SNARE complex assembly that controls synaptic vesicle fusion.

Westra et al. [9] performed confinement analyses on simulated random walks and trajectories that exhibited transient confined behavior by optimizing the parameters for various experimental conditions. The authors also developed a tool to visualize confinement areas in heat maps that allowed the spatial mapping of confinement hotspots relative to subcellular markers. Using these optimized tools, they reliably detected the subdiffusive behavior of different membrane components and revealed the different confinement properties of the two types of glutamate receptors in neurons. Their results provide a systematic analysis of the influence of different parameters used for detecting temporal confinement in single-molecule trajectories, and a visualization tool for mapping confinement zones in the cellular context.

Selikhanov et al. [10] proposed a rational design approach for genetically engineering the photosynthetic reaction center in *Cereibacter sphaeroides* by introducing disulfide bonds between its α-helices. This modification increased the thermal stability and functional activity of some mutants of photosynthetic pigment–protein complexes, while effectively maintaining the photochemical charge separation ability of the genetically modified reaction centers.

Radka [11] discussed the biochemical and biophysical advances in understanding how interfacial enzymes involved in exogenous fatty acid (eFA) metabolism interact with the membrane. The author aimed to provide a molecular mechanistic understanding of how peripheral membrane proteins use conformational changes to precisely regulate their activation, localization, and integration into the membrane, and how these protein–lipid interactions contribute to enzyme catalysis.

Beltran et al. [12] revealed the functional properties shared between borate transporter Bor1p in *Saccharomyces cerevisiae* and SLC4, including lipid-promoted dimerization, sensitivity to stilbene disulfonate-derived inhibitors, the requirements of an acidic residue at the solute binding site, and the conservation of disease-causing mutations.

Jasni et al. [13] aimed to identify and understand the protein structure, function, and interaction of the biological membrane of *Entamoeba histolytica* upon the immune response, which could contribute to vaccine development. Furthermore, they reviewed the current development of vaccines targeting amoebiasis.

Collectively, the studies included in this Special Issue update our current knowledge on the function, structure, and dynamics of membrane proteins. The technical approaches presented here, such as single-particle tracking, structural biology, and MD simulations, will help us understand the spatiotemporal dynamics and physical properties of membrane proteins. These studies provide new insights into the fundamental principles underlying the physiological functions of membrane proteins, such as GPCRs in cells and tissues, and could lead to the development of new drug designs.

Author Contributions: Writing—original draft preparation, Y.S.; writing—review and editing, S.S. All authors have read and agreed to the published version of the manuscript.

Funding: This research received no funding.

Acknowledgments: The authors of this Editorial are grateful to all the authors contributing to this Special Issue.

Conflicts of Interest: The authors declare no conflict of interest.

References

1. Senju, Y.; Suetsugu, S. Interaction of proteins with biomembranes. *Membranes* **2022**, *12*, 181. [CrossRef] [PubMed]
2. Kim, H.; Kim, E.; Lee, B.C. Investigation of phosphatidylserine-transporting activity of human TMEM16C isoforms. *Membranes* **2022**, *12*, 1005. [CrossRef] [PubMed]
3. Li, L.; Li, J. Dimerization of transmembrane proteins in cancer immunotherapy. *Membranes* **2023**, *13*, 393. [CrossRef] [PubMed]
4. Anosov, A.; Borisova, E.; Smirnova, E.; Korepanova, E.; Osipov, A. Effect of cytochrome C on the conductance of asolectin membranes and the occurrence of through pores at different pHs. *Membranes* **2023**, *13*, 268. [CrossRef] [PubMed]
5. Rodrigues, I.C.; Rodrigues, S.C.; Duarte, F.V.; Costa, P.M.D. The role of outer membrane proteins in UPEC antimicrobial resistance: A systematic review. *Membranes* **2022**, *12*, 981. [CrossRef] [PubMed]
6. Dengler, D.G.; Harikumar, K.G.; Yen, A.; Sergienko, E.A.; Miller, L.J. Mechanism of action and structure–activity relationships of tetracyclic small molecules acting as universal positive allosteric modulators of the cholecystokinin receptor. *Membranes* **2023**, *13*, 150. [CrossRef] [PubMed]
7. Isu, U.H.; Badiee, S.A.; Khodadadi, E.; Moradi, M. Cholesterol in Class C GPCRs: Role, relevance, and localization. *Membranes* **2023**, *13*, 301. [CrossRef] [PubMed]
8. Bykhovskaia, M. Molecular Dynamics simulations of the proteins regulating synaptic vesicle fusion. *Membranes* **2023**, *13*, 307. [CrossRef] [PubMed]
9. Westra, M.; MacGillavry, H.D. Precise detection and visualization of nanoscale temporal confinement in single-molecule tracking analysis. *Membranes* **2022**, *12*, 650. [CrossRef] [PubMed]
10. Selikhanov, G.; Atamas, A.; Yukhimchuk, D.; Fufina, T.; Vasilieva, L.; Gabdulkhakov, A. Stabilization of *Cereibacter sphaeroides* photosynthetic reaction center by the introduction of disulfide bonds. *Membranes* **2023**, *13*, 154. [CrossRef] [PubMed]
11. Radka, C.D. Interfacial enzymes enable gram-positive microbes to eat fatty acids. *Membranes* **2023**, *13*, 423. [CrossRef] [PubMed]
12. Beltran, J.L.; McGrath, L.G.; Caruso, S.; Bain, R.K.; Hendrix, C.E.; Kamran, H.; Johnston, H.G.; Collings, R.M.; Henry, M.N.; Abera, T.L.; et al. Borate transporters and SLC4 bicarbonate transporters share key functional properties. *Membranes* **2023**, *13*, 235. [CrossRef] [PubMed]
13. Jasni, N.; Saidin, S.; Kin, W.W.; Arifin, N.; Othman, N. *Entamoeba histolytica*: Membrane and non-membrane protein structure, function, immune response interaction, and vaccine development. *Membranes* **2022**, *12*, 1079. [CrossRef] [PubMed]

membranes

MDPI

Article

Investigation of Phosphatidylserine-Transporting Activity of Human TMEM16C Isoforms

Hanggu Kim, Eunyoung Kim and Byoung-Cheol Lee *

Neurovascular Unit Research Group, Korea Brain Research Institute (KBRI), Daegu 41068, Korea
* Correspondence: lbc0523@kbri.re.kr

Abstract: Lipid scrambling is a rapid process that dissipates the asymmetrical distribution of phospholipids in the plasma membrane. It is involved in various physiological functions such as blood coagulation and apoptosis. Many TMEM16 members are recognized as Ca^{2+}-activated phospholipid scramblases, which transport phospholipids between the two leaflets of the plasma membrane nonspecifically and bidirectionally; among these, TMEM16C is abundant in the brain, especially in neuronal cells. We investigated the scrambling activity of three human TMEM16C isoforms with different N-terminus lengths. After optimizing conditions to minimize endogenous scrambling activity, an annexin V-based imaging assay was used to detect phosphatidylserine (PS) scrambling in 293T cells. Unlike previous results, our data showed that human TMEM16C isoform 1 and isoform 3 exposed PS to the cell surface. A surface biotinylation assay showed that the surface expression of isoform 2, which did not show scrambling activity, was ~5 times lower than the other isoforms. In contrast to other TMEM16 proteins, flux assays and electrophysiology recording showed TMEM16C does not possess ion-transporting activity. We conclude that the N-terminus of TMEM16C determines whether TMEM16C can translocate to the plasma membrane and facilitate scrambling activity; membrane-localized TMEM16C isoforms 1 and 3 transport PS to the outer leaflet.

Keywords: membrane proteins; lipid transport; scramblases; ion channels

Citation: Kim, H.; Kim, E.; Lee, B.-C. Investigation of Phosphatidylserine-Transporting Activity of Human TMEM16C Isoforms. *Membranes* **2022**, *12*, 1005. https://doi.org/10.3390/membranes12101005

Academic Editors: Yosuke Senju and Shiro Suetsugu

Received: 18 September 2022
Accepted: 14 October 2022
Published: 17 October 2022

1. Introduction

Lipid scrambling is a process that changes the asymmetric distribution of phospholipids in the outer- and inner leaflets of cell membranes [1]. Whereas phosphatidylcholine and sphingomyelin largely reside in the outer leaflet, the inner leaflet is composed primarily of phosphatidylserine (PS) and phosphatidylethanolamine. This asymmetry is created and maintained by two ATP-driven pumps called flippase and floppase [1,2]; the enzyme that scrambles or mixes lipids in the membrane is called "scramblase". Lipid scrambling is used as a signal by cells [2], the most well-known being the "eat me" signal from apoptotic cells. In normal cells, PS is dominant in the inner leaflet; however, lipid scrambling causes dying cells to expose PS on the outer membrane, a change recognized by phagocytic cells, which then remove apoptotic cells [3]. Lipid scrambling is involved in various other physiological processes such as blood coagulation, synaptic pruning, viral infection, fertilization, and myoblast fusion [4].

There are three types of lipid scramblases in cell membranes, which are classified by their activation mechanism. The first, Ca^{2+}-activated lipid scramblases, are activated by increased intracellular Ca^{2+} [5]. The second type, the Xk-related (Xkr) family, are activated by caspase signals; they contain caspase recognition domains and their cleavage by caspase is essential for scrambling activity [6]. The final class, constitutively active or non-regulated lipid scramblases, includes G protein-coupled receptors such as Rhodopsin [7,8]. In recent decades, several studies have tried to determine the molecular identity of Ca^{2+}-activated lipid scramblases, with members of the TMEM16 family now accepted as belonging to this class [9].

TMEM16 proteins are composed of 10 family members and were initially reported as Ca^{2+}-activated Cl^- channels [10–12]. Despite the similarity in their genetic sequence, TMEM16 proteins are functionally divergent. For instance, TMEM16A and TMEM16B are recognized as Ca^{2+}-activated Cl^- channels, but TMEM16C, TMEM16D, TMEM16E, TMEM16F, TMEM16G, TMEM16J, and TMEM16K have Ca^{2+}-activated lipid-scrambling activity [13–16]. Among these, TMEM16E, TMEM16F, and TMEM16K are dual-function, non-selective ion channels and lipid scramblases [14–16]. Structural studies show they are dimeric proteins with 10 transmembrane domains [15,17–20]. TMEM16 proteins have a primary Ca^{2+} binding site within the transmembrane region and an additional binding site near the dimeric interface. Among TMEM16 family members, TMEM16C is mainly found in the brain [13,21,22], especially in neuronal cells [23,24]. TMEM16C does not have ion- or PS-transporting activity, but does transport phosphatidylcholine and galactosylceramide [13]. TMEM16C also regulates pain-related signals in rat dorsal root ganglia by modulating the activity of sodium-activated potassium channels [24,25]; although the mechanism is not well understood, changes in lipid distribution by the lipid-scrambling activity of TMEM16C may be involved.

TMEM16C proteins are associated with many neuronal diseases, including febrile seizure [26] and autosomal-dominant craniocervical dystonia [22]. TMEM16C is also associated with late-onset Alzheimer's disease [27]. Interestingly, transcriptional analysis shows that TMEM16C is downregulated in patients suffering from Alzheimer's disease compared with healthy controls [28]. These results suggest that TMEM16C could be a new target for the treatment of brain disease.

In this study, we investigated the lipid-scrambling and ion-transporting activity of human TMEM16C isoforms. Among three human isoforms, isoform 1 and 3 transported PS to the membrane outer leaflet, while isoform 2 did not. We confirmed that these results were due to differences in the surface expression level of each isoform. We also found that GFP tagging on isoform 1 severely reduced the scrambling activity of TMEM16C. Ion channel activity of the TMEM16C isoforms was not observed in either flux assays or in the electrophysiological recording. These results suggest that human TMEM16C isoforms that present on the surface membrane are PS-transporting lipid scramblases but not ion channels.

2. Materials and Methods

2.1. Construction and Expression of TMEM16C Isoforms

A human, cDNA ORF clone of TMEM16C isoform 1 (NCBI, NP_001300655.1) was purchased from GenScript (Clone ID, Ohu71184). Full-length DNAs of isoform 2 (NP_113606.2) and isoform 3 (NP_001300656.1) were generated by PCR and cloned into a pCEP4 vector (Thermo). For GFP tagging of each DNA construct, pCAG-GFP (Addgene, #11150) vectors were used. To estimate transcription efficiency, pCAG-GFP virgin vectors were co-transfected with TMEM16C constructs at a molar ratio of 1:100 (GFP:TMEM16C). To confirm the presence of TMEM16C isoforms in the human brain, whole brain QUICK-Clone cDNA (TaKaRa) and PCR were used with isoform 1 and isoform 2-specific primers, namely isoform 1 forward primer: ATGTCAGTTTTAAAATTTGAACTG; isoform 2 forward primer: ATGGTCCACCATTCAGGCTCCATT; isoform 3 forward primer: GCAATGAAGGATTCCAAATGCAGCTG; and universal reverse primer: TCCCCCCGGGGAGGCCATTCATGGTG.

2.2. Scrambling Assay

24 to 48 h after transfection using X-tremeGENE HP (Roche), scrambling activity was measured in 293T cells. A 0.5% Alexa Fluor 568-tagged annexin V (Invitrogen) solution was prepared by making a scrambling solution (140 mM NaCl, 10 mM $CaCl_2$, 10 mM HEPES, pH 7.4). To activate phospholipid scramblases, an equal volume of scrambling solution containing 4 μM of the Ca^{2+} ionophore A23187 (Sigma) was added to reach a final concentration of 2 μM. As an unstimulated control, 0.1% DMSO was added to cells in annexin V-binding solution. Exposure of PS on the outer cell membrane was measured

by the accumulation of annexin V–Alexa Fluor 568 and captured with time-lapse imaging and a super-sensitive, high-resolution confocal laser scanning microscope TCS SP (Leica) for 20 min after the addition of the ionophore. Image acquisition was controlled by LAS X software (Leica).

2.3. *Surface Biotinylation and Western Blot*

For the surface biotinylation of TMEM16C isoforms, 293T cells were seeded at a density of 6×10^5 cells in a 6-well plate. Cells were transfected with the DNA of TMEM16C isoforms and incubated for 48 h. During the experiment, the plate was fixed on ice. 293T cells were washed three times with PBS and incubated with 2 mL of biotinylation solution containing EZ-LinkTM-sulfo-NHS-SS-Biotin (Thermo) (0.35 mg/mL in PBS) for 20 min. Next, an ice-cold PBS buffer containing 50 mM glycine was added to stop the reaction. After an additional incubation of 5 min, cells were collected by centrifugation at $4000\times g$ for 1 min. Cells were washed with PBS and lysed with lysis buffer containing 1% Triton X-100 and protease inhibitor cocktail by inverting for 1 h at 4 °C. The supernatant was collected and biotinylated proteins were bound to Streptavidin Plus UltraLinkTM Resin (Thermo) for 3 h at 4 °C. After this incubation period, samples were spun down and washed three times with 1 mL of lysis buffer. The pellet of streptavidin beads was resuspended in 40 μL of 3× LDS buffer for 10 min. Protein samples were then loaded onto the protein gel and a western blot was conducted with target antibodies, TMEM16C (Human protein atlas), actin (Cell signaling) and transferrin receptor (abcam). Relative surface expression ratios were calculated by normalizing the band intensity of TMEM16C in the surface to the band intensity of TMEM16C in the total (surface/total).

2.4. *Halide-Quenching Flux Assay*

To measure the ion-transporting activity of TMEM16C isoforms using a cell-based assay, YFP-H148Q/I152L-stable cell lines were used. YFP-H148Q/I152L in pcDNA3.1 was transfected into 293 cells and stable cells were selected using G418 antibiotics. YFP-expressing 293 cells were seeded into 96-well plates and TMEM16C isoforms and human TMEM16A were transfected with X-tremeGENE HP (Roche). Human TMEM16A (NCBI, XM_011545127.3) was used as a positive control. Prior to measurement of the fluorescence signal, the medium was changed to 50 μL basal buffer (137 mM NaCl, 2.7 mM KCl, 2 mM $CaCl_2$, 1 mM $MgCl_2$, 10 mM HEPES, pH 7.4). The signal was acquired using the Flexstation 3 (Molecular device) at an excitation and emission wavelength of 513 nm and 527 nm, respectively. After stabilization of the signal for 60 s, 100 μL of iodide-containing buffer (137 mM NaI, 2.7 mM KCl, 2 mM $CaCl_2$, 1 mM $MgCl_2$, 10 mM HEPES, pH 7.4) was added to each well to monitor iodide-transporting activity. Iodide buffers both with and without the Ca^{2+} ionophore (10 μM) were prepared.

2.5. *Electrophysiology*

The macroscopic current of TMEM16C isoforms was recorded by whole-cell configuration. Patch pipettes were fabricated from borosilicate glass and each resistance was between 2–4 Mohm. The currents were amplified using an Axopatch 200B (Molecular Device) and filtered at 2 kHz with a lowpass Bessel filter. Signals were digitized at a rate of 5 kHz using an Axon Digidata 1550B digitizer (Molecular Device). Ionic currents were evoked by voltage stimulus delivered from a holding potential of 0 mV to test voltages ranging from −100 to 100 mV. For the activation of TMEM16C isoforms, recording conditions described previously were used [29]. The intracellular solution contained 146 mM CsCl, 2 mM $MgCl_2$, 5 mM EGTA, 10 mM Sucrose, 10 mM HEPES, pH 7.3. Based on the MaxChealator (https://somapp.ucdmc.ucdavis.edu/ (accessed on 3 May 2022), free Ca^{2+} concentration was calculated and the required amount of $CaCl_2$ was added. The extracellular solution contained 140 mM NaCl, 5 mM KCl, 2 mM $CaCl_2$, 1 mM $MgCl_2$, 15 mM Glucose, 10 mM HEPES, pH 7.4. The pH of both solutions was titrated with NMDG. As a positive control, TMEM16F (NCBI, NM_001025356.3) isoform 1 was used to measure the macroscopic currents activated by intracellular Ca^{2+}. For each data set, the significance of the difference was tested using a paired sample t-test using Origin software (Originlab). In all cases, $p < 0.05$ was considered significant.

3. Results

3.1. Human TMEM16C Isoforms Contain Different N-Terminuses

Most studies on TMEM16C are conducted using mouse [13,24] and rat TMEM16C [30]. Mouse and rat isoform 1 are almost identical (Figure 1A,B) and correspond to human TMEM16C isoform 2. In humans, there are two further isoforms of TMEM16C: isoform 1 is a novel TMEM16C isoform that contains a longer N-terminal segment than mouse and rat TMEM16C (Figure 1A); isoform 3 contains the shortest N-terminus and corresponds to mouse isoform 2 (Figure 1A,B). To identify whether isoform 1 is present in the human brain at the transcriptional level, we obtained commercially available cDNA from human brain. PCR demonstrated that isoform 1 of human TMEM16C could be amplified using human isoform 1-specific primers and the amount of amplified DNA was less than that of isoform 2 (Figure 1C, left). Since the CDS (Coding Sequence) of isoform 3 is perfectly matched to that of isoform 1, isoform 3-specific primer was designed in the UTR (Untranslated Region) of isoform 3. Finally, we could observe isoform 3-specific PCR fragments in the gel (Figure 1C, right). These results suggest that all three human TMEM16C isoforms exist in the human brain at the transcriptional level.

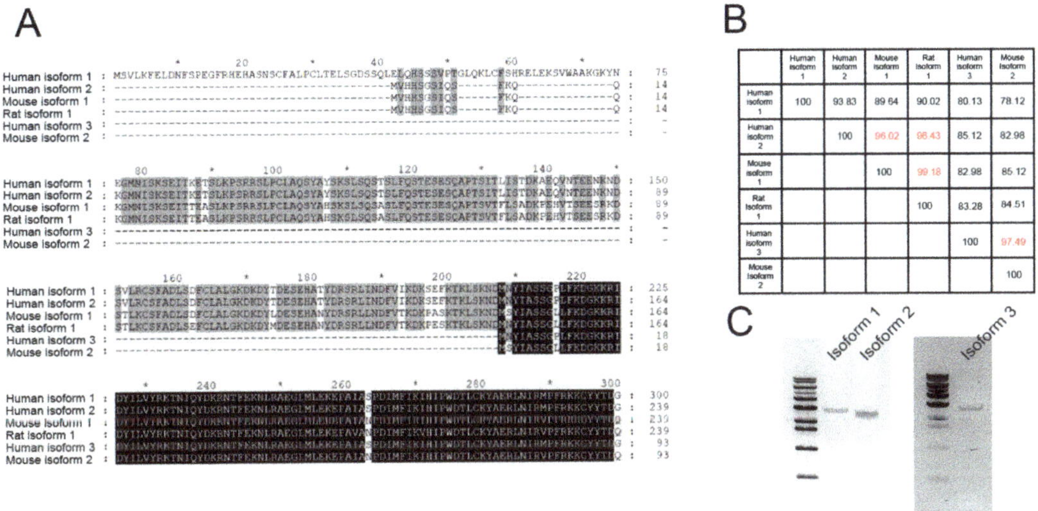

	Human isoform 1	Human isoform 2	Mouse isoform 1	Rat isoform 1	Human isoform 3	Mouse isoform 2
Human isoform 1	100	93.83	89.64	90.02	80.13	78.12
Human isoform 2		100	96.02	98.43	85.12	82.98
Mouse isoform 1			100	99.18	82.98	85.12
Rat isoform 1				100	83.28	84.51
Human isoform 3					100	97.49
Mouse isoform 2						100

Figure 1. Three human TMEM16C isoforms and their similarity to other orthologues. (**A**) Sequence alignment of the N-terminal regions of human, mouse and rat TMEM16C isoforms. Alignment was conducted using Clustal Omega. (**B**) Sequence similarity between human TMEM16C isoforms and orthologues; similarity was calculated using the Ident and Sim program from the Sequence Manipulation Suite. (**C**) PCR results for human TMEM16C isoforms using cDNA from human brain. The first column in the gels is 1 kb DNA ladder to estimate the length of DNA.

3.2. Endogenous Scrambling Activity of the 293T Cell and the Effects of GFP Tagging on the Scrambling Activity

To measure scrambling activity, we conducted an optical imaging, cell-based assay to monitor exposure of PS on cell membranes. In unstimulated conditions, most PS exists in the inner leaflet of the plasma membrane (Figure 2A, left). Treatment with the Ca^{2+} ionophore A23187 increases intracellular Ca^{2+}, allowing Ca^{2+}-activated lipid scramblases to be activated for transporting PS to outer leatlet. Since annexin V could bind to PS specifically,

PS exposed on the outer leaflet can then be visualized by fluorophore-conjugated annexin V (Figure 2A, right). Thus, accumulation of the fluorescent signal can be monitored using real-time imaging.

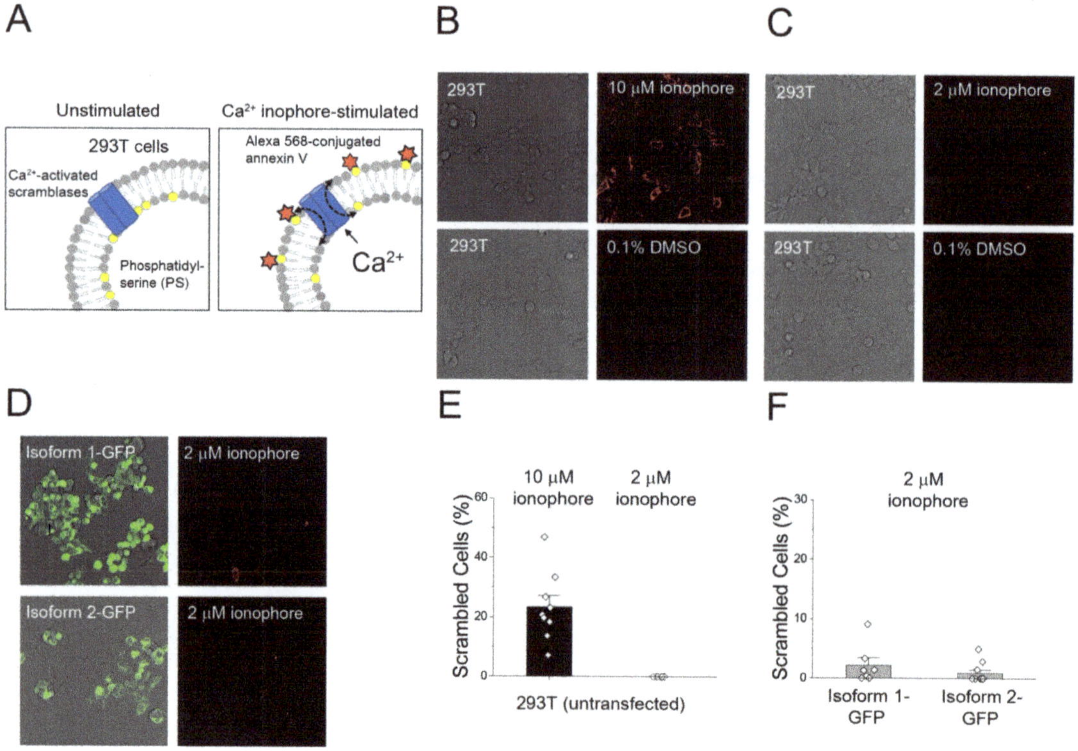

Figure 2. Scrambling activity of endogenous TMEM16F in 293T cells and the effect of GFP tagging on human TMEM16C. (**A**) Schematic diagram of cell-based imaging using fluorophore-labeled annexin V. Phosphatidylserine (PS) exposed by Ca^{2+}-activated scramblase activity upon treatment with the Ca^{2+} ionophore A23187 was stained with Alexa 568-conjugated annexin V. Accumulation of red signal was captured by microscope imaging. (**B**) Endogenous scrambling activity of TMEM16F in 293T cells. 10 μM ionophore was added to 293T cells to increase intracellular Ca^{2+} concentration. DMSO (0.1%) was used as a control. (**C**) Optimization of imaging conditions to minimize endogenous lipid-scrambling activity in 293T cells. 293T cells treated with 2 μM ionophore showed minimal scrambling activity, equivalent to the control treatment. (**D**) Lipid-transporting activity of GFP-tagged human TMEM16C isoforms 1 and 2. 2 μM ionophore was added to transfected 293T cells to increase intracellular Ca^{2+} and stimulate TMEM16C activity. (**E**) Quantification of lipid-scrambling activity in 293T cells upon treatment with 10 μM and 2 μM ionophore. Data are presented as mean + standard error of the mean (SEM); n = 9 for 10 μM ionophore and n = 8 for 2 μM ionophore. (**F**) Quantification of lipid-scrambling activity of GFP-tagged TMEM16C isoform-transfected 293T cells after treatment with 2 μM ionophore. Data are presented as mean + SEM; n = 7 for isoform 1-GFP and n = 10 for isoform 2-GFP.

Prior to measuring the scrambling activity of TMEM16C isoforms, the endogenous scrambling activity of 293T cells was investigated. As reported previously [31], 293T cells showed scrambling activity after 10 μM ionophore treatment (Figure 2B), with 23.3% of cells showing scrambling activity (Figure 2E). The control treatment, 0.1% DMSO, did not affect the scrambling process in 293T cells (Figure 2B). To find the optimal conditions to prevent endogenous TMEM16F activity from interfering with the measurement of TMEM16C activity, we tested the effect of various concentrations of a Ca^{2+} ionophore on the scrambling activity of 293T cells. We concluded that no scrambling activity was observed in the presence of 10 mM $CaCl_2$ and 2 μM ionophore (Figure 2C,E).

To confirm transfection with our genes of interest, we used a GFP-tagged construct, which has been used previous in studies measuring the scrambling activity of TMEM16F [16,32]. However, in the presence of 2 μM ionophore, scrambling activity was only observed in 2.2% and 0.9% of cells expressing isoform 1-GFP and isoform 2-GFP, respectively (Figure 2D,F). These results suggest that a very small number of cells expressing GFP-tagged TMEM16C proteins responded to an increment of intracellular Ca^{2+}, consistent with a previous result that showed that tagged mouse TMEM16C displayed minimal PS scrambling activity [13].

3.3. Scrambling Activity of Three TMEM16C Isoforms

Since the ion-and lipid-transporting activity of nhTMEM16, a fungal homologue of TMEM16, is inhibited by GFP tagging [33], we next probed the effect of GFP tagging of the scrambling activity of TMEM16C isoforms. We generated constructs without a GFP tag and repeated the cell-based imaging assay. To validate the transfection of the genes and expression of the TMEM16C, DNA expressing GFP was co-transfected with TMEM16C isoforms at a molar ratio of 1:100; we assumed that cells expressing GFP were also expressing TMEM16C isoforms. In each case, transfection efficiency was similar in ionophore- (red bars) and DMSO-treated (control; black bars) cells (Figure 3A–C): 41.1% and 43.7% for isoform 1, 33.8% and 31.9% for isoform 2, and 34.9% and 40.1% for isoform 3, respectively (Figure 3D).

Unlike with the GFP-tagged protein, 23.5% of cells expressing isoform 1 without a GFP tag presented PS on the outer leaflet in the presence of 2 μM ionophore (Figure 3A,E). These results suggest that GFP tagging was inhibiting the scrambling activity of isoform 1, as shown in a previous study which was performed with fungal TMEM16 homologues, nhTMEM16 [33]. Consistent with a previous study [13], isoform 2, which corresponds to mouse isoform 1, showed very low activity, with 2.8% of cells displaying PS on the outer leaflet (Figure 3B). Isoform 3, the shortest TMEM16C isoform, also transported PS to the outer membrane (Figure 3C). Upon the increment of intracellular Ca^{2+}, 10.3% of 293T cells expressing TMEM16C isoform 3 transported PS (Figure 3E). All cells were also exposed to a control condition (0.1% DMSO) without Ca^{2+} addition. For isoform 2 and 3 of TMEM16C, 0.1% DMSO did not stimulate any scrambling activity of 293T cells; however, in cells expressing isoform 1, 0.1% DMSO resulted in the scrambling activity in 0.5% of cells (black bars, Figure 3E). This may result from the expression of isoform 1 affecting cell viability, causing cells to generate the apoptotic "eat me" signal which would also be stained with the annexin V-based assay.

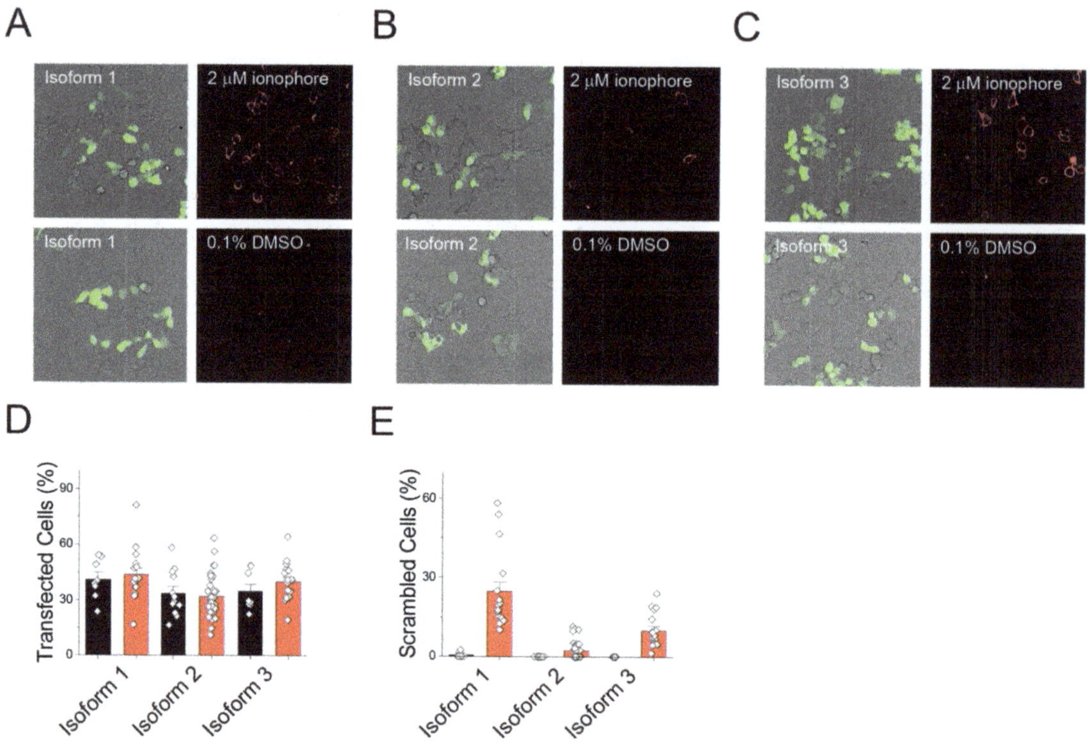

Figure 3. Scrambling activity of three human TMEM16C isoforms without GFP tagging. Monitoring of the lipid-transporting activity of TMEM16C isoform 1 (**A**), isoform 2 (**B**), and isoform 3 (**C**). A total of 293T cells were transfected with each isoform, without a GFP tag. To confirm transfection, a pCAG-GFP construct was co-transfected with TMEM16C at a ratio of 100:1 (TMEM16C:GFP). 2 μM ionophore was used to increase intracellular Ca^{2+} concentration; DMSO (0.1%) was used as a control. (**D**) Transfection efficiency of each isoform. The percentage of transfected cells was calculated for ionophore- (red bars) and DMSO-treated cells (black bars) by counting the total cells in brightfield images and GFP-expressing cells in fluorescence images. (**E**) Quantification of scrambling activity of human TMEM16C isoforms after treatment with a control (DMSO) or 2 μM ionophore. Data are presented as mean + SEM. Isoform 1: DMSO, n = 9 and ionophore, n = 16; isoform 2: DMSO, n = 12 and ionophore, n = 34; isoform 3: DMSO, n = 8 and ionophore, n = 17.

3.4. Surface Expression of Human TMEM16C Isoforms

These results suggested that the isoforms with the longest (isoform 1) and shortest (isoform 3) N-terminuses can transport PS, while the isoform of intermediate length N-terminus (isoform 2) cannot. Thus, we next investigated the surface expression level of each isoform in 293T cells by conducting a surface biotinylation assay. No differences in the expression of each isoform was observed when the expression levels of the isoforms were compared using the immunoblotting of total cell lysates, (Figure 4A). However, a surface biotinylation assay showed much lower expression of isoform 2 on the cell surface than isoform 1 or 3: the relative surface expression ratios of isoform 1 and 3 were 0.46 and 0.45, respectively, while that of isoform 2 was 0.10 (Figure 4B,C). The actin and transferrin receptors were used as a control protein for cytosolic- and membrane proteins, respectively. Immunoblotting results for the transferrin receptor and actin showed that only proteins on the cell surface are biotinylated (Figure 4B). These results suggest that the N-terminal region of TMEM16C is critical for the translocation of TMEM16C to the plasma membrane. Additionally, we tested the surface expression of the GFP tagged TMEM16C isoforms by conducting surface biotinyaltion assays with a GFP tagged construct. Unlike untagged isoforms, GFP tagged isoform 3 showed somewhat lower expression in total fraction (Figure 4C). The relative surface expression ratios of GFP tagged isoform 1, 2, and 3 were 1.63, 0.52, and 0.52, respectively. These results suggest that GFP tagged TMEM16C isoform 1 and 2 are more abundant in the cell surface than untagged constructs.

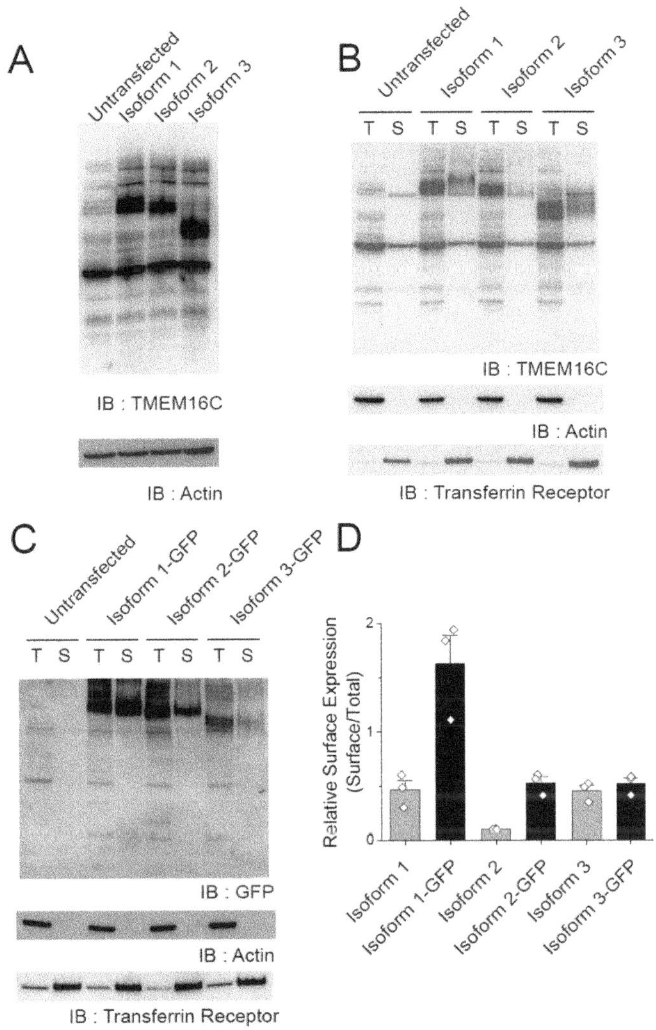

Figure 4. Surface expression of human TMEM16C isoforms. (**A**) Immunoblotting of cell lysates expressing human TMEM16C isoforms. Anti-TMEM16C (upper image) and anti-actin (lower image) antibodies were used. (**B**) Immunoblotting of surface biotinylated TMEM16C proteins. Surface fraction (S) samples were loaded at a 10-fold higher concentration than the total fraction (T). Anti-TMEM16C (upper image), anti-actin (middle image) and anti-transferrin receptor (lower image) antibodies were used to validate the biotinylation of surface-expressed proteins. (**C**) Immunoblotting of surface biotinylated GFP tagged TMEM16C proteins. (**D**) Quantification of relative surface expression of each isoform. Relative surface expression ratios were calculated by normalizing the band intensity of TMEM16C in the surface to the band intensity of TMEM16C in the total. Mean + SEM; n = 3 for each isoform.

3.5. *Ion Channel Activity of Human TMEM16C Scramblases*

Next, we studied the ion-transporting activity of TMEM16C isoforms. Previous studies suggest that the heterologous expression of mouse TMEM16C isoform 1, which corresponds to human isoform 2, does not result in ion-transporting activity [13,24]. Prior to conducting electrophysiology studies, we measured ion transportation using a YFP-based halide ion-quenching assay. Since fluorescent signals from YFP(H148Q/I152L) could be quenched by the halide ion, this assay was largely used in the study on the anion transporting activity of ion channels. After establishing YFP (H148Q/I152L)-expressing stable cell lines, cells were transfected with DNA for each TMEM16C isoform. In all cases, the iodide treatment of TMEM16C-expressing cells in the presence of 10 μM ionophore did not decrease the fluorescent signal compared with untransfected cells (Figure 5A). However, cells expressing human TMEM16A, a well-known Ca^{2+}-activated Cl^- channel, showed the decrease in the YFP signal after treatment of Ca^{2+} ionophore and iodide (Figure 5A). To measure the electrical activity of human TMEM16C directly, electrophysiological recordings were performed using whole-cell configuration. At 2 μM intracellular Ca^{2+}, no TMEM16C isoforms showed any changes in the ionic current compared to untransfected cells (Figure 5B). Since the purpose of patch clamp recording was to determine whether TMEM16C isoforms have a channel function or not, a high concentration of free Ca^{2+} (200 μM) was also tested in the intracellular solution, as reported in other studies [29,31]. Compared with untransfected cells, TMEM16C isoform-expressing cells did not show any significant changes in the whole-cell current upon Ca^{2+} and voltage stimulus (Figure 5C). As a positive control for validating our assay system, human TMEM16F isoform 1 was also transfected and their electrical currents were measured in the presence of 200 μM Ca^{2+}. The cells expressing human TMEM16F showed large macroscopic currents (Figure 5D). All current traces recorded from human TMEM16C isoforms were almost similar to the endogenous currents from untransfected 293T cells. We could only observe the endogenous outward membrane currents. These currents were activated with voltage-dependent activation kinetics upon higher positive voltage stimulus (Figure 5B,C). These activation kinetics over time were also shown in the current traces of TMEM16F transfected cells. The current density from all isoforms was comparable with that of untransfected 293T cells (Figure 5D), and these values are statistically insignificant. These results suggest that TMEM16C does not have ion-conducting activity.

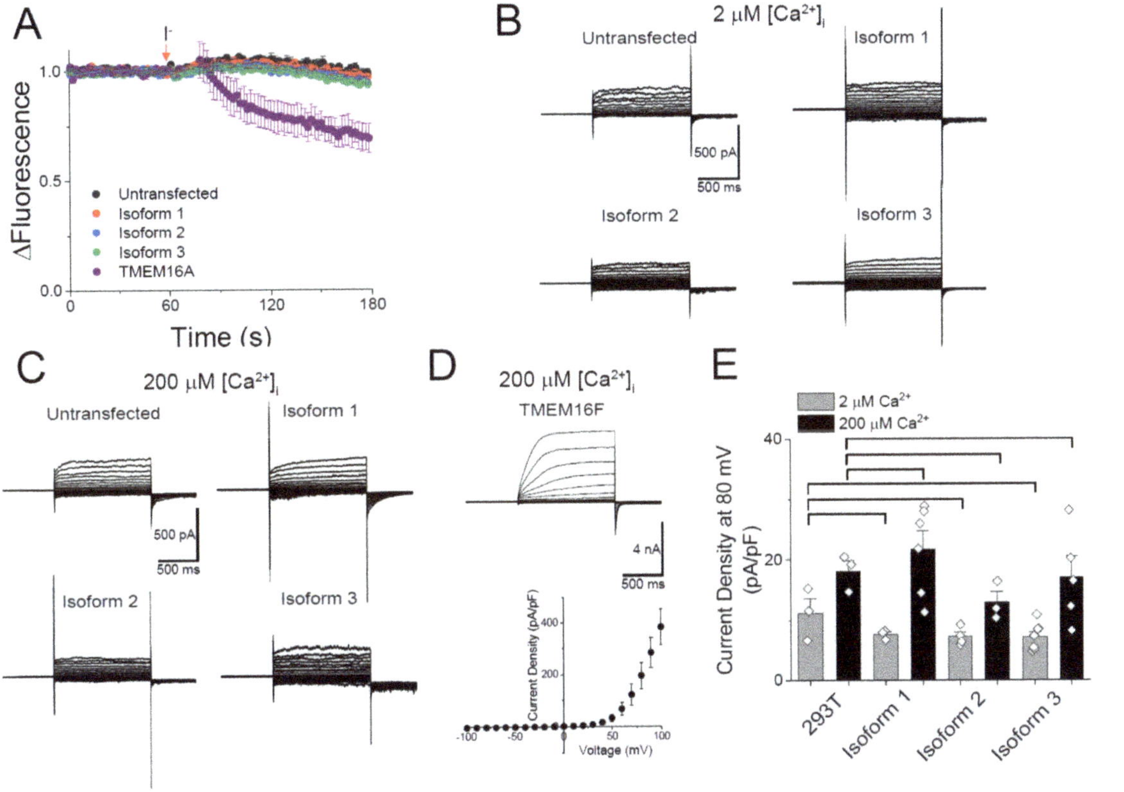

Figure 5. Ion-transporting activity of human TMEM16C isoforms. (**A**) Measurement of ion-transporting activity using an iodide-quenching assay. Each TMEM16C isoform and human TMEM16A were transfected into 293T cells stably expressing YFP-H148Q/I152L. Mean + SEM; n = 3. (**B**,**C**) Representative whole-cell current recordings of cells transiently transfected with human TMEM16C isoforms. 2 μM (**B**) and 200 μM (**C**) intracellular Ca^{2+} was added and ionic currents were evoked with voltage steps ranging from −100 mV to +100 mV in 10 mV increments. (**D**) Whole-cell recording of human TMEM16F expressing cells. The macroscopic current was measured in the presence of 200 μM intracellular Ca^{2+} (upper). The representative I-V relationship of TMEM16F (lower). (**E**) Average current densities in untransfected and TMEM16C isoform-transfected 293T cells. Current density was measured at +80 mV. Data are plotted as mean + SEM (n = 3–6).

4. Discussion

To date, many studies measuring ion-transporting and/or the lipid-scrambling activity of the TMEM16C protein have used tagged mouse or rat TMEM16C isoform 1. For instance, the Nagata group used mouse isoform 1 with C-terminal FLAG tagging [13], and the Jan group measured the activity of mouse isoform 1 with N-terminal HA tagging and C-terminal GFP tagging [24]. Several studies have used GFP or YFP-tagged constructs to investigate other members of the TMEM16 family, such as TMEM16A and TMEM16F [12,16,17,32], without observing an effect of the fluorescent tagging on the protein's function. However, we previously examined the effect of GFP tagging on the activity of nhTMEM16, a fungal homologue of TMEM16 protein, and found that GFP tagging inhibited both its ion- and lipid-transporting activity [33]. Likewise, the present study suggests that GFP tagging of human TMEM16C isoform 1 severely reduces PS-transporting activity. After removing the GFP tag from the isoform 1 construct, the percentage of scrambled cells increased from 2.2% to 23.5%; for isoform 2, a slight increase from 0.9% to 2.3% scrambling was observed after GFP tag removal. To determine whether the reduction of scrambling activity of GFP tagged TMEM16C is caused by the inhibition of protein function itself or due to the defects on the surface expression of the protein, we conducted a surface biotinylation assay by using GFP tagged constructs. We found that more TMEM16C isoform 1 and 2 exist in the cell surface after adding the GFP tag. These results strongly suggest that GFP tagging on isoform 1 could inhibit the PS transporting activity without inhibiting the translocation of isoform 1 to the cell surface. For isoform 2, even though the scrambling activity was slightly increased after cleavage of the GFP tag, GFP tagging inhibited the PS transporting function not by inhibiting its translocation to the cell surface.

Previous findings from studies of mouse TMEM16C show that this protein did not transport PS in the presence of the Ca^{2+} increment [13]. Consistent with this result, our experiments showed that human isoform 2, corresponding to mouse isoform 1, did not transport PS to the cell surface because its surface expression was significantly lower than that of other isoforms; however, human isoforms 1 and 3 transported PS, similar to TMEM16F. The exposure PS to the outer leaflet is a representative consequence of lipid scrambling. Since the most basic feature of lipid scramblases is non-selective lipid transport, the result that showed that TMEM16C discriminated between PS and other lipids, phosphatidylcholine and ceramide was unexpected [13]. For this reason, interest in TMEM16C as a lipid scramblase has declined despite its involvement in neurological diseases. Thus, our finding provides evidence that certain TMEM16C isoforms possess the predominant feature of lipid scramblases, PS-transporting activity. Furthermore, we measured the PS-transporting activity of TMEM16C in the presence of a lower Ca^{2+} concentration than that required to activate endogenous human TMEM16F in 293T cells. These results suggest that TMEM16C could have a role in cell surface PS exposure in response to small increases in intracellular Ca^{2+}.

In summary, we investigated the lipid- and ion-transporting activity of three human TMEM16C isoforms using optical imaging, a fluorophore-quenching flux assay, and electrophysiological recording. We showed that isoform 1 and isoform 3 could transport PS to the outer leaflet, while isoform 2 could not, due to significantly lower isoform 2 surface expression. We could not detect macroscopic ionic currents from surface-expressed TMEM16C isoforms 1 and 3. Taken together, these results suggest that human TMEM16C present on cell surface membranes is a PS-transporting lipid scramblase but not an ion channel.

Author Contributions: Conceptualization, H.K. and B.-C.L.; Methodology, H.K., E.K. and B.-C.L.; Software, H.K., E.K. and B.-C.L.; Validation, H.K., E.K. and B.-C.L.; Formal analysis, H.K., E.K. and B.-C.L.; Investigation, H.K., E.K. and B.-C.L.; Resources, B.-C.L.; Data curation, H.K., E.K. and B.-C.L.; Writing-Original Draft Preparation, E.K. and B.-C.L. Writing-Review & Editing, H.K., E.K. and B.-C.L.; Visualization, H.K., E.K. and B.-C.L.; Supervision, B.-C.L.; Project administration, B.-C.L.; Funding acquisition, B.-C.L. All authors have read and agreed to the published version of the manuscript.

Funding: This work was supported by the KBRI Basic Research Program funded by the Ministry of Science and ICT of the Republic of Korea (22-BR-01-02 to B.L.) and a National Research Foundation (NRF) of Korea grant funded by the Ministry of Science and ICT (MIST) of the Republic of Korea (2019R1C1C1002699 to B.L.).

Institutional Review Board Statement: Ethical review and approval for this study were waived by the Institutional Review Board of Korea Brain Research Institute (No. KBRI-202103-BR-001). The materials (human brain cDNA from Takara) used in this study are separated and processed from human materials for the use by general public without direct collecting human biological materials.

Informed Consent Statement: Not applicable.

Data Availability Statement: The data that support the findings of this study are available on request from the corresponding author, B.L.

Acknowledgments: The imaging data from TCS SP and fluorescent signal from Flexstation 3 were supported by Brain Research Core Facilities in KBRI. The DNA for human TMEM16A was kindly provided by Joo Hyun Nam (Dongguk Univ., Republic of Korea).

Conflicts of Interest: The authors declare that they have no conflicting interests.

References

1. Pomorski, T.; Menon, A. Lipid flippases and their biological functions. *Experientia* **2006**, *63*, 2908–2921. [CrossRef]
2. Bevers, E.M.; Williamson, P.L. Phospholipid scramblase: An update. *FEBS Lett.* **2010**, *584*, 2724–2730. [CrossRef] [PubMed]
3. Balasubramanian, K.; Schroit, A.J. Aminophospholipid Asymmetry: A Matter of Life and Death. *Annu. Rev. Physiol.* **2003**, *65*, 701–734. [CrossRef] [PubMed]
4. Bevers, E.M.; Williamson, P.L. Getting to the Outer Leaflet: Physiology of Phosphatidylserine Exposure at the Plasma Membrane. *Physiol. Rev.* **2016**, *96*, 605–645. [CrossRef] [PubMed]
5. Pedemonte, N.; Galietta, L.J. Structure and Function of TMEM16 Proteins (Anoctamins). *Physiol. Rev.* **2014**, *94*, 419–459. [CrossRef] [PubMed]
6. Suzuki, J.; Imanishi, E.; Nagata, S. Exposure of Phosphatidylserine by Xk-related Protein Family Members during Apoptosis. *J. Biol. Chem.* **2014**, *289*, 30257–30267. [CrossRef]
7. Goren, M.A.; Morizumi, T.; Menon, I.; Joseph, J.S.; Dittman, J.S.; Cherezov, V.; Stevens, R.C.; Ernst, O.P.; Menon, A.K. Constitutive phospholipid scramblase activity of a G protein-coupled receptor. *Nat. Commun.* **2014**, *5*, 5115. [CrossRef] [PubMed]
8. Menon, I.; Huber, T.; Sanyal, S.; Banerjee, S.; Barré, P.; Canis, S.; Warren, J.D.; Hwa, J.; Sakmar, T.P.; Menon, A.K. Opsin Is a Phospholipid Flippase. *Curr. Biol.* **2011**, *21*, 149–153. [CrossRef]
9. Ehlen, H.W.; Chinenkova, M.; Moser, M.; Munter, H.-M.; Krause, Y.; Gross, S.; Brachvogel, B.; Wuelling, M.; Kornak, U.; Vortkamp, A. Inactivation of anoctamin-6/Tmem16f, a regulator of phosphatidylserine scrambling in osteoblasts, leads to decreased mineral deposition in skeletal tissues. *J. Bone Miner. Res.* **2012**, *28*, 246–259. [CrossRef]
10. Caputo, A.; Caci, E.; Ferrera, L.; Pedemonte, N.; Barsanti, C.; Sondo, E.; Pfeffer, U.; Ravazzolo, R.; Zegarra-Moran, O.; Galietta, L.J.V. TMEM16A, A Membrane Protein Associated with Calcium-Dependent Chloride Channel Activity. *Science* **2008**, *322*, 590–594. [CrossRef]
11. Schroeder, B.C.; Cheng, T.; Jan, Y.N.; Jan, L.Y. Expression Cloning of TMEM16A as a Calcium-Activated Chloride Channel Subunit. *Cell* **2008**, *134*, 1019–1029. [CrossRef] [PubMed]
12. Yang, Y.D.; Cho, H.; Koo, J.Y.; Tak, M.H.; Cho, Y.; Shim, W.-S.; Park, S.P.; Lee, J.; Lee, B.; Kim, B.-M.; et al. TMEM16A confers receptor-activated calcium-dependent chloride conductance. *Nature* **2008**, *455*, 1210–1215. [CrossRef]
13. Suzuki, J.; Fujii, T.; Imao, T.; Ishihara, K.; Kuba, H.; Nagata, S. Calcium-dependent Phospholipid Scramblase Activity of TMEM16 Protein Family Members. *J. Biol. Chem.* **2013**, *288*, 13305–13316. [CrossRef]
14. Gyobu, S.; Miyata, H.; Ikawa, M.; Yamazaki, D.; Takeshima, H.; Suzuki, J.; Nagata, S. A Role of TMEM16E Carrying a Scrambling Domain in Sperm Motility. *Mol. Cell. Biol.* **2016**, *36*, 645–659. [CrossRef] [PubMed]
15. Bushell, S.R.; Pike, A.C.W.; Falzone, M.E.; Rorsman, N.J.G.; Ta, C.M.; Corey, R.A.; Newport, T.D.; Christianson, J.C.; Scofano, L.F.; Shintre, C.A.; et al. The structural basis of lipid scrambling and inactivation in the endoplasmic reticulum scramblase TMEM16K. *Nat. Commun.* **2019**, *10*, 3956. [CrossRef] [PubMed]
16. Yu, K.; Whitlock, J.M.; Lee, K.; Ortlund, E.A.; Cui, Y.Y.; Hartzell, H.C. Identification of a lipid scrambling domain in ANO6/TMEM16F. *eLife* **2015**, *4*, e06901. [CrossRef]
17. Brunner, J.D.; Lim, N.K.; Schenck, S.; Duerst, A.; Dutzler, R. X-ray structure of a calcium-activated TMEM16 lipid scramblase. *Nature* **2014**, *516*, 207–212. [CrossRef]
18. Dang, S.; Feng, S.; Tien, J.; Peters, C.; Bulkley, D.; Lolicato, M.; Zhao, J.; Zuberbühler, K.; Ye, W.; Qi, L.; et al. Cryo-EM structures of the TMEM16A calcium-activated chloride channel. *Nature* **2017**, *552*, 426–429. [CrossRef]
19. Alvadia, C.; Lim, N.K.; Mosina, V.C.; Oostergetel, G.T.; Dutzler, R.; Paulino, C. Cryo-EM structures and functional characterization of the murine lipid scramblase TMEM16F. *eLife* **2019**, *8*, e44365. [CrossRef]
20. Falzone, E.M.; Rheinberger, J.; Lee, B.-C.; Peyear, T.; Sasset, L.; Raczkowski, A.M.; Eng, E.T.; Di Lorenzo, A.; Andersen, O.S.; Nimigean, C.M.; et al. Structural basis of Ca2+-dependent activation and lipid transport by a TMEM16 scramblase. *eLife* **2019**, *8*, e43229. [CrossRef]
21. Uhlén, M.; Fagerberg, L.; Hallström, B.M.; Lindskog, C.; Oksvold, P.; Mardinoglu, A.; Sivertsson, Å.; Kampf, C.; Sjöstedt, E.; Asplund, A.; et al. Proteomics. Tissue-Based Map of the Human Proteome. *Science* **2015**, *347*, 1260419. [CrossRef] [PubMed]
22. Charlesworth, G.; Plagnol, V.; Holmström, K.M.; Bras, J.; Sheerin, U.-M.; Preza, E.; Rubio-Agusti, I.; Ryten, M.; Schneider, S.A.; Stamelou, M.; et al. Mutations in ANO3 Cause Dominant Craniocervical Dystonia: Ion Channel Implicated in Pathogenesis. *Am. J. Hum. Genet.* **2012**, *91*, 1041–1050. [CrossRef] [PubMed]
23. Karlsson, M.; Zhang, C.; Méar, L.; Zhong, W.; Digre, A.; Katona, B.; Sjöstedt, E.; Butler, L.; Odeberg, J.; Dusart, P.; et al. A single–cell type transcriptomics map of human tissues. *Sci. Adv.* **2021**, *7*, eabh2169. [CrossRef]
24. Huang, F.; Wang, X.; Ostertag, E.M.; Nuwal, T.; Huang, B.; Jan, Y.-N.; Basbaum, A.I.; Jan, L.Y. TMEM16C facilitates Na+-activated K+ currents in rat sensory neurons and regulates pain processing. *Nat. Neurosci.* **2013**, *16*, 1284–1290. [CrossRef]
25. Gadotti, V.M.; Zamponi, G.W. TMEM16C cuts pain no SLACK. *Nat. Neurosci.* **2013**, *16*, 1165–1166. [CrossRef]
26. Feenstra, B.; Pasternak, B.; Geller, F.; Carstensen, L.; Wang, T.; Huang, F.; Eitson, J.L.; Hollegaard, M.V.; Svanström, H.; Vestergaard, M.; et al. Common variants associated with general and MMR vaccine–related febrile seizures. *Nat. Genet.* **2014**, *46*, 1274–1282. [CrossRef] [PubMed]

27. Briones, N.; Dinu, V. Data mining of high density genomic variant data for prediction of Alzheimer's disease risk. *BMC Med. Genet.* **2012**, *13*, 7. [CrossRef]
28. Hargis, E.K.; Blalock, E.M. Transcriptional signatures of brain aging and Alzheimer's disease: What are our rodent models telling us? *Behav. Brain Res.* **2016**, *322*, 311–328. [CrossRef]
29. Jiang, T.; Yu, K.; Hartzell, H.C.; Tajkhorshid, E. Lipids and ions traverse the membrane by the same physical pathway in the nhTMEM16 scramblase. *eLife* **2017**, *6*, e28671. [CrossRef] [PubMed]
30. Li, Y.; Zhang, L.; Li, J.; Wang, C.; Chen, Y.; Yuan, Y.; Xie, K.; Wang, G.; Yu, Y. A Role for Transmembrane Protein 16C/Slack Impairment in Excitatory Nociceptive Synaptic Plasticity in the Pathogenesis of Remifentanil-induced Hyperalgesia in Rats. *Neurosci. Bull.* **2021**, *37*, 669–683. [CrossRef] [PubMed]
31. Le, T.; Le, S.C.; Yang, H. Drosophila Subdued is a moonlighting transmembrane protein 16 (TMEM16) that transports ions and phospholipids. *J. Biol. Chem.* **2019**, *294*, 4529–4537. [CrossRef]
32. Le, T.; Jia, Z.; Le, S.C.; Zhang, Y.; Chen, J.; Yang, H. An inner activation gate controls TMEM16F phospholipid scrambling. *Nat. Commun.* **2019**, *10*, 1846. [CrossRef]
33. Lee, B.-C.; Menon, A.K.; Accardi, A. The nhTMEM16 Scramblase Is Also a Nonselective Ion Channel. *Biophys. J.* **2016**, *111*, 1919–1924. [CrossRef]

membranes MDPI

Review

Dimerization of Transmembrane Proteins in Cancer Immunotherapy

Lei Li [1] and Jingying Li [1,2,*]

1 College of Chemistry, Fuzhou University, Fuzhou 350108, China
2 College of Biological Science and Engineering, Fuzhou University, Fuzhou 350108, China
* Correspondence: lijingying@fzu.edu.cn

Abstract: Transmembrane proteins (TMEMs) are integrated membrane proteins that span the entire lipid bilayer and are permanently anchored to it. TMEMs participate in various cellular processes. Some TMEMs usually exist and perform their physiological functions as dimers rather than monomers. TMEM dimerization is associated with various physiological functions, such as the regulation of enzyme activity, signal transduction, and cancer immunotherapy. In this review, we focus on the dimerization of transmembrane proteins in cancer immunotherapy. This review is divided into three parts. First, the structures and functions of several TMEMs related to tumor immunity are introduced. Second, the characteristics and functions of several typical TMEM dimerization processes are analyzed. Finally, the application of the regulation of TMEM dimerization in cancer immunotherapy is introduced.

Keywords: transmembrane proteins; dimerization; cancer immunotherapy; small-molecule drugs; dimerization regulation

Citation: Li, L.; Li, J. Dimerization of Transmembrane Proteins in Cancer Immunotherapy. *Membranes* **2023**, *13*, 393. https://doi.org/10.3390/membranes13040393

Academic Editors: Yosuke Senju and Shiro Suetsugu

Received: 2 February 2023
Revised: 24 March 2023
Accepted: 26 March 2023
Published: 30 March 2023

1. Introduction

Membrane proteins are essential for the physiological functions of the cell membrane. They can be classified into two types based on their interaction with the membrane: peripheral membrane proteins, which bind to the membrane through non-covalent interactions, and integrated membrane proteins, which more firmly bind to the membrane through hydrophobic interactions [1]. TMEMs are integrated membrane proteins that span the entire lipid bilayer and are permanently anchored in it [2]. Based on their different structures, TMEMs can be divided into two categories: α-helical proteins and β-barrel proteins (Figure 1). These two structures are the dominant structures of all transmembrane proteins, with α-helical proteins being mostly found in the cytoplasm and subcellular septum, while β-barrel proteins are mostly found in chloroplasts, bacteria, and mitochondrial membranes [3].

A variety of TMEMs play a regulatory role in cancer immunotherapy. Within the TMEM family, immune checkpoint proteins are called immune system regulators; they can regulate the activity of various immune cells and play important roles in the process of maintaining the body's immune balance [4]. Cancer cells can use inhibitory immune checkpoint proteins to generate tumor immune escape [5]. Traditional immune checkpoint proteins mainly target T cells, with Programmed death ligand 1 (PD-L1) and Toll-Like Receptor-4 (TLR-4) being well-known immune checkpoint proteins related to T-cell activation [5,6]. The dimerization of PD-L1 or TLR-4 can cause changes in T-cell activity, making them a potential target for cancer immunotherapy. For example, the PD-L1/Programmed death 1 (PD-1) signaling pathway can inhibit T-cell activity, allowing cancer cells to escape the killing action performed by immune cells [7]. BMS-8, a compound by Bristol-Myers Squibb (BMS), binds to PD-L1 and causes it to form a homodimer, ultimately blocking the interaction between PD-L1 and PD-1, and leading to T-cell activation [8]. Rhamnogalacturonan II (RG-II) could effectively induce the dimerization of TLR-4, activate myeloid differentiation factor 88 (MyD88)-independent and MyD88-dependent pathways, promote the maturation and differentiation of dendritic cells (DCs), produce a series of cytokines to regulate inflammatory responses, and affect the activity of T cells [9].

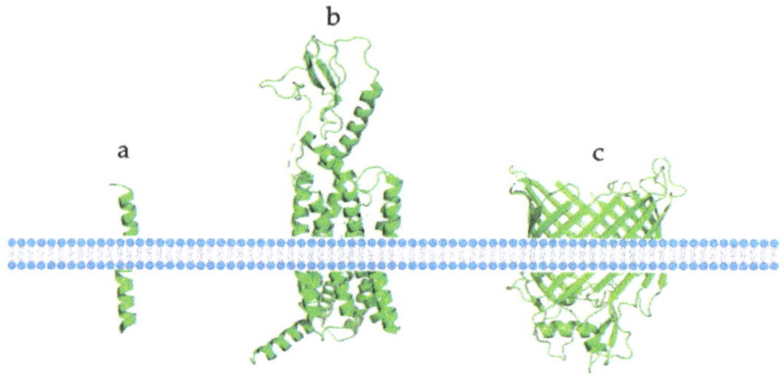

Figure 1. Different types of TMEMs found in the cell membrane. (**a**) A single transmembrane α-helix protein. (**b**) An α-helix protein that crosses the membrane multiple times. (**c**) A transmembrane β-barrel protein.

Recent studies have demonstrated the potential of immune checkpoint proteins in natural killer (NK) cells for cancer immunotherapy [10]. Current immune checkpoint proteins associated with NK-cell activity include natural killer Group 2 member A (NKG2A), leukocyte immunoglobulin-like receptors (LIRS), human leukocyte antigen G (HLA-G), and Cluster of differentiation 96 (CD96) [10–14]. HLA-G is a major histocompatibility complex molecule that binds to inhibitory receptors on white blood cells and protects the fetus from attack by the mother's immune cells [15]. Because of its immunosuppressive effects, HLA-G has been extensively studied for its role in cancer immunotherapy. For example, after homodimer formation, HLA-G can bind to its ligand, Ig-like transcript 2 (ILT2), with a higher affinity to regulate the cytotoxic activity of T cells and NK cells [16]. NKG2A is one of the key immune checkpoint proteins of NK cells and is involved in the regulation of NK-cell activity [17]. NKG2A can form a heterodimer with Cluster of differentiation 94 (CD94), which results in the inhibition of the activity of NK cells [17]. Therefore, blocking the interaction between NKG2A and CD94 can activate NK-cell activity, further inhibiting the immune escape of cancer cells.

Four immune checkpoint proteins, PD-L1, TLR4, HLA-G, and NKG2A, belong to the TMEM family and can form homodimers or heterodimers to regulate cancer immunotherapy. Therefore, this review focuses on these four immune checkpoint proteins. Firstly, we introduce the structures and functions of these proteins; then, we analyze the binding properties and functions of these immune checkpoint proteins and their receptors. Finally, we discuss the regulation of the dimerization of these immune checkpoint proteins and their potential as targets for cancer immunotherapy through dimerization regulation. Understanding the mechanisms of TMEM dimerization regulation can provide insights into developing novel anti-tumor drugs.

2. Structures and Functions of Immune Checkpoint Proteins in Cancer Immunotherapy

The structures and physiological functions of immune checkpoint proteins are crucial to the development of new drugs for tumor immunotherapy. This section mainly introduces the structures and physiological functions of several immune checkpoint proteins, including PD-L1, TLR4, HLA-G, and NKG2A.

2.1. PD-L1

PD-L1 is a transmembrane protein expressed in macrophages, activated B cells, T cells, and many solid tumor cells at higher levels than in normal tissues [18]. Encoded by the PD-L gene, PD-L1 belongs to the cluster of differentiation 28 (CD28)/B7/cytotoxic T lymphocyte antigen-4 (CTLA-4) family [19,20]. The gene encoding the PD-L1 protein contains seven exons, and the PD-L1 protein contains 290 amino acids and has a molecular mass of 40 kDa [21]. PD-L1 consists of an intracellular domain (30 amino acids), a transmembrane domain (hydrophobic), and two extracellular Ig-like domains (IgC- and IgV-like domains) [22–24].

PD-L1 plays a crucial role in tumor immunotherapy, as it inhibits T cells and NK cells when combined with PD-1 [25,26]. The PD-L1/PD-1 complex produces a signal that suppresses cytotoxic T cells, leading to T-cell depletion, which protects local tissues from immune-cell-induced inflammation [27]. The PD-1/PD-L1 signaling pathway is also involved in immune tolerance, which refers to the inability of immune cells to carry out a normal immune response under specific antigen stimulation [28]. Inhibiting the PD-1/PD-L1 signaling pathway enhances the immune response in two ways: it promotes the maturation and differentiation of immune cells, and it enhances the activity of immune cells [29,30]. However, the activation of immune cells can result in skin disorders, such as vitiligo and lichenoid dermatitis associated with renal metabolism, as well as the uncommon but difficult-to-treat psoriasis [29,30]. Therefore, understanding the structure and physiological function of PD-L1 is crucial to developing new drugs for tumor immunotherapy.

2.2. *TLR-4*

TLR-4 is a member of the Toll-like receptor (TLR) family that plays a critical role in regulating the balance of the immune system [31,32]. It is predominantly located on the cell membrane of adipose cells, as well as on the outer membrane of natural killer (NK) cells, macrophages, and monocytes [33]. Encoded by the TLR-4 gene, TLR-4 has a molecular mass of 69 KDa in humans. The full-length TLR-4 protein comprises three parts: an extracellular domain containing 608 amino acids, an intracellular domain containing 187 amino acids, and a transmembrane domain [34]. The extracellular domain is further divided into three sections based on its amino acid sequence: an N-terminal domain, a central domain, and a C-terminal domain [35].

TLR-4 plays an essential role in regulating the response to tissue damage and inflammation, which is crucial to maintaining tissue homeostasis [36]. For instance, TLR-4 binds to ligands such as a cluster of differentiation 14 (CD14) or myeloid differentiation protein 2 (MD2) to activate downstream signaling pathways that regulate inflammatory responses via two distinct pathways (myD88-dependent and myD88-independent pathways) [37]. These inflammatory signals link appetite with the mesolimbic dopamine (DA) system and the hypothalamus, significantly affecting appetite [38]. The link between chronic inflammation and cancer development was first discovered in the 19th century [39,40]. Until recently, it was thought that TLR-4-induced inflammation might have two distinct effects on tumor therapy [41]. While chronic inflammation promotes tumor cell growth, inducing acute inflammation can effectively kill tumor cells [41]. By inducing TLR-4 dimerization with small-molecule agonists, downstream signaling pathways can be activated to generate a strong pro-inflammatory response to kill tumor cells, which may have important implications in anti-tumor immunity [41]. Thus, TLR-4 agonists have been widely studied as potential agents for cancer immunotherapy.

2.3. *HLA-G*

HLA-G is a non-classical major histocompatibility complex-I (MHC-I) molecule that plays a crucial role in fetal and maternal immune tolerance [42]. While primarily expressed in the placental trophoblast, it is also expressed in various cancer cells, suggesting its involvement in regulating cancer development [43–46]. The gene responsible for HLA-G has eight exons and seven introns, with the extracellular domains being encoded by exons 2, 3, and 4; the transmembrane and cytoplasmic domains being encoded by exons 5 and 6; and exon 1 encoding signal peptides [47]. HLA-G has a molecular mass of 37–39 kDa and consists of seven isomers (HLA-G1 to -G7) that differ in the heavy-chain α1, α2, and α3 domains of the extracellular domain; β2 microglobulin; and transmembrane and cytoplasmic domains [43]. HLA-G1 and HLA-G5, both of which contain complete α1-α2-α3 domains, peptides, and β2m in their extracellular domains, are the most highly expressed members among all isomers [48–50]. Other isomers, such as HLA-G2, HLA-G3, and HLA-G4, lack one or both α-domains and are expressed in a membrane-anchored form containing a single α1 domain, α1-α3 domains, and α1-α2 domains, respectively. The remaining isomers are expressed in a non-membrane-anchored form [48–50].

HLA-G has been well-studied due to its role in maternal and infant immune tolerance [51,52]. During pregnancy, HLA-G mediates immune tolerance and supports fetal growth by binding to immune cells and exerting its immunosuppressive effects [53]. As an immunosuppressive molecule, HLA-G regulates immune cell activity by inhibiting immune cell maturation and cytotoxicity, inducing immune cell apoptosis, and activating downstream signaling pathways [54]. HLA-G binds to inhibitory receptors located on T cells, resulting in changes in T-cell function and the inhibition of T-cell maturation and proliferation, as well as changes in the cellular activity of $CD8^+$T cells and $CD4^+$T cells [55]. HLA-G also has an inhibitory effect on DCs maturation and differentiation when binding to Ig-like transcript 4 (ILT4), a ligand expressed on DCs, inhibiting antigen presentation and impeding signal transmission between other immune cells and DCs [56,57].

2.4. NKG2A

Killer-cell immunoglobulin-like receptors (KIRs) can be classified as stimulatory or inhibitory receptors based on their function [58]. NKG2A, an inhibitory receptor in the KIR family, is commonly expressed in NK cells and $CD8^+$T cells [59,60]. Located in the NK complex, the NKG2A gene is composed of seven exons [61]. Recombinant human NKG2A, made up of 159 amino acids, has a molecular mass of 26 KDa and comprises three components: an extracellular lectin-like domain, a transmembrane domain, and an intracellular domain [62,63]. The intracellular domain of NKG2A contains two immunoreceptor tyrosine-based inhibitory motifs (ITIMs), which are associated with Src homology region 2 domain-containing phosphatase-1 (SHP-1) and can bind to ligands to cause ITIM tyrosine phosphorylation [64]. The SHP-1 tyrosine phosphatase is then recruited intracellularly, and tyrosine residues on the activated cascade signaling molecules are dephosphorylated, thus transmitting the inhibitory signal [65].

NKG2A plays a critical role in regulating immune cell function and is commonly expressed in NK cells in human peripheral blood [66]. After heterodimerization with CD94, NKG2A can bind to the human leukocyte antigen E (HLA-E) ligand to activate downstream signaling pathways, thereby inhibiting NK-cell function; when the binding of NKG2A to HLA-E is blocked, NK cells can resume cytotoxic activity [67–69]. Additionally, NKG2A is expressed in some T cells, such as $CD8^+$T cells, but unlike its expression in NK cells, it is only expressed in $CD8^+$T cells in patients [70,71]. For example, the expression level of NKG2A is increased in cancer patients and patients with chronic viral infections, indirectly indicating the potential of NKG2A as a cancer therapeutic target [70,71]. NKG2A is highly expressed in lymphocytes in different tumor microenvironments [72,73]. For instance, NKG2A is overexpressed in human cervical cancer cells, and this mechanism is dependent on interleukin -15 (IL-15) to upregulate its expression in $CD8^+$T cells, thus inhibiting the cytotoxicity of lymphocytes and rendering them ineffective in killing tumors [74]. Blocking NKG2A-mediated signaling pathways can improve NK-cell dysfunction. Monalizumab, a novel anti-NKG2A antibody drug, has therapeutic effects on chronic lymphocytic leukemia (CLL) when used [75].

3. Binding Characteristics and Function of Transmembrane Protein-Receptor Dimers

Immune checkpoint proteins perform various physiological functions by interacting with receptors. Understanding the dimer structures and functions of these proteins, as well as the ways in which they interact with ligands, is important in the development of small-molecule drugs that regulate immune checkpoint protein dimerization.

3.1. PD-L1 Dimerization

Zak et al. reported the crystal structure diffraction data of the PD-L1 dimer (Figure 2a) [76]. The human PD-L1 dimer consists of two asymmetric PD-L1 molecules with a rotation angle of 30° between them, and the force between the two PD-L1 molecules is not strong, with a contact surface of 814.8 Å [77]. Both PD-1 and PD-L1 have an extracellular IgV-like domain through which they interact and activate downstream signaling pathways, ultimately inhibiting immune T-cell activity [78]. When PD-L1 dimerization occurs, the inhibitory signal of the downstream pathway is blocked, and the inhibition of T cells is removed [79].

PD-1 is an inhibitory receptor on T cells and a member of the CD28 family. It is generally expressed in regulatory T cells (Tregs, $CD4^+$, and $Foxp3^+$) and NK cells. PD-1 is also expressed in other immune cells, such as macrophages and B cells [80]. The full-length PD-1 protein consists of three parts containing a total of 268 amino acids: a cytoplasmic part (94 amino acids), a transmembrane structure (27 amino acids), and an extracellular part (147 amino acids) [81]. The cytoplasmic tail of PD-1 contains an immunoreceptor tyrosine-based switch motif (ITSM) and an ITIM [82]. After the interaction between PD-L1 and the extracellular IgV-like domain of PD-1, the tyrosine residues of these two domains are phosphorylated [81]. The activation of the downstream signaling pathway leads to the activation of SHP-1/-2 and the dephosphorylation of CD28. Various co-inhibitory receptors on T cells inhibit signal transduction, thereby inducing apoptosis and inhibiting cytokine secretion and cell differentiation [82–85].

A cluster of differentiation 80 (CD80) belongs to the transmembrane protein family, and full-length CD80 is composed of 254 amino acids [86]. Similar to PD-L1, CD80 also has an extracellular IgV-like domain, enabling it to bind to the IgV-like domain of PD-L1, thereby blocking the interaction between PD-1 and PD-L1 [86]. The interaction between CD80 and CTLA-4 can produce signals that inhibit the activity of T cells, and the heterodimerization of PD-L1/CD80 can effectively reduce the affinity of CTLA-4/CD80 interaction. Therefore, inducing the heterodimerization of PD-L1 and CD80 is an effective method to activate T cells [87]. Additionally, both CD80 and PD-1 act by binding to the IgV-like domain of PD-L1, so the heterodimerization of CD80/PD-L1 can effectively reduce the binding affinity between PD-L1 and PD-1, thereby removing the inhibitory effect on T cells [88].

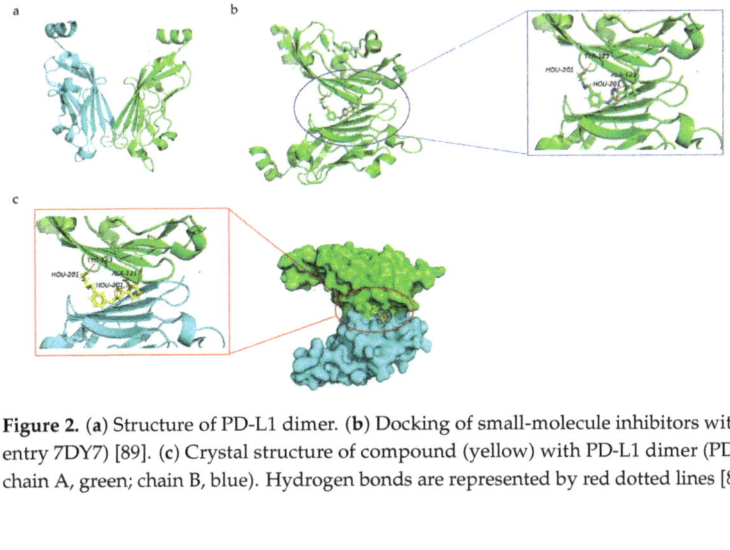

Figure 2. (**a**) Structure of PD-L1 dimer. (**b**) Docking of small-molecule inhibitors with PD-L1 (PDB entry 7DY7) [89]. (**c**) Crystal structure of compound (yellow) with PD-L1 dimer (PDB entry 7DY7; chain A, green; chain B, blue). Hydrogen bonds are represented by red dotted lines [89].

3.2. TLR-4 Dimerization

X-ray crystallography and experimental studies have demonstrated that TLRs are capable of forming homodimer signaling complexes, and small-molecule agonists can induce the homodimerization of TLR-4 to initiate downstream signaling pathways (Figure 3a) [90,91]. For example, Vladimir et al. designed a library of cell-permeating decoy peptides, each of which represents a nonfragmented patch of the TLR4 TIR surface [91]. They tested these peptides for the ability to inhibit early cytokine mRNA expression and mitogen-activated protein kinase (MAPK) activation in lipopolysaccharide (LPS)-stimulated primary murine macrophages. Five peptides-4R1, 4R3, 4BB, 4R9, and 4aE-potently inhibited all manifestations of TLR4, but not TLR2 signaling. These findings suggest that the area between the BB loop of TLR4 and its fifth helical region mediates TLR4 TIR dimerization [91]. The transmembrane domain of TLR4 plays a key role in its dimerization process, and the sequences that constitute the transmembrane domain, as well as their secondary and tertiary structures formed by the insertion into the membrane, affect dimerization [92]. For instance, 25% of the amino acids in the TLR-4 transmembrane domain (^{632}TIIGVSVLSVLVVSVVAVLVY^{652}) constitute polar residues. The ^{637}SxxS^{640} motif (x = any amino acid) among these polar residues is believed to contribute to the stability of homologous dimer complexes [93]. Once dimerization occurs, TLR-4 activates various downstream signaling pathways, inducing inflammatory responses, and activating monocytes (macrophages and DCs) and neutrophils, thereby enhancing the immune response and promoting tumor killing [41,94,95].

Lipopolysaccharide (LPS), an immune system agonist, activates monocytes and neutrophils via the TLR-4 signaling pathway, ultimately activating the body's innate immune system [41]. The process whereby LPS activates the TLR-4 signaling pathway is as follows: First, LPS forms a complex with its binding protein, LBP (LPS-binding protein). This complex then binds to a cluster of differentiation 14 (CD14) and transfers LPS to myeloid differential protein-2 (MD2) [96]. Upon interaction with LPS, the structure of MD2 changes, and the TLR-4 receptor binds to the MD2/TLR-4 complex to form a homodimer (Figure 3b) [96]. Activated TLR-4-MD2-LPS homodimer complexes perform downstream signal transduction to initiate innate immune responses [34].

High Mobility Group Box 1 (HMGB1) is typically expressed in the nucleus, where it can enhance transcription after binding to DNA [97–100]. HMGB1 is also released into the cytoplasm during apoptosis, injury, or death, and it is secreted in cancer cells and immune cells as a response mechanism to external stimuli [97–100]. The full-length HMGB1 protein contains 215 amino acids, and its structure is highly conserved [101]. It is mainly composed of three parts: the first part, called Box A, comprises amino acid residues 9–79; the second part, Box B, contains amino acid residues 95–163; and the last part is a C-terminal tail containing amino acid residues 186–215 [101]. The B box has been identified as a functional domain recognized by TLR-4, and amino acid residues 89–108 are the site of interaction with TLR-4 [101]. Upon interaction with TLR-4, downstream signaling pathways can be activated to regulate the inflammatory response and innate immunity [102,103]. For example, TLR-4-deficient animals are protected from ischemia–reperfusion injury in the liver, kidney, and heart, indicating that TLR-4 plays a critical role in aseptic inflammation [104].

Fibronectin (FN), typically expressed in the extracellular matrix, is a protein with multiple domains that is overexpressed in many types of cancer cells and plays a crucial role in tumor growth and metastasis [105,106]. FN molecules contain two distinct chains with molecular weights of 220 and 250 kDa, which are connected at their C ends by two disulfide bonds to form FN molecules [107,108]. The two chains can be divided into three parts according to the amino acid sequence: first, there are 12 modules composed of 40 amino acid residues; then, 2 modules contain 50 amino acid residues; finally, there are 15 to 17 modules made up of 90-to-100 amino acid residues [107,108]. It is worth mentioning that FNIII EDA (extra domains A) and FNIII EDB are the key structures for FN to play its physiological functions, and FNIII EDA is the site where FN interacts with

TLR-4. After FN combines with TLR-4, it can activate the downstream (MyD88-dependent) signaling pathway and, finally, activate the innate immune system of the body [109].

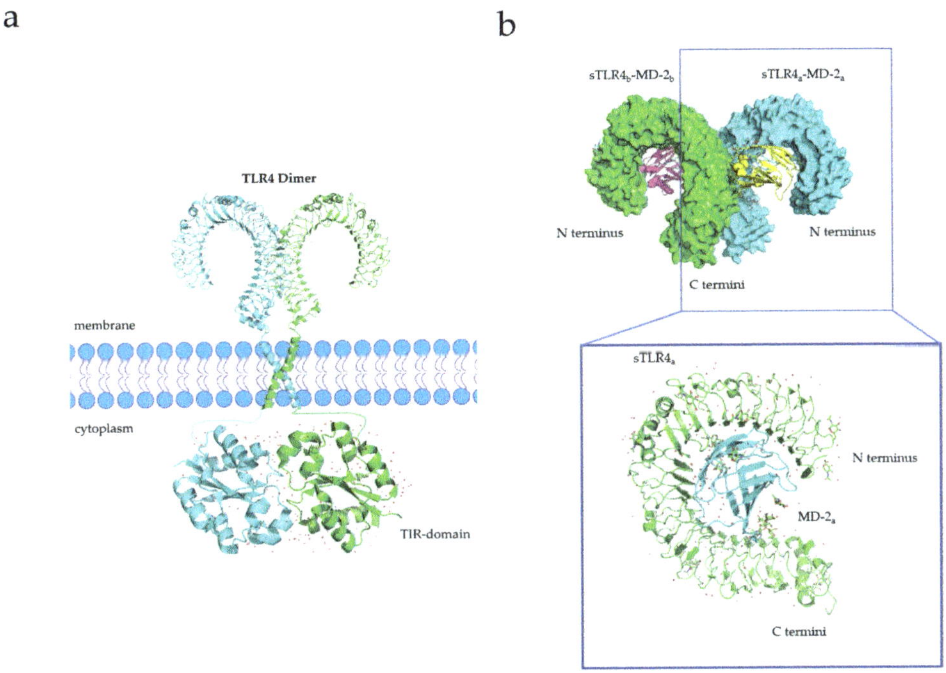

Figure 3. (**a**) Transmembrane structure of full-length TLR4 dimer, constructed based on the X-ray structure of TIR domain and reported NMR data [90]. (**b**) Crystal structure of sTLR4/MD-2/LPS complex and structural basis of sTLR4 interaction with MD-2 (PDB entry 3VQ2) [110].

3.3. *HLA-G Dimerization*

Boyson and Shiroishi et al. reported the crystal structure of the HLA-G dimer, showing that the soluble form of HLA-G can form dimers with intermolecular Cys^{42}-Cys^{42} disulfide bonds (Figure 4a) [111,112]. The Cys^{42} residue, located in the center of the $\alpha 1$ helical structure, contributes to the stability of the HLA-G dimer by interacting with the Cys^{42} residue of the other HLA-G protein [111]. The two exposed binding sites above the HLA-G dimer bind to CD8 and LILRB-1/-2, respectively. Shiroishi et al. proposed a complex model of HLA-G binding to its receptor, ILT2 (HLA-G dimer: receptor = 1:2) (Figure 4b). These findings suggest that the stability of the HLA-G dimer is dependent on its structural orientation [111]. Interestingly, HLA-G dimers are associated with T-cell activation and disease occurrence [113]. Therefore, HLA-G dimerization has great potential in cancer immunotherapy and may become a new target for cancer therapy.

Ig-like transcripts (ILTs) are members of the immunomodulatory receptor family and are commonly expressed on various immune cells [114]. Ig-like transcripts 4 (ILT4) and 2 (ILT2), which belong to the immunoglobulin-like receptor (LILR) family, are expressed on leukocytes [115]. ILT2 is frequently expressed on dendritic cells, T cells, and NK cells, while ILT4 is mainly expressed on DCs and monocytes [115]. The extracellular domains of ILT2 and ILT4 consist of four parts (immunoglobulin domains D1–D4), as well as a cytoplasmic tail and transmembrane domain [116]. The binding sites of HLA-G are the D1 and D2 domains located outside the cell (Figure 4b) [117–119]. The binding of ILT to HLA-G can inhibit the cytokine secretion, cell differentiation, and cell proliferation of immune cells, and induce cytotoxicity and apoptosis [115]. For example, after ILT4 binds to HLA-G, it can recruit SHP-1/-2 and activate the nuclear factor-κ-gene binding (NF-κB) pathway, leading to the increase in Interleukin-6 (IL-6) level and then the activation of the STAT3 pathway, which affects the function and maturation of DCs and leads to impaired innate immunity [120].

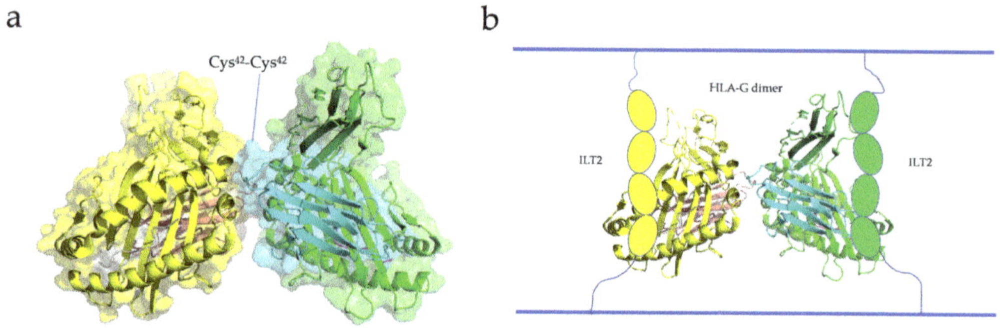

Figure 4. (**a**) Crystal structure of a dimer of a disulfide-bonded HLA-G-peptide complex. HLA-G heavy chains and β2m are shown in cartoon models, wrapped in translucent surfaces (yellow and green) [111]. (**b**) ILT2 and HLA-G dimer binding model. The HLA-G dimer is presented in a cartoon model, and the four Ig domains of ILT2 are circular (PDB entry 2D31) [15].

KIR2DL4 is a member of the killer-cell immunoglobulin-like receptor (KIR) and is commonly expressed on killer cells; it consists of two domains outside the cell, a positively charged transmembrane arginine residue, and an intracellular ITIM tail [121]. KIR2DL4 contains two distinct signaling domains (activation and inhibition), so Attia et al. hypothesized that it has both functions [57]. To demonstrate this, they created a chimeric receptor with both activation and inhibition domains, and the results showed that both KIR2DL4 signaling domains were active, suggesting that under different conditions, KIR2DL4 can inhibit or activate the activity of NK cells [57]. The interaction between KIR2DL4 and HLA-G can activate the downstream signaling pathway, thus regulating the activity of NK cells [122]. The structure of KIR2DL4 is unique, and its role in cancer will likely be the focus of future research.

3.4. NKG2A Dimerization

CD94, also known as Kp43, is a transmembrane protein expressed on the surface of most freshly isolated natural killer (NK) cells, as well as some subpopulations of γδT and αβT cells, although at different levels [123,124]. CD94 belongs to the Type II transmembrane protein family, and its carbohydrate recognition domain (CRD) is located outside the cell and contains two glycosylation sites [123,124]. The intracellular domain consists of only seven amino acids. CD94 can form heterodimers with NKG2A to perform its physiological functions [123,124]. The NKG2A/CD94 heterodimer can bind to the HLA-E expressed on immune T and NK cells to inhibit their activity and function [125]. Brett K. Kaiser and co-workers determined the crystal structure of the HLAE/NKG2A/CD94 protein complex using X-ray crystallography (Figure 5) [126].

HLA-E is a major histocompatibility complex (MHC) Class I molecule highly expressed in many solid tumor cells [127]. Generally, HLA-E forms trimers on the outer membrane of the cell, consisting of an α-heavy chain with a molecular weight of 45 kDa and a light chain (beta-2 microglobulin) encoded by chromosome 15 [128,129]. The binding of HLA-E to ligands has high specificity and only peptides with specific amino acid sequences can bind to HLA-E and be stably expressed [128,129]. For example, HLA-E can specifically bind to NKG2A/CD94 receptors on $CD8^+$ T and NK cells and activate downstream signaling pathways that lead to the inhibition of target-cell function and activity [128,129]. When formed in a complex with CD94/NKG2A, HLA-E can generate inhibitory signals, reduce the secretion of cytokines, and directly inhibit the killing effect of immune cells on cancer cells. This signaling pathway is often used by tumor cells to evade immune-cell-mediated killing [75].

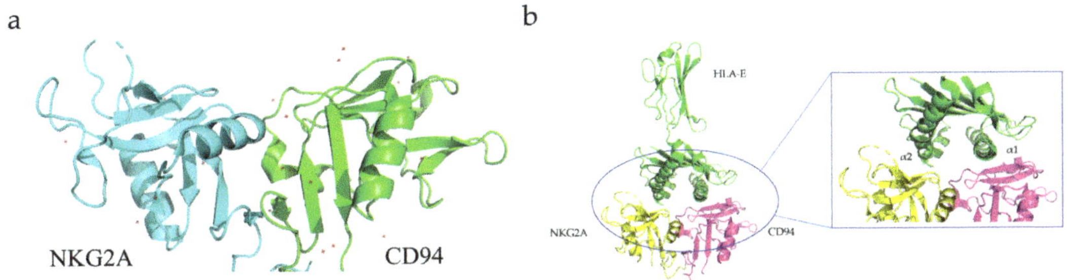

Figure 5. (**a**) Crystal structure of NKG2A/CD94 complex (PDB entry 3BDW). (**b**) Co-crystal structure of human HLA-E/NKG2A/CD94 (PDB entry 3CDG) and amplified structure of HLA-E binding interface with NKG2A/CD94 [86].

4. Application of Regulation of TMEM Dimerization in Anti-Tumor Immunity

Several TMEMs play important physiological roles as dimers. Among them, immune checkpoint proteins are crucial to tumor immunotherapy, and small-molecule drugs that regulate their dimerization have been extensively studied.

PD-L1, which exists in a monomeric form, interacts with PD-1 to produce signals that suppress immune cells, protecting cancer cells from being killed by the immune system [77]. Therefore, inhibiting the interaction between PD-L1 and PD-1 using small-molecule inhibitors is a promising direction in cancer immunotherapy. Two major classes of synthetic small-molecule inhibitors have been reported to date, one that directly blocks the PD-L1 binding site to PD-1 and another that induces PD-L1 dimerization, thus blocking the PD-1 binding site [130]. The drug-induced formation of PD-L1 homodimers can prevent PD-L1/PD-1 binding, thereby blocking downstream signaling pathways [77]. For example, BMS company has reported that the compound with (2-methyl-3-biphenyl) methanol as a scaffold can effectively induce the dimerization of PD-L1 and interact with the hydrophobic tunnel formed by two PD-L1 molecules (Figure 2b) [76,131]. These compounds can form hydrogen bonds with the amino acid residues (such as Ala121 or Tyr123) of PD-L1 (Figure 2c), making the formed PD-L1 dimer more stable and thus preventing the binding of PD-L1 to its ligands (such as PD-1) [132]. (S)-BMS-200, a compound with core scaffolds of 2,3-dihydro-1,4-benzodioxinyl, can stably bind to the PD-L1 homodimer and induce remarkable conformational changes in the key residues on the dimer, thus accelerating compact interactions [133]. Moreover, intermolecular interaction studies revealed that compound N-[2-(aminocarbonyl) phenyl] [1,1′-biphenyl]-4-carboxamide (APBC) can bind to the hydrophobic pocket formed by two PD-L1 monomers, especially anchoring residues Y56, M115, and A121. APBC forms a key hydrogen bond with critical residue D122, thereby stabilizing the structure of the dimer. APBC can bind and block the binding interface of PD-L1, thus blocking the binding of PD-1 [134]. In a mouse tumor model with high PD-L1 expression, the administration of APBC resulted in enhanced infiltration of $CD8^+$T cells and increased cytokine levels in the tumor microenvironment [134]. Compared with the control group, the tumor growth and survival rate of mice in the APBC group were significantly improved, and the tumor growth inhibition rate reached 62.1% after administering 10 mg/kg APBC [134]. In conclusion, small-molecule drugs that induce PD-L1 dimerization generally bind to PD-L1 and induce its dimerization, thus blocking the binding of PD-1 to PD-L1 and achieving the objective of immunotherapy. Therefore, these small-molecule drugs may become potential candidates for new anti-cancer drugs.

The dimerization of TLR-4 is crucial to regulating both innate and adaptive immune systems. Immunomax®, a TLR-4 agonist, binds to TLR-4 and promotes TLR-4 homodimerization, which in turn triggers the formation of intracellular signal transduction complexes and activates downstream signaling pathways [135,136]. In a 4T1 breast cancer mouse model, Immunomax® significantly improved survival; inhibited tumor growth and metastasis; increased the percentage of NK cells, CD4$^+$T cells, and CD8$^+$T cells in the mouse spleen; and significantly reduced the percentage of bone marrow-derived suppressor cells (MDSCs), making it a potential anti-cancer drug candidate [136]. Huang et al. established a sarcoma-containing C57BL/10J mouse model to validate the anti-cancer effects of cationic polymers such as cationic dextran (C-dextran) and polyethylenimine (PEI) [137]. These cationic polymers activated the TLR-4 signaling pathway, resulting in TLR-4 dimerization, acute inflammation, and direct tumor cell killing, thereby prolonging the survival time of mice, and inhibiting tumor growth and metastasis [137]. Park et al. found that rhamnogalacturonan II (RG-II) could activate the downstream signaling pathway of TLR-4. Further studies showed that RG-II could effectively induce the dimerization of TLR-4, activate MyD88-independent and MyD88-dependent pathways, promote the maturation and differentiation of DCs, and produce a series of cytokines to regulate inflammatory responses [9]. The effect of RG-II on mice carrying lymphoma C57BL6 was further verified, and the results showed that the activity of CD8$^+$T cells was significantly enhanced after the administration of RG-II and that the tumor growth and metastasis of mice were effectively alleviated [9]. In conclusion, inducing TLR-4 dimerization is an emerging cancer immunotherapy approach. Therefore, TLR-4 agonists have been widely studied by the scientific community, and investigations are still in progress.

HLA-G is typically expressed on specific immune cells or as a soluble dimer (sHLA-G) in the blood [138]. Upon the formation of sHLA-G dimers, the activity of HLA-G molecules is increased, enhancing their binding affinity for ligand ILT-2/ILT-4, which is expressed by tumor-associated white blood cells [138,139]. The binding of sHLA-G dimers to their ligand activates the downstream signaling pathway, leading to the inhibition of DC function and maturation and weakened activity of NK cells and CD8$^+$T cells, ultimately inhibiting the immune response [138,139]. In a study by Nathalie et al., HLA-G dimers were identified in Fon$^+$ cell lines from melanoma patients, and the formation of HLA-G dimers inhibited NK-cell activity, protecting Fon$^+$ cells from NK-cell-mediated killing and promoting rapid tumor growth and metastasis, potentially leading to patient death [140]. Singer et al. found different levels of HLA-G5 homodimers in malignant effusions of four ovarian cancer patients, and disease severity was positively correlated with the level of dimers [141,142]. These findings suggest that high levels of sHLA-G dimers are associated with advanced disease and a poor prognosis. Currently, there is limited knowledge regarding the regulation of the HLA-G dimer, which is typically regulated at the gene level [143]. For instance, the 3′-UTR of HLA-G can interact with multiple microRNAs (miR-152, miR-133a, and miR-148a) to reduce HLA-G expression and further downregulate the level of sHLA-G dimers in the blood [143]. In addition to microRNAs, Reches et al. identified an RNA-binding protein, HNRNPR (RBP HNRNPR), that interacts with the 3′-UTR of HLA-G and regulates its expression [144]. In summary, reducing the expression level of the HLA-G protein can reduce the formation of HLA-G dimers, thereby reducing the binding affinity of HLA-G and its receptors, and enhancing the immune response. This class of drugs is currently under development and has great potential for cancer immunotherapy.

NKG2A forms a heterodimer with CD94 in the majority of NK and CD8$^+$T cells; this heterodimer binds to its ligand, HLA-E, to suppress NK- and CD8$^+$T-cell activity. Tumor cells exploit this mechanism to evade the immune response [145,146], making it essential to block the interaction of HLA-E and NKG2A to enhance anti-tumor immune responses [86]. NKG2A and CD94 can only perform their functions after forming heterodimers. For instance, in klrd1-knockout mice, NKG2A could not form a heterodimer with CD94 due to the lack of the gene encoding CD94, resulting in the absence of inhibition of NK-cell maturation and development, as well as no reduction in the number and activity of NK cells [147]. Likewise, in DBA/2J mice with spontaneous mutations in the Klrd1 gene, CD94 expression was prevented, yet normal NK-cell development was maintained following the induction of NKG2A expression, without significant immunosuppression [148]. CD94 was expressed in both $CD8^+$T cells and $CD4^+$T cells, whereas NKG2A was only expressed in $CD8^+$T cells. Thus, inducing CD94/NKG2A expression in mixed-culture lymphocyte populations resulted in impaired $CD8^+$T-cell activity, which was restored upon the addition of anti-CD94 antibodies [149]. These findings suggest that blocking the formation of NKG2A/CD94 heterodimers can prevent downstream signaling pathways, thereby eliminating the inhibitory effect on immune cells [75]. In conclusion, inhibiting the formation of NKG2A/CD94 heterodimers can block the immunosuppressive signal of NKG2A, which has significant implications for the development of anti-tumor drugs.

5. Summary and Outlook

Cancer has long been one of the leading causes of death in humans, and the introduction of immune checkpoint inhibitors has opened new opportunities for cancer therapy. However, a significant proportion of patients do not respond to immune checkpoint inhibitors, and some solid tumors have primary resistance to immune checkpoint inhibitors [150]. Therefore, the search for more effective immunotherapy is urgent. A variety of TMEMs play important roles in cancer immunotherapy, and activating their downstream signaling pathways can regulate the activity of immune cells and play a role in killing tumor cells. The dimerization of TMEMs is often postulated to be the initial step in controlling their downstream signaling pathways [151].

Drugs (e.g., monoclonal antibodies and small-molecule immune checkpoint inhibitors) that regulate the dimerization of several TMEMs are under development, and some of these compounds show good biological activity [77]. A breakthrough has been made in the research of antibody immune checkpoint inhibitors; the U.S. Food and Drug Administration (FDA) has approved seven types of monoclonal antibodies, but these drugs directly act on immune checkpoint proteins and their ligands, which can cause multiple, nonspecific toxic effects and side effects in some patients, which are expensive to treat [152,153]. In contrast, research on small-molecule immune checkpoint inhibitors is still in its infancy. Small-molecule inhibitors have the advantages of weak immune-related adverse reactions, good penetration in tumor cells, and high bioavailability [154,155]. Despite many exciting advantages, the regulation of the dimerization of transmembrane proteins using small-molecule inhibitors still faces several challenges. First, although protein dimerization is thought to act as an on–off switch for cascade signaling, small-molecule drug-regulatory protein dimerization is not yet universally applicable, and most current research targets are immune checkpoint proteins. Secondly, current research on these small-molecule drugs is still in the initial stage, and the mechanism and characteristics of the drugs have not been fully understood, so the development cost is high; the yield is low; and the time span is long. Third, because these small-molecule inhibitors inhibit signaling pathways through indirect action, their effects are generally weaker than those of antibodies, and there are challenges to increasing their efficacy to antibody levels.

Indeed, cancer immune escape and immune tolerance might overpower drug monotherapy; then, based on biomarker analysis, targeted combinational treatments might be needed to achieve a meaningful immune response. Several promising novel immunomodulatory therapies and their optimal combinations are currently under clinical investigation. However, severe immune-related side effects may be the result of disinhibiting the brakes of the immune system, and their management must be contemplated. In conclusion, the patient-specific configuration of immune-system–cancer-cell interactions and the specific immune escape mechanism will need to be understood to guide personalized treatment options for cancer immunotherapy.

Author Contributions: L.L. conceived and designed the review; J.L. contributed new ideas; L.L. and J.L. wrote the draft of the manuscript. All authors have read and agreed to the published version of the manuscript.

Funding: This research was funded by National Key R&D Program of China (2020YFA0210800), National Natural Science Foundation of China (Nos. 21974022 and 22027805), Major Project of Science and Technology of Fujian Province (2020HZ06006).

Institutional Review Board Statement: Not applicable.

Informed Consent Statement: Not applicable.

Data Availability Statement: Not applicable.

Conflicts of Interest: There are no conflicts of interest to declare. The funders had no role in the design or implementation of the research.

References

1. Smith, S.M. Strategies for the purification of membrane proteins. In *Protein Chromatography*; Springer: Berlin/Heidelberg, Germany, 2011; pp. 485–496. [CrossRef]
2. Schmit, K.; Michiels, C. TMEM proteins in cancer: A review. *Front. Pharmacol.* **2018**, *9*, 1345. [CrossRef] [PubMed]
3. Vinothkumar, K.R.; Henderson, R. Structures of membrane proteins. *Q. Rev. Biophys.* **2010**, *43*, 65–158. [CrossRef] [PubMed]
4. Huang, Y.; Liang, D.; Liu, J.; Zeng, J.; Zeng, Y. The Breakthroughs in Cancer Immune Checkpoint Based Therapy: A Review of Development in Immune Checkpoint Study and its Application. *Comb. Chem. High Throughput Screen.* **2017**, *20*, 430–439. [CrossRef] [PubMed]
5. Fan, Y.; Zhang, C.; Jin, S.; Gao, Z.; Cao, J.; Wang, A.; Li, D.; Wang, Q.; Sun, X.; Bai, D. Progress of immune checkpoint therapy in the clinic (Review). *Oncol. Rep.* **2019**, *41*, 3–14. [CrossRef]
6. Litak, J.; Grochowski, C.; Litak, J.; Osuchowska, I.; Gosik, K.; Radzikowska, E.; Kamieniak, P.; Rolinski, J. TLR-4 Signaling vs. Immune Checkpoints, miRNAs Molecules, Cancer Stem Cells, and Wingless-Signaling Interplay in Glioblastoma Multiforme-Future Perspectives. *Int. J. Mol. Sci.* **2020**, *21*, 3114. [CrossRef]
7. Ying, H.; Zhang, X.; Duan, Y.; Lao, M.; Xu, J.; Yang, H.; Liang, T.; Bai, X. Non-cytomembrane PD-L1: An atypical target for cancer. *Pharmacol. Res.* **2021**, *170*, 105741. [CrossRef]
8. Sugiura, D.; Maruhashi, T.; Okazaki, I.M.; Shimizu, K.; Maeda, T.K.; Takemoto, T.; Okazaki, T. Restriction of PD-1 function by cis-PD-L1/CD80 interactions is required for optimal T cell responses. *Science* **2019**, *364*, 558–566. [CrossRef]
9. Park, S.N.; Noh, K.T.; Jeong, Y.I.; Jung, I.D.; Kang, H.K.; Cha, G.S.; Lee, S.J.; Seo, J.K.; Kang, D.H.; Hwang, T.H.; et al. Rhamnogalacturonan II is a Toll-like receptor 4 agonist that inhibits tumor growth by activating dendritic cell-mediated $CD8^+$ T cells. *Exp. Mol. Med.* **2013**, *45*, e8. [CrossRef]
10. Khan, M.; Arooj, S.; Wang, H. NK Cell-Based Immune Checkpoint Inhibition. *Front. Immunol.* **2020**, *11*, 167. [CrossRef]
11. Godal, R.; Bachanova, V.; Gleason, M.; McCullar, V.; Yun, G.H.; Cooley, S.; Verneris, M.R.; McGlave, P.B.; Miller, J.S. Natural killer cell killing of acute myelogenous leukemia and acute lymphoblastic leukemia blasts by killer cell immunoglobulin-like receptor-negative natural killer cells after NKG2A and LIR-1 blockade. *Biol. Blood Marrow Transplant. J. Am. Soc. Blood Marrow Transplant.* **2010**, *16*, 612–621. [CrossRef]
12. Moretta, A.; Bottino, C.; Vitale, M.; Pende, D.; Biassoni, R.; Mingari, M.C.; Moretta, L. Receptors for HLA class-I molecules in human natural killer cells. *Annu. Rev. Immunol.* **1996**, *14*, 619–648. [CrossRef] [PubMed]
13. Moretta, L.; Mingari, M.C.; Pende, D.; Bottino, C.; Biassoni, R.; Moretta, A. The molecular basis of natural killer (NK) cell recognition and function. *J. Clin. Immunol.* **1996**, *16*, 243–253. [CrossRef]
14. Wagtmann, N.; Rajagopalan, S.; Winter, C.C.; Peruzzi, M.; Long, E.O. Killer cell inhibitory receptors specific for HLA-C and HLA-B identified by direct binding and by functional transfer. *Immunity* **1995**, *3*, 801–809. [CrossRef]
15. Clements, C.S.; Kjer-Nielsen, L.; McCluskey, J.; Rossjohn, J. Structural studies on HLA-G: Implications for ligand and receptor binding. *Hum. Immunol.* **2007**, *68*, 220–226. [CrossRef]
16. Hò, G.T.; Celik, A.A.; Huyton, T.; Hiemisch, W.; Blasczyk, R.; Simper, G.S.; Bade-Doeding, C. NKG2A/CD94 Is a New Immune Receptor for HLA-G and Distinguishes Amino Acid Differences in the HLA-G Heavy Chain. *Int. J. Mol. Sci.* **2020**, *21*, 4362. [CrossRef] [PubMed]
17. Wang, X.; Xiong, H.; Ning, Z. Implications of NKG2A in immunity and immune-mediated diseases. *Front. Immunol.* **2022**, *13*, 960852. [CrossRef]
18. Brown, J.A.; Dorfman, D.M.; Ma, F.R.; Sullivan, E.L.; Munoz, O.; Wood, C.R.; Greenfield, E.A.; Freeman, G.J. Blockade of programmed death-1 ligands on dendritic cells enhances T cell activation and cytokine production. *J. Immunol.* **2003**, *170*, 1257–1266. [CrossRef]
19. Greenwald, R.J.; Freeman, G.J.; Sharpe, A.H. The B7 family revisited. *Annu. Rev. Immunol.* **2005**, *23*, 515. [CrossRef] [PubMed]
20. Nishimura, H.; Honjo, T. PD-1: An inhibitory immunoreceptor involved in peripheral tolerance. *Trends Immunol.* **2001**, *22*, 265–268. [CrossRef]
21. Dong, H.; Zhu, G.; Tamada, K.; Chen, L. B7-H1, a third member of the B7 family, co-stimulates T-cell proliferation and interleukin-10 secretion. *Nat. Med.* **1999**, *5*, 1365–1369. [CrossRef]

22. Lin, D.Y.-W.; Tanaka, Y.; Iwasaki, M.; Gittis, A.G.; Su, H.-P.; Mikami, B.; Okazaki, T.; Honjo, T.; Minato, N.; Garboczi, D.N. The PD-1/PD-L1 complex resembles the antigen-binding Fv domains of antibodies and T cell receptors. *Proc. Natl. Acad. Sci. USA* **2008**, *105*, 3011–3016. [CrossRef] [PubMed]
23. Shi, D.; Zhou, S.; Liu, X.; Zhao, C.; Liu, H.; Yao, X. Understanding the structural and energetic basis of PD-1 and monoclonal antibodies bound to PD-L1: A molecular modeling perspective. *Biochim. et Biophys. Acta (BBA)-Gen. Subj.* **2018**, *1862*, 576–588. [CrossRef] [PubMed]
24. Zak, K.M.; Kitel, R.; Przetocka, S.; Golik, P.; Guzik, K.; Musielak, B.; Dömling, A.; Dubin, G.; Holak, T.A.J.S. Structure of the complex of human programmed death 1, PD-1, and its ligand PD-L1. *Structure* **2015**, *23*, 2341–2348. [CrossRef] [PubMed]
25. Alvarez, M.; Simonetta, F.; Baker, J.; Morrison, A.R.; Wenokur, A.S.; Pierini, A.; Berraondo, P.; Negrin, R.S. Indirect impact of PD-1/PD-L1 blockade on a murine model of NK cell exhaustion. *Front. Immunol.* **2020**, *11*, 7. [CrossRef] [PubMed]
26. Wherry, E.J.; Kurachi, M.J.N.R.I. Molecular and cellular insights into T cell exhaustion. *Nat. Rev. Immunol.* **2015**, *15*, 486–499. [CrossRef]
27. Swaika, A.; Hammond, W.A.; Joseph, R.W. Current state of anti-PD-L1 and anti-PD-1 agents in cancer therapy. *Mol. Immunol.* **2015**, *67*, 4–17. [CrossRef]
28. Xing, Y.; Hogquist, K. T-cell tolerance: Central and peripheral. *Cold Spring Harb. Perspect. Biol.* **2012**, *4*, a006957. [CrossRef]
29. Ramos-Casals, M.; Brahmer, J.R.; Callahan, M.K.; Flores-Chávez, A.; Keegan, N.; Khamashta, M.A.; Lambotte, O.; Mariette, X.; Prat, A.; Suárez-Almazor, M.E. Immune-related adverse events of checkpoint inhibitors. *Nat. Rev. Dis. Prim.* **2020**, *6*, 38. [CrossRef]
30. Sibaud, V. Dermatologic reactions to immune checkpoint inhibitors. *Am. J. Clin. Dermatol.* **2018**, *19*, 345–361. [CrossRef]
31. Hua, Z.; Hou, B. TLR signaling in B-cell development and activation. *Cell. Mol. Immunol.* **2013**, *10*, 103–106. [CrossRef]
32. Shi, Y.-J.; Gong, H.-F.; Zhao, Q.-Q.; Liu, X.-S.; Liu, C.; Wang, H.J.T.L. Critical role of toll-like receptor 4 (TLR4) in dextran sulfate sodium (DSS)-Induced intestinal injury and repair. *Toxicol. Lett.* **2019**, *315*, 23–30. [CrossRef] [PubMed]
33. Pahwa, R.; Devaraj, S.; Jialal, I. The effect of the accessory proteins, soluble CD14 and lipopolysaccharide-binding protein on Toll-like receptor 4 activity in human monocytes and adipocytes. *Int. J. Obes.* **2016**, *40*, 907–911. [CrossRef] [PubMed]
34. Mishra, V.; Pathak, C. Human Toll-Like Receptor 4 (hTLR4): Structural and functional dynamics in cancer. *Int. J. Biol. Macromol.* **2019**, *122*, 425–451. [CrossRef] [PubMed]
35. Kim, H.M.; Park, B.S.; Kim, J.-I.; Kim, S.E.; Lee, J.; Oh, S.C.; Enkhbayar, P.; Matsushima, N.; Lee, H.; Yoo, O.J.J.C. Crystal structure of the TLR4-MD-2 complex with bound endotoxin antagonist Eritoran. *Cell* **2007**, *130*, 906–917. [CrossRef] [PubMed]
36. He, L.; Zhang, L.; Li, Z.; Zhang, Q. The roles of toll-like receptors in carcinogenesis and cancer immunotherapy. *Chin.-Ger. J. Clin. Oncol.* **2010**, *9*, 118–120. [CrossRef]
37. Yirmiya, R.; Goshen, I.J.B. Immune modulation of learning, memory, neural plasticity and neurogenesis. *Brain Behav. Immun.* **2011**, *25*, 181–213. [CrossRef] [PubMed]
38. Hotamisligil, G.S.J.N. Inflammation, metaflammation and immunometabolic disorders. *Nature* **2017**, *542*, 177–185. [CrossRef]
39. Kluwe, J.; Mencin, A.; Schwabe, R.F. Toll-like receptors, wound healing, and carcinogenesis. *J. Mol. Med.* **2009**, *87*, 125–138. [CrossRef]
40. Maruyama, K.; Selmani, Z.; Ishii, H.; Yamaguchi, K. Innate immunity and cancer therapy. *Int. Immunopharmacol.* **2011**, *11*, 350–357. [CrossRef]
41. Shetab Boushehri, M.A.; Lamprecht, A. TLR4-based immunotherapeutics in cancer: A review of the achievements and shortcomings. *Mol. Pharm.* **2018**, *15*, 4777–4800. [CrossRef]
42. Foroni, I.; Couto, A.R.; Bettencourt, B.F.; Santos, M.; Lima, M.; Bruges-Armas, J.J.H.; Diseases, A.I. HLA-E, HLA-F and HLA-G—The non-classical side of the MHC cluster. In *HLA and Associated Important Diseases*; IntechOpen: London, UK, 2014; Volume 3, pp. 61–109. [CrossRef]
43. Carosella, E.D.; Rouas-Freiss, N.; Tronik-Le Roux, D.; Moreau, P.; LeMaoult, J. HLA-G: An immune checkpoint molecule. *Adv. Immunol.* **2015**, *127*, 33–144. [CrossRef]
44. de Kruijf, E.M.; Sajet, A.; van Nes, J.G.; Natanov, R.; Putter, H.; Smit, V.T.; Liefers, G.J.; van den Elsen, P.J.; van de Velde, C.J.; Kuppen, P.J.K. HLA-E and HLA-G expression in classical HLA class I-negative tumors is of prognostic value for clinical outcome of early breast cancer patients. *J. Immunol.* **2010**, *185*, 7452–7459. [CrossRef]
45. Shen, X.; Wang, P.; Dai, P.; Jin, B.; Tong, Y.; Lin, H.; Shi, G. Correlation between human leukocyte antigen-G expression and clinical parameters in oral squamous cell carcinoma. *Indian J. Cancer* **2018**, *55*, 340. [CrossRef] [PubMed]
46. Zeestraten, E.; Reimers, M.; Saadatmand, S.; Dekker, J.T.; Liefers, G.; Van Den Elsen, P.; Van De Velde, C.; Kuppen, P.J.K. Combined analysis of HLA class I, HLA-E and HLA-G predicts prognosis in colon cancer patients. *Br. J. Cancer* **2014**, *110*, 459–468. [CrossRef] [PubMed]
47. Arnaiz-Villena, A.; Juarez, I.; Suarez-Trujillo, F.; López-Nares, A.; Vaquero, C.; Palacio-Gruber, J.; Martin-Villa, J.M. HLA-G: Function, polymorphisms and pathology. *Int. J. Immunogenet.* **2021**, *48*, 172–192. [CrossRef] [PubMed]
48. Diehl, M.; Münz, C.; Keilholz, W.; Stevanović, S.; Holmes, N.; Loke, Y.W.; Rammensee, H.-G.J.C.B. Nonclassical HLA-G molecules are classical peptide presenters. *Curr. Biol.* **1996**, *6*, 305–314. [CrossRef]
49. Paul, P.; Cabestre, F.A.; Lefebvre, S.; Khalil-Daher, I.; Vazeux, G.; Quiles, R.M.M.; Bermond, F.; Dausset, J.; Carosella, E.D. Identification of HLA-G7 as a new splice variant of the HLA-G mRNA and expression of soluble HLA-G5,-G6, and-G7 transcripts in human transfected cells. *Hum. Immunol.* **2000**, *61*, 1138–1149. [CrossRef] [PubMed]

50. Takahashi, A.; Kuroki, K.; Okabe, Y.; Kasai, Y.; Matsumoto, N.; Yamada, C.; Takai, T.; Ose, T.; Kon, S.; Matsuda, T. The immunosuppressive effect of domain-deleted dimer of HLA-G2 isoform in collagen-induced arthritis mice. *Hum. Immunol.* **2016**, *77*, 754–759. [CrossRef] [PubMed]
51. Lin, X.-X.; Xie, Y.-M.; Zhao, S.-J.; Liu, C.-Y.; Mor, G.; Liao, A.-H. Human leukocyte antigens: The unique expression in trophoblasts and their crosstalk with local immune cells. *Int. J. Biol. Sci.* **2022**, *18*, 4043–4052. [CrossRef] [PubMed]
52. Moffett, A.; Colucci, F. Co-evolution of NK receptors and HLA ligands in humans is driven by reproduction. *Immunol. Rev.* **2015**, *267*, 283–297. [CrossRef] [PubMed]
53. Xu, X.; Zhou, Y.; Wei, H. Roles of HLA-G in the maternal-fetal immune microenvironment. *Front. Immunol.* **2020**, *11*, 592010. [CrossRef] [PubMed]
54. Saurabh, A.; Chakraborty, S.; Kumar, P.; Mohan, A.; Bhatnagar, A.K.; Rishi, N.; Mitra, D.K.J.T. Inhibiting HLA-G restores IFN-γ and TNF-α producing T cell in pleural tuberculosis. *Tuberculosis* **2018**, *109*, 69–79. [CrossRef] [PubMed]
55. Martín-Villa, J.M.; Vaquero-Yuste, C.; Molina-Alejandre, M.; Juarez, I.; Suárez-Trujillo, F.; López-Nares, A.; Palacio-Gruber, J.; Barrera-Gutiérrez, L.; Fernández-Cruz, E.; Rodríguez-Sainz, C. HLA-G: Too Much or Too Little? Role in Cancer and Autoimmune Disease. *Front. Immunol.* **2022**, *67*, 796054. [CrossRef] [PubMed]
56. Ristich, V.; Liang, S.; Zhang, W.; Wu, J.; Horuzsko, A. Tolerization of dendritic cells by HLA-G. *Eur. J. Immunol.* **2005**, *35*, 1133–1142. [CrossRef] [PubMed]
57. Attia, J.V.; Dessens, C.E.; van de Water, R.; Houvast, R.D.; Kuppen, P.J.; Krijgsman, D. The molecular and functional characteristics of HLA-G and the interaction with its receptors: Where to intervene for cancer immunotherapy? *Int. J. Mol. Sci.* **2020**, *21*, 8678. [CrossRef] [PubMed]
58. Muntasell, A.; Ochoa, M.C.; Cordeiro, L.; Berraondo, P.; de Cerio, A.L.-D.; Cabo, M.; López-Botet, M.; Melero, I. Targeting NK-cell checkpoints for cancer immunotherapy. *Curr. Opin. Immunol.* **2017**, *45*, 73–81. [CrossRef]
59. Li, F.; Wei, H.; Wei, H.; Gao, Y.; Xu, L.; Yin, W.; Sun, R.; Tian, Z. Blocking the Natural Killer Cell Inhibitory Receptor NKG2A Increases Activity of Human Natural Killer Cells and Clears Hepatitis B Virus Infection in Mice. *Gastroenterology* **2013**, *144*, 392–401. [CrossRef]
60. Rapaport, A.S.; Schriewer, J.; Gilfillan, S.; Hembrador, E.; Crump, R.; Plougastel, B.F.; Wang, Y.; Le Friec, G.; Gao, J.; Cella, M.; et al. The Inhibitory Receptor NKG2A Sustains Virus-Specific $CD8^+$ T Cells in Response to a Lethal Poxvirus Infection. *Immunity* **2015**, *43*, 1112–1124. [CrossRef]
61. Plougastel, B.; Jones, T.; Trowsdale, J. Genomic structure, chromosome location, and alternative splicing of the humanNKG2A gene. *Immunogenetics* **1996**, *44*, 286–291. [CrossRef]
62. Iwaszko, M.; Bogunia-Kubik, K. Clinical Significance of the HLA-E and CD94/NKG2 Interaction. *Arch. Immunol. et Ther. Exp.* **2011**, *59*, 353–367. [CrossRef]
63. Mingari, M.C.; Pietra, G.; Moretta, L. Immune Checkpoint Inhibitors: Anti-NKG2A Antibodies on Board. *Trends Immunol.* **2019**, *40*, 83–85. [CrossRef]
64. Tang, L.; Bai, J.; Chung, C.-S.; Lomas-Neira, J.; Chen, Y.; Huang, X.; Ayala, A. Programmed Cell Death Receptor Ligand 1 Modulates the Regulatory T Cells' Capacity to Repress Shock/Sepsis-Induced Indirect Acute Lung Injury by Recruiting Phosphatase Src Homology Region 2 Domain-Containing Phosphatase 1. *Shock* **2015**, *43*, 47–54. [CrossRef]
65. Wen, L.-Z.; Ding, K.; Wang, Z.-R.; Ding, C.-H.; Lei, S.-J.; Liu, J.-P.; Yin, C.; Hu, P.-F.; Ding, J.; Chen, W.-S.; et al. SHP-1 Acts as a Tumor Suppressor in Hepatocarcinogenesis and HCC Progression. *Cancer Res.* **2018**, *78*, 4680–4691. [CrossRef]
66. Mahapatra, S.; Mace, E.M.; Minard, C.G.; Forbes, L.R.; Vargas-Hernandez, A.; Duryea, T.K.; Makedonas, G.; Banerjee, P.P.; Shearer, W.; Orange, J.S. High-resolution phenotyping identifies NK cell subsets that distinguish healthy children from adults. *PLoS ONE* **2017**, *12*, e0181134. [CrossRef]
67. Kamiya, T.; Seow, S.V.; Wong, D.; Robinson, M.; Campana, D. Blocking expression of inhibitory receptor NKG2A overcomes tumor resistance to NK cells. *J. Clin. Investig.* **2019**, *129*, 2094–2106. [CrossRef]
68. Wang, X.; Cui, Y.; Luo, G.; Wang, Q.; Hu, J.; He, W.; Yuan, J.; Zhou, J.; Wu, Y.; Sun, X.; et al. Activated mouse $CD4^+Foxp3-$ T cells facilitate melanoma metastasis via Qa-1-dependent suppression of NK-cell cytotoxicity. *Cell Res.* **2012**, *22*, 1696–1706. [CrossRef] [PubMed]
69. Zhang, C.; Wang, X.-M.; Li, S.-R.; Twelkmeyer, T.; Wang, W.-H.; Zhang, S.-Y.; Wang, S.-F.; Chen, J.-Z.; Jin, X.; Wu, Y.-Z.; et al. NKG2A is a NK cell exhaustion checkpoint for HCV persistence. *Nat. Commun.* **2019**, *10*, 1507. [CrossRef] [PubMed]
70. Borst, L.; van der Burg, S.H.; van Hall, T. The NKG2A–HLA-E Axis as a Novel Checkpoint in the Tumor Microenvironment. *Clin. Cancer Res.* **2020**, *26*, 5549–5556. [CrossRef] [PubMed]
71. Van Hall, T.; André, P.; Horowitz, A.; Ruan, D.F.; Borst, L.; Zerbib, R.; Narni-Mancinelli, E.; Van Der Burg, S.H.; Vivier, E. Monalizumab: Inhibiting the novel immune checkpoint NKG2A. *J. Immunother. Cancer* **2019**, *7*, 263. [CrossRef]
72. Li, Q.; Cai, S.; Li, M.; Zhou, X.; Wu, G.; Kang, K.; Yuan, J.; Wang, R.; Huyan, T.; Zhang, W. Natural killer cell exhaustion in lung cancer. *Int. Immunopharmacol.* **2021**, *96*, 107764. [CrossRef]
73. van Montfoort, N.; Borst, L.; Korrer, M.J.; Sluijter, M.; Marijt, K.A.; Santegoets, S.J.; van Ham, V.J.; Ehsan, I.; Charoentong, P.; André, P.; et al. NKG2A Blockade Potentiates CD8 T Cell Immunity Induced by Cancer Vaccines. *Cell* **2018**, *175*, 1744–1755.e15. [CrossRef]

74. Sheu, B.-C.; Chiou, S.-H.; Lin, H.-H.; Chow, S.-N.; Huang, S.-C.; Ho, H.-N.; Hsu, S.-M. Up-regulation of Inhibitory Natural Killer Receptors CD94/NKG2A with Suppressed Intracellular Perforin Expression of Tumor-Infiltrating CD8+ T Lymphocytes in Human Cervical Carcinoma. *Cancer Res.* **2005**, *65*, 2921–2929. [CrossRef] [PubMed]
75. McWilliams, E.M.; Mele, J.M.; Cheney, C.; Timmerman, E.A.; Fiazuddin, F.; Strattan, E.J.; Mo, X.; Byrd, J.C.; Muthusamy, N.; Awan, F.T. Therapeutic CD94/NKG2A blockade improves natural killer cell dysfunction in chronic lymphocytic leukemia. *OncoImmunology* **2016**, *5*, e1226720. [CrossRef]
76. Zak, K.M.; Grudnik, P.; Guzik, K.; Zieba, B.J.; Musielak, B.; Dömling, A.; Dubin, G.; Holak, T.A. Structural basis for small molecule targeting of the programmed death ligand 1 (PD-L1). *Oncotarget* **2016**, *7*, 30323–30335. [CrossRef]
77. Bailly, C.; Vergoten, G. Protein homodimer sequestration with small molecules: Focus on PD-L1. *Biochem. Pharmacol.* **2020**, *174*, 113821. [CrossRef] [PubMed]
78. Zak, K.M.; Grudnik, P.; Magiera-Mularz, K.; Dömling, A.; Dubin, G.; Holak, T.A. Structural Biology of the Immune Checkpoint Receptor PD-1 and Its Ligands PD-L1/PD-L2. *Structure* **2017**, *25*, 1163–1174. [CrossRef] [PubMed]
79. Perry, E.; Mills, J.J.; Zhao, B.; Wang, F.; Sun, Q.; Christov, P.P.; Tarr, J.C.; Rietz, T.A.; Olejniczak, E.T.; Lee, T.; et al. Fragment-based screening of programmed death ligand 1 (PD-L1). *Bioorganic Med. Chem. Lett.* **2019**, *29*, 786–790. [CrossRef]
80. Okazaki, T.; Honjo, T. PD-1 and PD-1 ligands: From discovery to clinical application. *Int. Immunol.* **2007**, *19*, 813–824. [CrossRef]
81. Schöniger, S.; Jasani, B. The PD-1/PD-L1 Pathway: A Perspective on Comparative Immuno-Oncology. *Animals* **2022**, *12*, 2661. [CrossRef]
82. Wang, Q.; Bardhan, K.; Boussiotis, V.A.; Patsoukis, N. The PD-1 Interactome. *Adv. Biol.* **2021**, *5*, e2100758. [CrossRef]
83. Hui, E.; Cheung, J.; Zhu, J.; Su, X.; Taylor, M.J.; Wallweber, H.A.; Sasmal, D.K.; Huang, J.; Kim, J.M.; Mellman, I.; et al. T cell costimulatory receptor CD28 is a primary target for PD-1-mediated inhibition. *Science* **2017**, *355*, 1428–1433. [CrossRef] [PubMed]
84. Sheppard, K.-A.; Fitz, L.J.; Lee, J.M.; Benander, C.; George, J.A.; Wooters, J.; Qiu, Y.; Jussif, J.M.; Carter, L.L.; Wood, C.R.; et al. PD-1 inhibits T-cell receptor induced phosphorylation of the ZAP70/CD3ζ signalosome and downstream signaling to PKCθ. *FEBS Lett.* **2004**, *574*, 37–41. [CrossRef] [PubMed]
85. Yokosuka, T.; Takamatsu, M.; Kobayashi-Imanishi, W.; Hashimoto-Tane, A.; Azuma, M.; Saito, T. Programmed cell death 1 forms negative costimulatory microclusters that directly inhibit T cell receptor signaling by recruiting phosphatase SHP2. *J. Exp. Med.* **2012**, *209*, 1201–1217. [CrossRef] [PubMed]
86. Huang, R.Y.-C.; Wang, Y.; Jhatakia, A.D.; Deng, A.X.; Bee, C.; Deshpande, S.; Rangan, V.S.; Bezman, N.; Gudmundsson, O.; Chen, G. Higher-Order Structure Characterization of NKG2A/CD94 Protein Complex and Anti-NKG2A Antibody Binding Epitopes by Mass Spectrometry-Based Protein Footprinting Strategies. *J. Am. Soc. Mass Spectrom.* **2021**, *32*, 1567–1574. [CrossRef]
87. Kang-Pettinger, T.; Walker, K.; Brown, R.; Cowan, R.; Wright, H.; Baravalle, R.; Waters, L.C.; Muskett, F.W.; Bowler, M.W.; Sawmynaden, K.; et al. Identification, binding, and structural characterization of single domain anti-PD-L1 antibodies inhibitory of immune regulatory proteins PD-1 and CD80. *J. Biol. Chem.* **2022**, *299*, 102769. [CrossRef]
88. Zhao, Y.; Lee, C.K.; Lin, C.-H.; Gassen, R.B.; Xu, X.; Huang, Z.; Xiao, C.; Bonorino, C.; Lu, L.-F.; Bui, J.D.; et al. PD-L1:CD80 Cis-Heterodimer Triggers the Co-stimulatory Receptor CD28 While Repressing the Inhibitory PD-1 and CTLA-4 Pathways. *Immunity* **2019**, *51*, 1059–1073.e9. [CrossRef] [PubMed]
89. Wang, T.; Cai, S.; Cheng, Y.; Zhang, W.; Wang, M.; Sun, H.; Guo, B.; Li, Z.; Xiao, Y.; Jiang, S. Discovery of Small-Molecule Inhibitors of the PD-1/PD-L1 Axis That Promote PD-L1 Internalization and Degradation. *J. Med. Chem.* **2022**, *65*, 3879–3893. [CrossRef]
90. Mineev, K.S.; Goncharuk, S.A.; Goncharuk, M.V.; Volynsky, P.E.; Novikova, E.V.; Aresinev, A.S. Spatial structure of TLR4 transmembrane domain in bicelles provides the insight into the receptor activation mechanism. *Sci. Rep.* **2017**, *7*, 6864. [CrossRef]
91. Toshchakov, V.Y.; Szmacinski, H.; Couture, L.A.; Lakowicz, J.R.; Vogel, S.N. Targeting TLR4 Signaling by TLR4 Toll/IL-1 Receptor Domain-Derived Decoy Peptides: Identification of the TLR4 Toll/IL-1 Receptor Domain Dimerization Interface. *J. Immunol.* **2011**, *186*, 4819–4827. [CrossRef]
92. Reuven, E.M.; Fink, A.; Shai, Y. Regulation of innate immune responses by transmembrane interactions: Lessons from the TLR family. *Biochim. et Biophys. Acta (BBA)-Biomembr.* **2014**, *1838*, 1586–1593. [CrossRef]
93. Kargas, V.; Marzinek, J.K.; Holdbrook, D.A.; Yin, H.; Ford, R.C.; Bond, P.J. A polar SxxS motif drives assembly of the transmembrane domains of Toll-like receptor 4. *Biochim. et Biophys. Acta (BBA)-Biomembr.* **2017**, *1859*, 2086–2095. [CrossRef] [PubMed]
94. Kashani, B.; Zandi, Z.; Pourbagheri-Sigaroodi, A.; Bashash, D.; Ghaffari, S.H. The role of toll-like receptor 4 (TLR4) in cancer progression: A possible therapeutic target? *J. Cell. Physiol.* **2020**, *236*, 4121–4137. [CrossRef] [PubMed]
95. Krüger, C.L.; Zeuner, M.-T.; Cottrell, G.S.; Widera, D.; Heilemann, M. Quantitative single-molecule imaging of TLR4 reveals ligand-specific receptor dimerization. *Sci. Signal.* **2017**, *10*, eaan1308. [CrossRef]
96. Soares, J.-B.; Pimentel-Nunes, P.; Jr, R.R.-A.; Leite-Moreira, A. The role of lipopolysaccharide/toll-like receptor 4 signaling in chronic liver diseases. *Hepatol. Int.* **2010**, *4*, 659–672. [CrossRef]
97. Chiba, S.; Baghdadi, M.; Akiba, H.; Yoshiyama, H.; Kinoshita, I.; Dosaka-Akita, H.; Fujioka, Y.; Ohba, Y.; Gorman, J.V.; Colgan, J.D.; et al. Tumor-infiltrating DCs suppress nucleic acid–mediated innate immune responses through interactions between the receptor TIM-3 and the alarmin HMGB1. *Nat. Immunol.* **2012**, *13*, 832–842. [CrossRef] [PubMed]
98. He, S.-J.; Cheng, J.; Feng, X.; Yu, Y.; Tian, L.; Huang, Q. The dual role and therapeutic potential of high-mobility group box 1 in cancer. *Oncotarget* **2017**, *8*, 64534–64550. [CrossRef]

99. Kang, R.; Zhang, Q.; Zeh, H.J., 3rd; Lotze, M.T.; Tang, D. HMGB1 in Cancer: Good, Bad, or Both? *Clin. Cancer Res.* **2013**, *19*, 4046–4057. [CrossRef]
100. Krysko, O.; Løve Aaes, T.; Bachert, C.; Vandenabeele, P.; Krysko, D.V. Many faces of DAMPs in cancer therapy. *Cell Death Dis.* **2013**, *4*, e631. [CrossRef]
101. Ding, J.; Cui, X.; Liu, Q. Emerging role of HMGB1 in lung diseases: Friend or foe. *J. Cell. Mol. Med.* **2016**, *21*, 1046–1057. [CrossRef]
102. Lee, C.-H.; Yoon, S.-J.; Lee, S.-M. Chlorogenic Acid Attenuates High Mobility Group Box 1 (HMGB1) and Enhances Host Defense Mechanisms in Murine Sepsis. *Mol. Med.* **2012**, *18*, 1437–1448. [CrossRef]
103. Tadie, J.-M.; Bae, H.-B.; Deshane, J.S.; Bell, C.P.; Lazarowski, E.R.; Chaplin, D.D.; Thannickal, V.J.; Abraham, E.; Zmijewski, J.W. Toll-Like Receptor 4 Engagement Inhibits Adenosine 5′-Monophosphate–Activated Protein Kinase Activation through a High Mobility Group Box 1 Protein–Dependent Mechanism. *Mol. Med.* **2012**, *18*, 659–668. [CrossRef]
104. Poulain, L.; Richard, V.; Lévy, P.; Dematteis, M.; Arnaud, C. Toll-Like Receptor-4 Mediated Inflammation Is Involved in the Cardiometabolic Alterations Induced by Intermittent Hypoxia. *Mediat. Inflamm.* **2015**, *2015*, 620258. [CrossRef] [PubMed]
105. Insua-Rodríguez, J.; Oskarsson, T. The extracellular matrix in breast cancer. *Adv. Drug Deliv. Rev.* **2016**, *97*, 41–55. [CrossRef] [PubMed]
106. Tivari, S.; Lu, H.; Dasgupta, T.; De Lorenzo, M.S.; Wieder, R. Reawakening of dormant estrogen-dependent human breast cancer cells by bone marrow stroma secretory senescence. *Cell Commun. Signal.* **2018**, *16*, 48. [CrossRef] [PubMed]
107. Pankov, R.; Yamada, K.M. Fibronectin at a glance. *J. Cell Sci.* **2002**, *115*, 3861–3863. [CrossRef]
108. Wu, C. Roles of integrins in fibronectin matrix assembly. *Histol. Histopathol.* **1997**, *12*, 233–240. [PubMed]
109. Julier, Z.; Martino, M.M.; de Titta, A.; Jeanbart, L.; Hubbell, J.A. The TLR4 Agonist Fibronectin Extra Domain A is Cryptic, Exposed by Elastase-2; use in a fibrin matrix cancer vaccine. *Sci. Rep.* **2015**, *5*, 8569. [CrossRef] [PubMed]
110. Yoon, S.-I.; Hong, M.; Wilson, I.A. An unusual dimeric structure and assembly for TLR4 regulator RP105–MD-1. *Nat. Struct. Mol. Biol.* **2011**, *18*, 1028–1035. [CrossRef] [PubMed]
111. Shiroishi, M.; Kuroki, K.; Ose, T.; Rasubala, L.; Shiratori, I.; Arase, H.; Tsumoto, K.; Kumagai, I.; Kohda, D.; Maenaka, K. Efficient Leukocyte Ig-like Receptor Signaling and Crystal Structure of Disulfide-linked HLA-G Dimer. *J. Biol. Chem.* **2006**, *281*, 10439–10447. [CrossRef]
112. Boyson, J.E.; Erskine, R.; Whitman, M.C.; Chiu, M.; Lau, J.M.; Koopman, L.A.; Valter, M.M.; Angelisova, P.; Horejsí, V.; Strominger, J.L. Disulfide bond-mediated dimerization of HLA-G on the cell surface. *Proc. Natl. Acad. Sci. USA* **2002**, *99*, 16180–16185. [CrossRef]
113. Santos, S.; Powis, S.; Arosa, F.A. Misfolding of Major Histocompatibility Complex Class I Molecules in Activated T Cells Allows cis-Interactions with Receptors and Signaling Molecules and Is Associated with Tyrosine Phosphorylation. *J. Biol. Chem.* **2004**, *279*, 53062–53070. [CrossRef] [PubMed]
114. Kang, X.; Kim, J.; Deng, M.; John, S.; Chen, H.; Wu, G.; Phan, H.; Zhang, C.C. Inhibitory leukocyte immunoglobulin-like receptors: Immune checkpoint proteins and tumor sustaining factors. *Cell Cycle* **2016**, *15*, 25–40. [CrossRef] [PubMed]
115. Krijgsman, D.; Roelands, J.; Hendrickx, W.; Bedognetti, D.; Kuppen, P.J.K. HLA-G: A New Immune Checkpoint in Cancer? *Int. J. Mol. Sci.* **2020**, *21*, 4528. [CrossRef] [PubMed]
116. Hirayasu, K.; Arase, H. Functional and genetic diversity of leukocyte immunoglobulin-like receptor and implication for disease associations. *J. Hum. Genet.* **2015**, *60*, 703–708. [CrossRef] [PubMed]
117. Kuroki, K.; Matsubara, H.; Kanda, R.; Miyashita, N.; Shiroishi, M.; Fukunaga, Y.; Kamishikiryo, J.; Fukunaga, A.; Fukuhara, H.; Hirose, K.; et al. Structural and Functional Basis for LILRB Immune Checkpoint Receptor Recognition of HLA-G Isoforms. *J. Immunol.* **2019**, *203*, 3386–3394. [CrossRef] [PubMed]
118. Shiroishi, M.; Kuroki, K.; Rasubala, L.; Tsumoto, K.; Kumagai, I.; Kurimoto, E.; Kato, K.; Kohda, D.; Maenaka, K. Structural basis for recognition of the nonclassical MHC molecule HLA-G by the leukocyte Ig-like receptor B2 (LILRB2/LIR2/ILT4/CD85d). *Proc. Natl. Acad. Sci. USA* **2006**, *103*, 16412–16417. [CrossRef]
119. Wang, Q.; Song, H.; Cheng, H.; Qi, J.; Nam, G.; Tan, S.; Wang, J.; Fang, M.; Shi, Y.; Tian, Z.; et al. Structures of the four Ig-like domain LILRB2 and the four-domain LILRB1 and HLA-G1 complex. *Cell. Mol. Immunol.* **2019**, *17*, 966–975. [CrossRef]
120. Mandel, I.; Ziv, D.H.; Goldshtein, I.; Peretz, T.; Alishekevitz, D.; Dror, A.F.; Hakim, M.; Hashmueli, S.; Friedman, I.; Sapir, Y.; et al. BND-22, a first-in-class humanized ILT2-blocking antibody, promotes antitumor immunity and tumor regression. *J. Immunother. Cancer* **2022**, *10*, e004859. [CrossRef]
121. Lin, A.; Yan, W.-H. Heterogeneity of HLA-G Expression in Cancers: Facing the Challenges. *Front. Immunol.* **2018**, *9*, 2164. [CrossRef]
122. Zheng, G.; Jia, L.; Yang, A.-G. Roles of HLA-G/KIR2DL4 in Breast Cancer Immune Microenvironment. *Front. Immunol.* **2022**, *13*, 791975. [CrossRef]
123. Van Beneden, K.; De Creus, A.; Stevenaert, F.; Debacker, V.; Plum, J.; Leclercq, G. Expression of Inhibitory Receptors Ly49E and CD94/NKG2 on Fetal Thymic and Adult Epidermal TCR Vγ3 Lymphocytes. *J. Immunol.* **2002**, *168*, 3295–3302. [CrossRef] [PubMed]
124. Miller, J.S.; McCullar, V. Human natural killer cells with polyclonal lectin and immunoglobulinlike receptors develop from single hematopoietic stem cells with preferential expression of NKG2A and KIR2DL2/L3/S2. *Blood* **2001**, *98*, 705–713. [CrossRef] [PubMed]

125. Eugène, J.; Jouand, N.; Ducoin, K.; Dansette, D.; Oger, R.; Deleine, C.; Leveque, E.; Meurette, G.; Podevin, J.; Matysiak, T.; et al. The inhibitory receptor CD94/NKG2A on CD8+ tumor-infiltrating lymphocytes in colorectal cancer: A promising new druggable immune checkpoint in the context of HLAE/β2m overexpression. *Mod. Pathol.* **2020**, *33*, 468–482. [CrossRef] [PubMed]
126. Kaiser, B.K.; Pizarro, J.C.; Kerns, J.; Strong, R.K. Structural basis for NKG2A/CD94 recognition of HLA-E. *Proc. Natl. Acad. Sci. USA* **2008**, *105*, 6696–6701. [CrossRef] [PubMed]
127. Veuillen, C.; Aurran-Schleinitz, T.; Castellano, R.; Rey, J.; Mallet, F.; Orlanducci, F.; Pouyet, L.; Just-Landi, S.; Coso, D.; Ivanov, V.; et al. Primary B-CLL Resistance to NK Cell Cytotoxicity can be Overcome In Vitro and In Vivo by Priming NK Cells and Monoclonal Antibody Therapy. *J. Clin. Immunol.* **2012**, *32*, 632–646. [CrossRef]
128. Prašnikar, E.; Perdih, A.; Borišek, J. What a Difference an Amino Acid Makes: An All-Atom Simulation Study of Nonameric Peptides in Inhibitory HLA-E/NKG2A/CD94 Immune Complexes. *Front. Pharmacol.* **2022**, *13*, 925427. [CrossRef]
129. Ruibal, P.; Franken, K.L.M.C.; van Meijgaarden, K.E.; van Loon, J.J.F.; van der Steen, D.; Heemskerk, M.H.M.; Ottenhoff, T.H.M.; Joosten, S.A. Peptide Binding to HLA-E Molecules in Humans, Nonhuman Primates, and Mice Reveals Unique Binding Peptides but Remarkably Conserved Anchor Residues. *J. Immunol.* **2020**, *205*, 2861–2872. [CrossRef]
130. Chai, I.; Kornyeyev, D.; Hsieh, E.; Magombedze, G.; Stapleton, L.; Hung, M.; Kwon, H.J.; Stefanutti, E.; Belzile, J.; Czerwieniec, G.; et al. Effects of small molecule-induced dimerization on the programmed death ligand 1 protein life cycle. *Sci. Rep.* **2022**, *12*, 21286. [CrossRef]
131. Guzik, K.; Zak, K.M.; Grudnik, P.; Magiera, K.; Musielak, B.; Törner, R.; Skalniak, L.; Dömling, A.; Dubin, G.; Holak, T.A. Small-Molecule Inhibitors of the Programmed Cell Death-1/Programmed Death-Ligand 1 (PD-1/PD-L1) Interaction via Transiently Induced Protein States and Dimerization of PD-L1. *J. Med. Chem.* **2017**, *60*, 5857–5867. [CrossRef]
132. Liu, C.; Seeram, N.P.; Ma, H. Small molecule inhibitors against PD-1/PD-L1 immune checkpoints and current methodologies for their development: A review. *Cancer Cell Int.* **2021**, *21*, 239. [CrossRef]
133. Guo, Y.; Jin, Y.; Wang, B.; Liu, B. Molecular Mechanism of Small-Molecule Inhibitors in Blocking the PD-1/PD-L1 Pathway through PD-L1 Dimerization. *Int. J. Mol. Sci.* **2021**, *22*, 4766. [CrossRef] [PubMed]
134. Wang, F.; Ye, W.; Wang, S.; He, Y.; Zhong, H.; Wang, Y.; Zhu, Y.; Han, J.; Bing, Z.; Ji, S.; et al. Discovery of a new inhibitor targeting PD-L1 for cancer immunotherapy. *Neoplasia* **2021**, *23*, 281–293. [CrossRef] [PubMed]
135. Akira, S.; Takeda, K. Toll-like receptor signalling. *Nat. Rev. Immunol.* **2004**, *4*, 499–511. [CrossRef]
136. Ghochikyan, A.; Pichugin, A.; Bagaev, A.; Davtyan, A.; Hovakimyan, A.; Tukhvatulin, A.; Davtyan, H.; Shcheblyakov, D.; Logunov, D.; Chulkina, M.; et al. Targeting TLR-4 with a novel pharmaceutical grade plant derived agonist, Immunomax®, as a therapeutic strategy for metastatic breast cancer. *J. Transl. Med.* **2014**, *12*, 322. [CrossRef]
137. Huang, Z.; Yang, Y.; Jiang, Y.; Shao, J.; Sun, X.; Chen, J.; Dong, L.; Zhang, J. Anti-tumor immune responses of tumor-associated macrophages via toll-like receptor 4 triggered by cationic polymers. *Biomaterials* **2013**, *34*, 746–755. [CrossRef]
138. Ajith, A.; Portik-Dobos, V.; Nguyen-Lefebvre, A.T.; Callaway, C.; Horuzsko, D.D.; Kapoor, R.; Zayas, C.; Maenaka, K.; Mulloy, L.L.; Horuzsko, A. HLA-G dimer targets Granzyme B pathway to prolong human renal allograft survival. *FASEB J.* **2019**, *33*, 5220–5236. [CrossRef]
139. Zilberman, S.; Schenowitz, C.; Agaugué, S.; Favier, B.; Riteau, B.; Rouzier, R.; Carosella, E.D.; Rouas-Freiss, N.; Menier, C. HLA-G1 and HLA-G5 active dimers are present in malignant cells and effusions: The influence of the tumor microenvironment. *Eur. J. Immunol.* **2012**, *42*, 1599–1608. [CrossRef]
140. Rouas-Freiss, N.; Bruel, S.; Menier, C.; Marcou, C.; Moreau, P.; Carosella, E.D. Switch ofHLA-G alternative splicing in a melanoma cell line causes loss of HLA-G1 expression and sensitivity to NK lysis. *Int. J. Cancer* **2005**, *117*, 114–122. [CrossRef] [PubMed]
141. Davidson, B.; Elstrand, M.B.; McMaster, M.T.; Berner, A.; Kurman, R.J.; Risberg, B.; Trope, C.G.; Shih, I.-M. HLA-G expression in effusions is a possible marker of tumor susceptibility to chemotherapy in ovarian carcinoma. *Gynecol. Oncol.* **2005**, *96*, 42–47. [CrossRef] [PubMed]
142. Singer, G.; Rebmann, V.; Chen, Y.-C.; Liu, H.-T.; Ali, S.Z.; Reinsberg, J.; McMaster, M.T.; Pfeiffer, K.; Chan, D.W.; Wardelmann, E.; et al. HLA-G is a potential tumor marker in malignant ascites. *Clin. Cancer Res.* **2003**, *9*, 4460–4464.
143. Schwich, E.; Rebmann, V.; Michita, R.T.; Rohn, H.; Voncken, J.W.; Horn, P.A.; Kimmig, R.; Kasimir-Bauer, S.; Buderath, P. HLA-G 3′ untranslated region variants +3187G/G, +3196G/G and +3035T define diametrical clinical status and disease outcome in epithelial ovarian cancer. *Sci. Rep.* **2019**, *9*, 5407. [CrossRef] [PubMed]
144. Reches, A.; Nachmani, D.; Berhani, O.; Duev-Cohen, A.; Shreibman, D.; Ophir, Y.; Seliger, B.; Mandelboim, O. HNRNPR Regulates the Expression of Classical and Nonclassical MHC Class I Proteins. *J. Immunol.* **2016**, *196*, 4967–4976. [CrossRef] [PubMed]
145. Ducoin, K.; Oger, R.; Mutala, L.B.; Deleine, C.; Jouand, N.; Desfrançois, J.; Podevin, J.; Duchalais, E.; Cruard, J.; Benlalam, H.; et al. Targeting NKG2A to boost anti-tumor CD8 T-cell responses in human colorectal cancer. *Oncoimmunology* **2022**, *11*, 2046931. [CrossRef] [PubMed]
146. Ram, D.R.; Lucar, O.; Hueber, B.; Reeves, R.K. Simian Immunodeficiency Virus Infection Modulates CD94 + (KLRD1+) NK Cells in Rhesus Macaques. *J. Virol.* **2019**, *93*, e00731-19. [CrossRef] [PubMed]
147. Orr, M.T.; Wu, J.; Fang, M.; Sigal, L.J.; Spee, P.; Egebjerg, T.; Dissen, E.; Fossum, S.; Phillips, J.H.; Lanier, L.L. Development and Function of CD94-Deficient Natural Killer Cells. *PLoS ONE* **2010**, *5*, e15184. [CrossRef]
148. Vance, R.E.; Jamieson, A.M.; Cado, D.; Raulet, D.H. Implications of CD94 deficiency and monoallelic NKG2A expression for natural killer cell development and repertoire formation. *Proc. Natl. Acad. Sci. USA* **2002**, *99*, 868–873. [CrossRef]

149. Mingari, M.C.; Ponte, M.; Bertone, S.; Schiavetti, F.; Vitale, C.; Bellomo, R.; Moretta, A.; Moretta, L. HLA class I-specific inhibitory receptors in human T lymphocytes: Interleukin 15-induced expression of CD94/NKG2A in superantigen- or alloantigen-activated CD8 $^+$ T cells. *Proc. Natl. Acad. Sci. USA* **1998**, *95*, 1172–1177. [CrossRef]
150. Starzer, A.M.; Preusser, M.; Berghoff, A.S. Immune escape mechanisms and therapeutic approaches in cancer: The cancer-immunity cycle. *Ther. Adv. Med Oncol.* **2022**, *14*, 17588359221096219. [CrossRef]
151. Wang, L.; Liang, H.; Sun, J.; Liu, Y.; Li, J.; Li, J.; Li, J.; Yang, H. Bispecific Aptamer Induced Artificial Protein-Pairing: A Strategy for Selective Inhibition of Receptor Function. *J. Am. Chem. Soc.* **2019**, *141*, 12673–12681. [CrossRef]
152. Akinleye, A.; Rasool, Z. Immune checkpoint inhibitors of PD-L1 as cancer therapeutics. *J. Hematol. Oncol.* **2019**, *12*, 92. [CrossRef]
153. Tan, S.; Li, D.; Zhu, X. Cancer immunotherapy: Pros, cons and beyond. *Biomed. Pharmacother.* **2020**, *124*, 109821. [CrossRef] [PubMed]
154. Jiao, P.; Geng, Q.; Jin, P.; Su, G.; Teng, H.; Dong, J.; Yan, B. Small Molecules as PD-1/PD-L1 Pathway Modulators for Cancer Immunotherapy. *Curr. Pharm. Des.* **2019**, *24*, 4911–4920. [CrossRef] [PubMed]
155. Shaabani, S.; Huizinga, H.P.S.; Butera, R.; Kouchi, A.; Guzik, K.; Magieramularz, K.; Holak, T.A.; Domling, A. A patent review on PD-1/PD-L1 antagonists: Small molecules, peptides, and macrocycles (2015–2018). *Expert Opin. Ther. Pat.* **2018**, *28*, 665–678. [CrossRef] [PubMed]

membranes

MDPI

Article

Effect of Cytochrome C on the Conductance of Asolectin Membranes and the Occurrence of Through Pores at Different pHs

Andrey Anosov [1,2,*], Elizaveta Borisova [1], Elena Smirnova [1], Eugenia Korepanova [3] and Anatoly Osipov [3]

1 The Department of Medical and Biological Physics, Sechenov First Moscow State Medical University (Sechenov University), 119991 Moscow, Russia
2 Kotelnikov Institute of Radioengineering and Electronics of RAS, 125009 Moscow, Russia
3 The Department of General and Medical Biophysics, Pirogov Russian National Research Medical University, 117997 Moscow, Russia
* Correspondence: anosov_a_a@staff.sechenov.ru

Abstract: The study of the electrical parameters of asolectin bilayer lipid membranes in the presence of cytochrome c (cyt c) at various concentrations showed that an increase in the concentration of cyt c leads to an increase in the membrane conductance and the appearance of through pores. The studied membranes did not contain cardiolipin, which is commonly used in studying the effect of cyt c on membrane permeability. In the presence of cyt c, discrete current fluctuations were recorded. The occurrence of these fluctuations may be associated with the formation of through pores. The diameter of these pores was ~0.8 nm, which is smaller than the size of the cyt c globule (~3 nm). Measurements carried out at pH values from 6.4 to 8.4 showed that the concentration dependence of the membrane conductance increases with increasing pH. To assess the binding of cyt c to the bilayer, we measured the concentration and pH dependences of the difference in surface potentials induced by the unilateral addition of cyt c. The amount of bound cyt c at the same concentrations decreased with increasing pH, which did not correspond to the conductance trend. An analysis of conductance traces leads to the conclusion that an increase in the integral conductance of membranes is associated with an increase in the lifetime of pores. The formation of "long-lived" pores, of which the residence time in the open state is longer than in the closed state, was achieved at various combinations of pHs and cyt c concentrations: the higher the pH, the lower the concentration at which the long-lived pores appeared and, accordingly, a higher conductance was observed. The increase in conductance and the formation of transmembrane pores are not due to the electrostatic interaction between cyt c and the membrane. We hypothesize that an increase in pH leads to a weakening of hydrogen bonds between lipid heads, which allows cyt c molecules to penetrate into the membrane. This disrupts the order of the bilayer and leads to the occurrence of through pores.

Keywords: bilayer lipid membranes; cytochrome c; electrical conductance; lipid pores

Citation: Anosov, A.; Borisova, E.; Smirnova, E.; Korepanova, E.; Osipov, A. Effect of Cytochrome C on the Conductance of Asolectin Membranes and the Occurrence of Through Pores at Different pHs. *Membranes* **2023**, *13*, 268. https://doi.org/10.3390/membranes13030268

Academic Editors: Bernhard Schuster, Shiro Suetsugu and Yosuke Senju

Received: 21 January 2023
Revised: 6 February 2023
Accepted: 22 February 2023
Published: 24 February 2023

1. Introduction

The study of the interaction of cyt c with bilayer lipid membranes is of particular interest since similar processes occur in mitochondria when programmed cell death (apoptosis) is triggered. The exact mechanism of the interaction of fatty acids and acid phospholipids with cyt c is unclear. Cyt c interacts with lipid membranes via electrostatic interactions, hydrogen bonds, and hydrophobic effects [1–7]. Firstly, cyt c can be adsorbed on the membrane surface. This binding is determined by the electrostatic interaction of positively charged cyt c with negative charges on the membrane surface, which is stabilized by hydrogen bonds at low pH values [1]. Secondly, the hydrophobic interaction of fatty acid chains of lipids with the hydrophobic internal structure of cyt c is possible [2,6,8]. Gorbenko et al. [1] showed that at a physiological pH, a shallow cyt c location is assumed, whereas at a pH below 6.0, the protein tends to be inserted into the membrane core. In experiments, these effects manifest themselves when the membrane contains cardiolipin, which is the most likely candidate for the role of an anionic phospholipid that attaches cyt c to the inner mitochondrial membrane [1,8–10].

Various physical methods are used to study cyt c–lipid interactions. Tuominen et al. [8] used steady-state absorption and fluorescence spectroscopy to show the presence of the extended lipid anchorage of cyt c to phospholipid membranes. In this mechanism, one of the phospholipid acyl chains protrudes from the membrane and enters the hydrophobic channel in cyt c, whereas the other chains remain in the bilayer. Bernabeu et al. [9] used two-dimensional infrared correlation spectroscopy to study the interaction of cyt c with phospholipids under temperature changes. The presence of a specific interaction between the protein and vesicles containing cardiolipin was revealed. The adsorption of cyt c to anionic lipid bilayers was studied using atomic force microscopy [6]. It was found that cyt c inserts into the bilayer and resides in its hydrophobic core, changing the mechanical properties of the bilayer.

It is well known that the interaction of cyt c with phospholipid membranes containing cardiolipin leads to an increase in membrane permeability [10,11]. Bergstrom et al. [10] used confocal fluorescence microscopy to visualize the leakage of cyt c across the membranes of single giant unilamellar vesicles containing cardiolipin. The authors showed that cyt c leakage occurs only in cardiolipin-containing membranes and attributed the leakage to the opening of lipid pores formed by the cyt c–cardiolipin conjugate. The pore size was estimated from the permeability values by comparing results to those obtained for equinatoxin II, a member of the family of 20 kDa pore-forming toxins from sea anemones. Permeability was also determined by the calculation of diffusive transport. Two obtained approximate pore diameters of ~2 and ~20 nm were reasonably comparable to the diameter of cyt c (~3 nm). The authors suggested that the pore-forming process may be related to the ability of cyt c to induce a negative curvature stress upon binding to cardiolipin-containing membranes.

Kitt et al. [11] used optical-trapping confocal Raman microscopy to investigate the leakage of 3-nitrobenzenesulfonate from large unilamellar vesicles of dipalmitoylphosphatidylcholine and a dipalmitoylphosphatidylcholine–cardiolipin mixture. Cyt c-related leakage was observed only in vesicles containing cardiolipin. These results suggest that cyt c-induced permeability occurs due to the selective interaction of cyt c with cardiolipin. This leads to the unfolding of the protein, whereas the unfolded form interacts with the acyl chains of cardiolipin inside the bilayer, increasing the permeability of the membrane.

In [12,13], it was shown that the cyt c complex with cardiolipin is formed because of the attachment of the cyt c molecule to the membrane surface due to an electrostatic interaction. The subsequent insertion of one or two of the fatty acid chains of cardiolipin into the protein globule, due to hydrophobic interactions, leads to increased cyt c peroxidase activity, which in turn, increases membrane permeability [14]. It was shown in [15,16] that the addition of cyt c and hydrogen peroxide to asolectin membranes containing cardiolipin leads to the formation of pores, the size of which approximately corresponds to the size of a cyt c globule with a diameter of about 3 nm. Note that the peroxidase activity of cyt c leads to damage to the membrane structure (see, for example, [12–14,17]). We also note studies that investigated the toxicity of the interaction of cyt c with various nanostructures, such as for example, carbon nanotubes [18] or graphene nanosheets [19].

In [20], an effect of cyt c at different concentrations on the conductance of DIB (droplet interface bilayer) membranes of diphytanoylphosphocholine, cholesterol, and cardiolipin was investigated. Discrete current fluctuations of different amplitudes and durations were registered. The conductance of electrical spikes observed ranged from 1000 to 8000 pS, yielding pore diameter estimates ranging between 1.4 and 4.0 nm. The authors associated the occurrence of current spikes with the electrostatic interaction of cyt c with the membranes.

In [10,15,16,20], pores were registered in membranes containing cardiolipin. The outer mitochondrial membrane, through which cyt c passes during apoptosis, does not contain significant amounts of cardiolipin [21]. The asolectin used in our study is a natural mixture of phospholipids and is similar in composition to the phospholipids of the outer mitochondrial membrane. This makes it possible, to some extent, to simulate the interaction of cyt c with anionic phospholipids of the outer mitochondrial membrane.

In [22], the mitochondrial matrix and cytosolic pH were measured during staurosporine-induced apoptosis. Within an hour, the mitochondrial pH increased from 7.8 to 8.4, and the cytosolic pH decreased from 7.4 to 7.0. These findings indicate that the alteration of the intracellular pH may be an early event that regulates apoptosis.

In this paper, we investigated the influence of pH and cyt c on the electrical characteristics of asolectin bilayer lipid membranes. The acid phospholipids in asolectin can attach cyt c as cardiolipin does. We compared cyt c-induced conductances and membrane surface potential differences at various pH values to find out what are the mechanisms and processes that contribute to increased bilayer lipid membrane conductance.

2. Materials and Methods

2.1. Lipid, Cytochrome C, and Electrolytes

Asolectin (Avanti Polar Lipids, Alabaster, AL, USA) was used for the formation of planar lipid bilayer membranes (BLMs). Bulk solutions containing 0.1 M KCl and 5 mM Tris-HCl buffer (all reagents were of analytical grade) were used, and pH values were set to 6.4, 7.4, and 8.4. Cyt c from a bovine heart (Merck KGaA, Darmstadt, Germany) water solution of various concentrations (initial concentration being 5 mg/mL) was added to the bulk solutions.

2.2. Planar Lipid Bilayer Membranes

The BLMs were formed according to [23] over a 0.5 mm^2 circular hole in a 1 mm thick wall of a Teflon chamber at a room temperature of 21 ± 1 °C. The wall separated two compartments, each filled with 2.5 mL of the same electrolyte solution. The membrane-forming solution contained 30 mg of lipids dissolved in 1 mL of n-decane. Before each experiment, the vertical wall of the Teflon chamber was covered with a thin layer of dried membrane-forming solution. Once a small droplet (~0.1 μL) of lipid solution was placed below the hole, a bilayer was formed automatically in ~10 min. The formation of the bilayer was followed by capacitance measurements. To estimate the specific capacitance of the membrane, the area of the membrane formed on the hole was determined using a microscope. The specific capacitances of the studied membranes were in the range of 3–4 nF/mm^2.

Two methods of membrane formation were used in the measurements: a control membrane was formed, and every 10 min, cyt c was added, or a membrane was formed with cyt c previously added in the bulk solution.

The dependences of the conductance of asolectin membranes on cyt c concentration were measured for three pH values. Four control membranes and three membranes for each cyt c concentration were measured at pH 6.4; additionally, 9 and 5 membranes were measured at pH 7.4, and 5 and 3 membranes were measured at pH 8.4.

2.3. Electrical Measurements

Ag-AgCl STREF1 electrodes (OHAUS Corporation, Parsippany, NJ, USA) were placed into both compartments of the chamber. The membrane current was measured using a VA-10X amplifier (NPI Electronics GmbH, Tamm, Germany) with a feedback resistance of 5 GΩ and an integration constant of 20 ms. Current fluctuations were recorded with a sampling rate of 1 kHz in a 16-digit ADC (L-Card, Moscow, Russia). All the measurements were carried out at a room temperature of 21 ± 1 °C.

2.4. Measurement of Conductance and Estimation of the Radius of Pores in the Membrane

A constant voltage was applied to the membrane in the voltage clamp mode. The membrane current in the presence of cyt c shows quantized steps, which are usually associated with the formation of through pores. Considering these pores as cylinders and assuming that the electrolyte conductivity in the pore is equal to the conductivity in the solution, we estimated the pore radius r by using the well-known formula [24,25]:

$$r = \sqrt{\frac{Gh}{\pi g}}, \tag{1}$$

where h = 5 nm is the membrane thickness, G is the pore conductance, and g = 1.04 S/m is the conductivity of a 0.1 M KCl solution at room temperature.

2.5. Asymmetric Addition of Cyt C and Measurement of the Difference in Surface Potential

Surface potential is the difference between the membrane surface potential and bulk solution potential. For a symmetrical membrane in the absence of an external voltage, the difference between the surface potentials is zero. When charged molecules are adsorbed on one side of the membrane, the non-zero difference in the surface potential occurs. For homogeneous planar membranes with a homogeneous charge density, the surface potential is well described by the equations of Gouy–Chapman [26]. This model is used in studying the interaction of cyt c with membranes [1]. We used a method which was previously described in [27] for the measurements of the surface potential change with a one-sided cyt c addition to the membranes.

In the absence of cyt c, the surface potential depends on the number of the charged lipids in the membrane. The surface charge density is equal to

$$\sigma_0 = -\frac{eX}{A}, \tag{2}$$

where A is the area of a lipid molecule in the bilayer plane, e is the electron charge, and X is the molar fraction of charged lipids. The negative surface potential φ_0 can be calculated using the Gouy–Chapman equation

$$\sigma_0 = \sqrt{8\varepsilon\varepsilon_0 RTC_{KCl}}\sinh\left(\frac{F}{2RT}\varphi_0\right), \tag{3}$$

where F is the Faraday constant, R is the gas constant, T is the temperature, ε is the permittivity of water, ε_0 is the vacuum permittivity, and C_{KCl} is the concentration of KCl. The one-sided addition of positively charged cyt c leads to the increase in the surface charge by $\Delta\sigma$ and the surface potential by $\Delta\varphi$. In this case, the Gouy–Chapman equation is

$$\sigma_0 + \Delta\sigma = \sqrt{8\varepsilon\varepsilon_0 RTC_{KCl}}\sinh\left[\frac{F}{2RT}(\varphi_0 + \Delta\varphi)\right]. \tag{4}$$

The dependence between the surface charge $\Delta\sigma$ and the cyt c concentration C_{cyt} in the bulk solution is described by the Boltzmann equation

$$\Delta\sigma = zKC_{cyt}\exp\left[-\frac{F}{RT}(z\Delta\varphi + \varphi_0)\right], \tag{5}$$

where z is the number of charges in the cyt c molecule, and K is the binding constant. From Equations (3)–(5) we obtained the equation

$$\begin{aligned}&\sqrt{8\varepsilon\varepsilon_0 RTC_{KCl}}\left\{\sinh\left[\tfrac{F}{2RT}(\varphi_0 + \Delta\varphi)\right] - \sinh\left(\tfrac{F}{2RT}\varphi_0\right)\right\} = \\ &zKC_{cyt}\exp\left[-\tfrac{F}{RT}(z\Delta\varphi + \varphi_0)\right],\end{aligned} \tag{6}$$

whose numerical solution allows us to find the surface potential difference $\Delta\varphi$.

The measurement method of the surface potential exploits the nonmonotonic voltage dependence of membrane capacitance [28–30]. In [29], it was shown that the voltage corresponding to the minimum of the capacitance current could be uniquely calculated using current responses to upward and downward half-periods of the triangular voltage. With an asymmetric addition of cyt c, the resulting difference in the surface potentials can be compensated for by an external command voltage. A triangular alternating command voltage with amplitudes in the range of 100–200 mV and frequencies from 0.5 to 1 Hz was applied to the membrane and cyclic current-voltage characteristics were registered (Figure 1a). The half-difference of the current responses to upward and downward half-periods of the triangular voltage (Figure 1b) is proportional (if transitional processes are excluded) to the membrane capacitance [29]. To assess the difference in surface potentials for each concentration of cyt c and each pH, the voltage corresponding to the minimum half-difference of the current responses was recorded and compared with the corresponding data of the control membranes.

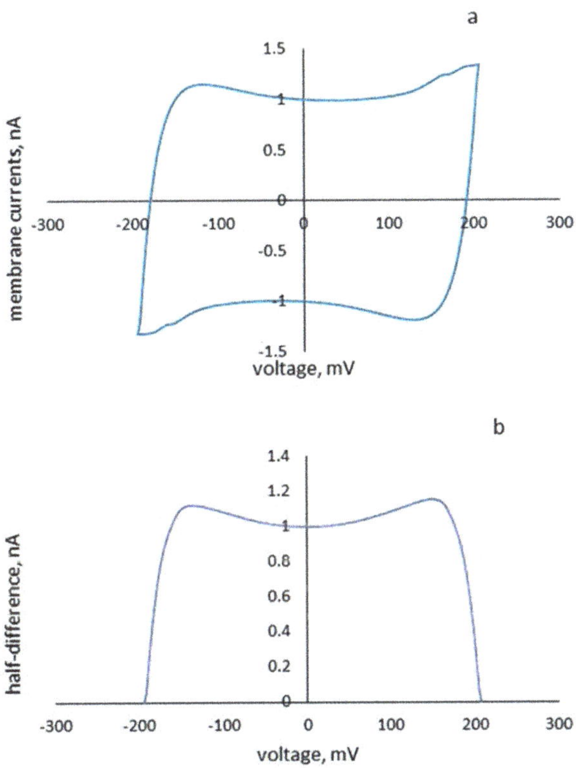

Figure 1. Cyclic current-voltage membrane characteristics (**a**) and the half-difference of the current responses to the upward and downward half-periods of the triangular voltage (**b**).

We added the same bulk solution volume to the opposite compartment to eliminate a differential pressure in the chamber caused by the cyt c addition.

3. Results

3.1. Conductance of Asolectin Membranes

It was found that membrane conductance increases with the addition of cyt c. Figure 2 shows an example of conductance traces when cyt c is added to a solution surrounding an asolectin membrane at pH 6.4. The control conductance of the membrane was 70 ± 10 pS; after the addition of cyt c, it increased to 240 ± 10 pS. After another 4 min, the basic conductance of the membrane decreased to about 230 ± 10 pS, and pores appeared in the membrane. Conductance traces and histograms are presented in Figures 2–4 at pH 6.4, 7.4, and 8.4, respectively.

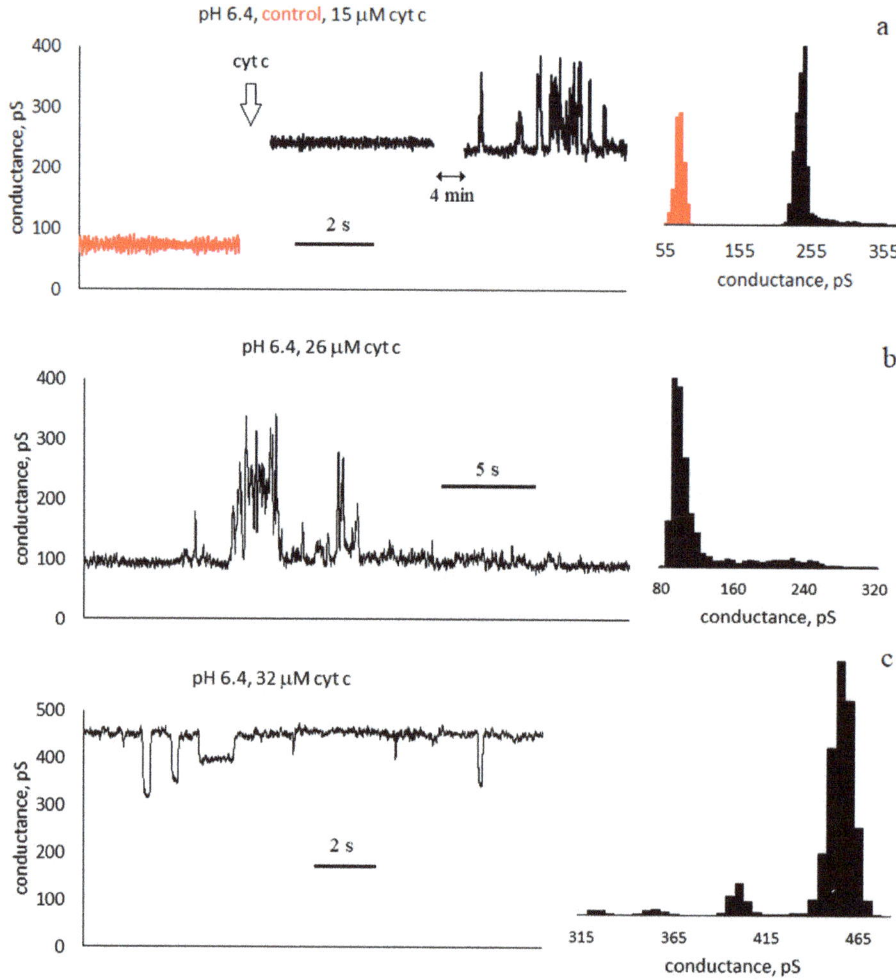

Figure 2. Conductance traces and histograms of asolectin membranes with symmetrical addition of cyt c at pH 6.4: (**a**) control (red) and conductance after addition of 15 µM cyt c, (**b**) 26 µM, and (**c**) 32 µM; 0.1 M KCl, 20 ms averaging.

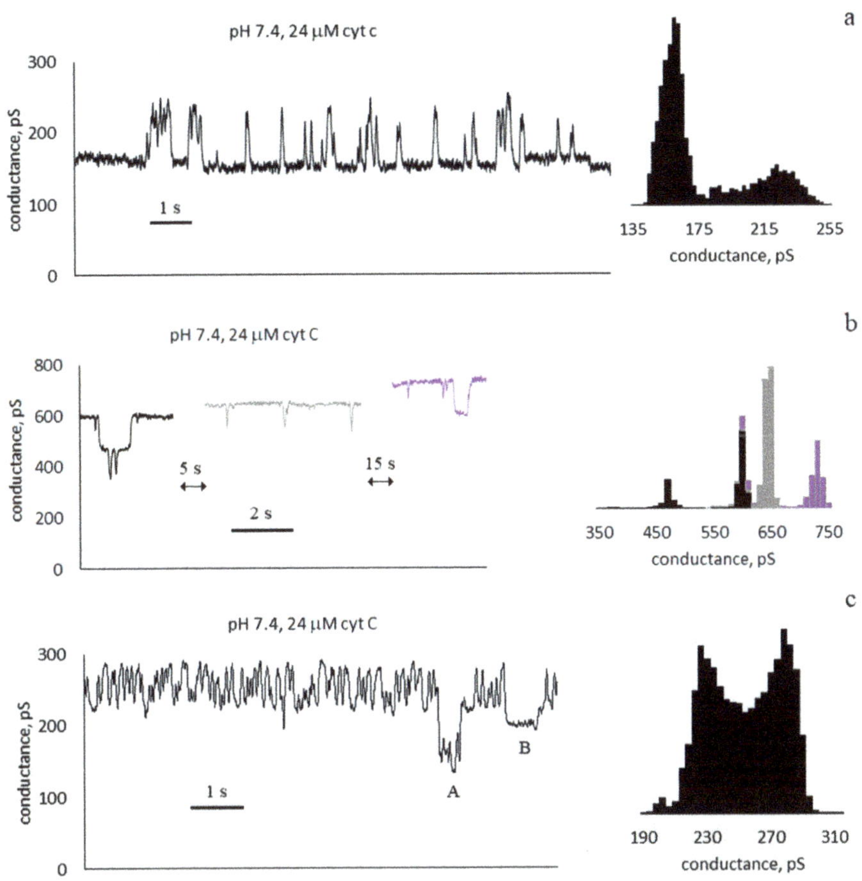

Figure 3. Conductance traces and histograms of asolectin membranes with symmetrical 24 μM addition of cyt c at pH 7.4: (**a**–**c**) are three different membranes; 0.1 M KCl, 20 ms averaging.

Figure 5 shows the dependences of the conductance of asolectin membranes on the concentration of cyt c in the surrounding solution at different pHs. The average conductance was measured at a given concentration one minute after the addition of cyt c. The results obtained show that the membrane conductance increases with an increase in the concentration of cyt c. At the same time, the conductance increases with an increase in pH from 6.4 to 8.4 at the same concentration of added cyt c. Figure 5 shows the standard errors of measurements carried out on different membranes. The variance of conductance from membrane to membrane under the same experimental conditions is significant and increases with an increase in the concentration of cyt c. For example, Figure 3 shows the traces of three membranes measured under the same conditions: pH 7.4 and 24μ M cyt c. The conductances of these three membranes are 170 ± 30, 630 ± 70, and 240 ± 30 pS (SDs are indicated after the ± signs). The average conductance for all three membranes is 350 ± 140 pS (SE is indicated after the ± sign).

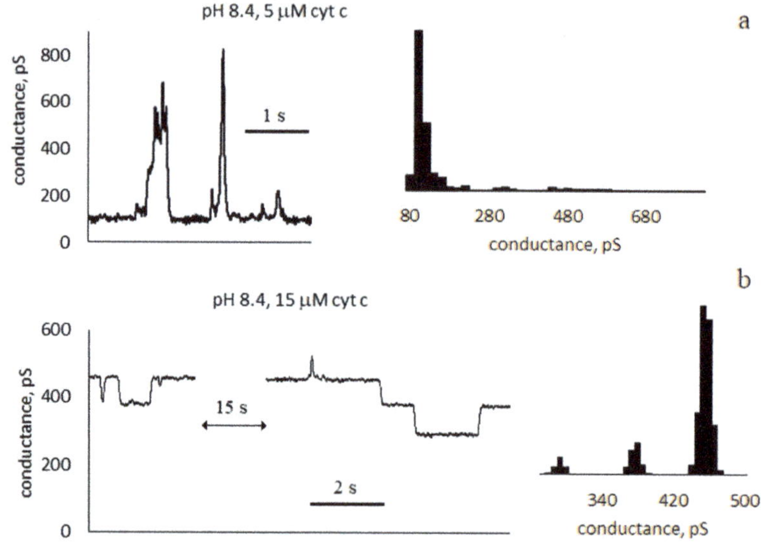

Figure 4. Conductance traces and histograms of asolectin membranes with symmetrical addition of cyt c at pH 8.4: (**a**) 5 µM, and (**b**) 15 µM; 0.1 M KCl, 20 ms averaging.

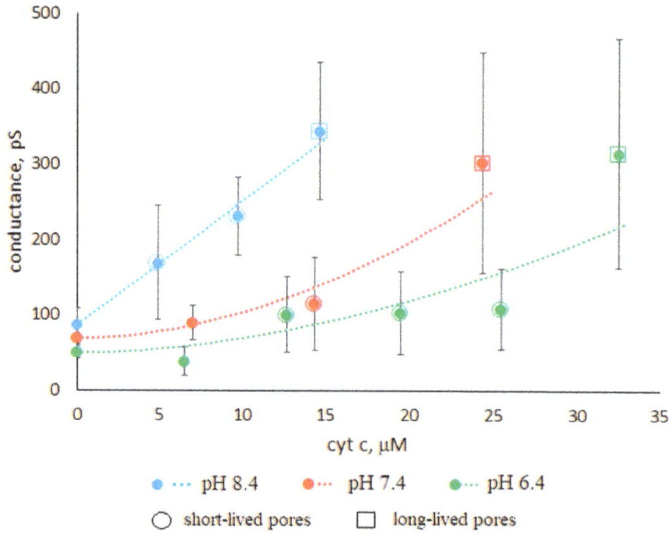

Figure 5. Dependence of the conductance of asolectin membranes on the cyt c concentration *C*. Cyt c additive is symmetrical, 0.1 M KCl. The approximating curves (dotted lines) show an increase in conductance with the addition of cyt c: 0.3 $C^{1.8}$ at pH 6.4, 0.4 $C^{1.9}$ at pH 7.4, and 16 $C^{1.0}$ at pH 8.4. Circles and squares show the appearance of short-lived and long-lived pores, respectively. Standard errors are shown.

We approximated the experimental dependences at different pHs by using the expression:

$$S = S_0 + aC^n,$$

where S_0 is the conductances of membranes at different pHs without cyt c, and a and n are the approximating coefficients. For linear approximation, we took the logarithm of the conductance increment $S - S_0$ (see Supplementary Materials, Figure S1). At pH 6.4, 7.4, and 8.4 we obtained the expressions $S - S_0 = (0.3 \pm 0.2)\ C^{1.8\pm0.4}$, $(0.4 \pm 0.3)C^{1.9\pm0.5}$, and $(16 \pm 9)\ C^{1.0\pm0.4}$ (see Figure 5, dotted lines), respectively.

The concentration of cyt c varied from 5 to 33 μM depending on pH. When the conductance reached ~300 pS, the membranes were destroyed. This occurred for the 15 μM cyt c concentration at pH 8.4, for 25 μM at pH 7.4, and for 32 μM at pH 6.4. We believe that the causes of membrane instability are associated with the formation of a large number of defects. This effect is similar to that of SDS at subsolubilizing concentrations on asolectin membranes [31].

3.2. *Characteristics of Through Pores*

The increase in conductance is accompanied by the appearance of discrete current fluctuations, which are usually associated with the appearance of through pores in the membrane (see Figures 2–4). The appearance of pores is a random event, and therefore, conductance fluctuations are not recorded in all membranes. At pH 6.4, with a cyt c concentration greater than 12 μM, pores appeared in 71% of membranes. At pH 7.4, with a cyt c concentration greater than 14 μM, pores appeared in 63% of membranes. At pH 8.4, pores appeared in 86% of membranes when cyt c was added. The registered current pulses are characterized by four parameters: shape, amplitude, duration τ, and frequency of occurrence f. Rectangular (Figures 2c, 3 and 4b) and non-rectangular (Figures 2b and 4a) pulses were obtained. The rectangular shape of the pulse corresponds to the formation of pores of a constant size. The radius of an equivalent cylindrical pore can be estimated by Formula (1). The observed rectangular pulses correspond to pores with a conductance of 100 $\pm$ 40 pS, which corresponds to pore radii in the range of 0.3–0.5 nm. Pulses with a short duration, close to the amplifier time constant of 20 ms, are distorted, have a triangular shape, and have a lower amplitude (Figures 2c, 3 and 4b). Note that the pore sizes are smaller than the size of the cyt c globule.

The temporal characteristics of the observed pulses differ greatly. Conventionally, they can be divided into three groups:

- Pulses with a duration much less than the interpulse interval $\tau \ll \frac{1}{f}$. For the trace in Figure 3a, the average pulse duration is τ = 100 $\pm$ 90 ms, and the frequency is $f \approx 2$ Hz).
- Pulses with a duration approximately equal to half of the interpulse interval. For the trace in Figure 3c, the average pulse duration is τ = 60 $\pm$ 40 ms, and the frequency is $f \approx 8$ Hz. In this case, in the conductance histogram there are two peaks of approximately the same size. The distributions of the durations of the opened and closed states are presented in the Supplementary Materials (Figure S2).
- Pulses with a duration slightly less than the interpulse interval $\tau \lesssim \frac{1}{f}$ (Figures 2c, 3b and 4). For the membrane in Figure 3b, the average pulse duration is τ = 1.5 $\pm$ 1.9 s, and the frequency is $f \approx 0.6$ Hz).

We will conditionally call the pores of the first and third groups "short-lived" and "long-lived", respectively. The concentrations of cyt c and membrane conductances at which short-lived and long-lived pores are observed are shown in Figure 5 by circles and squares, respectively. For example, at a pH of 7.4, a concentration of 14 μM, and a conductivity of 115 pS, a circle is drawn. This means that short-lived pores are registered in the membranes under these conditions. As can be seen from Figure 5, long-lived pores appear at higher cyt c concentrations. At all pH values, the presence of long-lived pores correlates with the high integral conductance of the membrane.

We also note that the appearance of pores is not a stationary process. This can be seen, for example, in Figure 2b: a burst (or two bursts) of conductance with a duration of ~10 s was registered, during which several (more than 10) pores appeared. However, neither before nor after that for 20 s were there any pores.

3.3. The Dependence of the Cyt C-Induced Difference in the Surface Potentials on pH

Figure 6 shows the experimental dependences of the difference in the surface potentials on the concentration of cyt c with the one-sided addition to the bulk solution at different pHs. An example of a cyt c-induced shift in the capacitive current minimum is presented in the Supplementary Materials (Figure S3). The results obtained show that the surface potential difference increased with an increasing concentration of cyt c. However, the cyt c-induced potential difference increased more at a lower pH.

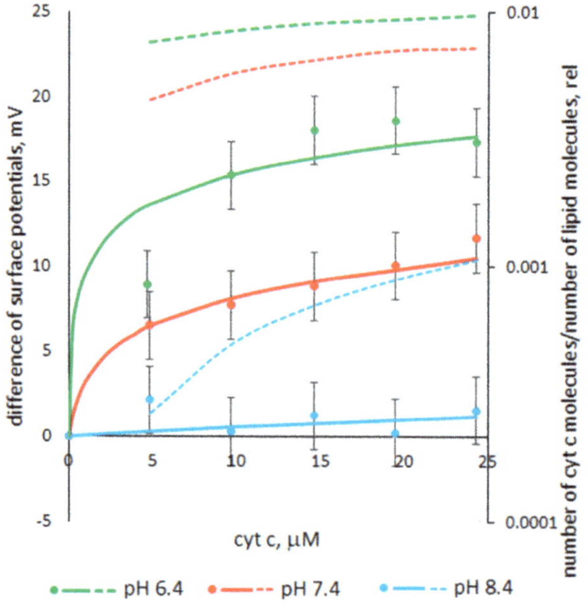

Figure 6. Experimental (markers) and calculated (curves) dependences of surface potential differences, as well as calculated dependences (dotted line) of the ratio of the number of cyt c molecules adsorbed on the membrane to the number of lipid molecules in the membrane, on the cyt c concentration at different pHs: 6.4 (green), 7.4 (red), and 8.4 (blue). Calculated parameters: for pH 6.4, the binding constant $K = 3000\ C{\cdot}m^{-2}{\cdot}M^{-1}$, and the number of charges of the cyt c molecule z = +8.5; for pH 7.4, $K = 110\ C{\cdot}m^{-2}{\cdot}M^{-1}$, z = +7.25; and for pH 8.4, $K = 1\ C{\cdot}m^{-2}{\cdot}M^{-1}$, z = +6. The bulk solution is 0.1 M KCl.

Calculating the surface potential in the absence of cyt c, we assumed that for an asolectin membrane at ~ pH 6, the fraction of charged lipids *X* is equal to 23% [30]. In the pH range 6–8.4, according to [32], phosphatidic acid (~3% in asolectin) changes its charge from −1.4 to −2. At pH 8.4, 4% of phosphatidylethanolamine molecules (~20% in asolectin) dissociate and become single-charged anions. In the pH range 6–8.4, phosphatidylinositol (~20% in asolectin) is a single-charged anion and the total charge of phosphatidylcholine (~50% in asolectin) is equal to zero. Thus, at pH 6.4 the percent of charged lipids is equal to *X* = 23.8%, whereas at pH 7.4 it is 24.9%, and at pH 8.4 it is 25.9%. The surface charges σ_0 calculated by Formula (2) are equal to –6.4 (pH 6.4), –6.6 (pH 7.4), and –6.9 (pH 8.4) $\mu C/cm^2$. In this calculation, the area of a lipid molecule in the bilayer plane was considered to be 0.6 nm^2 [33]. The surface potentials φ_0 calculated by Formula (3) are equal to –66 (pH 6.4), –68 (pH 7.4), and –69 (pH 8.4) mV.

Figure 6 also shows the dependences of the differences in surface potentials on the concentration of added cyt c calculated by Equation (6) at various pH values. The binding constant is used as a fitting parameter. The curves are in good agreement with the experimental data. Figure 6 also shows the dependences of the surface charge $\Delta\sigma$ on the concentration of cyt c calculated by Formula (5) at various pH values. The number of cyt c molecules per lipid molecule was determined as $\Delta\sigma A/ze$. The calculation shows that, at pH 6.4 and 7.4, cyt c molecules quite densely cover the membrane surface.

4. Discussion

Our measurements of cyt c-induced surface potential differences show that with a decrease in pH, the adsorption of cyt c molecules on the surface of asolectin membranes increases. Meanwhile, this does not lead to an increase in the conductance of the membrane, but on the contrary, the conductance increases with the increase in pH.

It is known that with an increase in pH from 6.4 to 8.4, the charge of cyt c decreases linearly from +8 to +6.5, or by 19% [34,35]. On the other hand, in this pH range, deprotonation of the polar heads of the acid lipids contained in asolectin occurs, which increases the negative surface charge of the membrane by ~8%. Thus, an increase in pH should lead to the weakening of the electrostatic interaction and to a decrease in the number of cyt c molecules adsorbed on the membrane surface (at a given cyt c concentration), which is observed in the experiment (Figure 6). However, the calculation shows that the binding constant K (fitting parameter) increases by about three orders of magnitude with the decrease in pH from 8.4 to 6.4. Such a difference in the values of the binding constants indicates the involvement of other interactions in the binding process. Since the number of protonated molecules sharply increases at a low pH, it can be assumed that the increase in binding between cyt c and lipids is due to an increase in the number of hydrogen bonds between them. This assumption is in line with [1] in which the enhancement of cyt c binding to liposomes containing cardiolipin was obtained at a lower pH. An increase in the number of protein-binding sites on the bilayer surface due to hydrogen bonds with protonated molecules was considered as a possible explanation for this dependence. In turn, such a modification is most likely to involve changes to the acid lipid protonation. Since the acid lipids contained in asolectin are known to exhibit peculiar protonation behavior [32,36], the increase in the amount of partially protonated species is most likely to account for the increased extent of cyt c binding to liposomes at a lower pH.

The data obtained in the experiment (Figure 5) show that the form of the dependence of the membrane conductance on the cyt c concentration strongly depends on pH. At the same concentration of cyt c, the greatest increase in conductance is observed at high pH values. This does not correlate with similar dependences of the surface potential difference and, accordingly, the number of bound cyt c molecules. It is possible that the pH-induced modification of the bilayer surface plays a key role not only in the observed enhancement of cyt c–membrane association at a low pH, but also in the change in permeability. The increase in pH leads to the destruction of hydrogen bonds between lipid heads. It is known that the increase in pH leads to a significant change in the lateral properties of the anion lipid bilayers, where the temperature of the phase transitions from the liquid state to the gel phase decreases, the fluidity rises, and the molecular packing density decreases [32,36–38].

Studies performed on monolayers have shown that alteration of the lateral structure of the bilayer at a high pH can lead to increased cyt c-induced changes in the lipid bilayer, facilitating pore formation [39,40]. Cyt c molecules penetrate into the monolayer for low surface pressures and are reversibly squeezed out at higher pressures [41]. A compressibility study showed that the adsorption or intermolecular aggregation of cyt c molecules on the lipid monolayer changes the fluidity of the membrane [7].

The hydrophobic interaction of proteins with membrane lipids is associated with anchoring the penetration of fatty acid chains to the protein thickness. This interaction leads to a change in the conformation of both cyt c and in the properties of the lipid layer. In this case, anchoring can occur in the presence [2] and in the absence [3] of cardiolipin in the membrane. Both electrostatic and hydrophobic interactions lead to the binding of cyt c on the membrane surface, and in our experiment, it is not possible to distinguish one from the other. In any case, the adsorption of cyt c on the surface cannot explain the increase in conductance and the appearance of pores.

If the protein remains on the membrane surface or anchors, then the cyt c molecule can create a heterogeneous structure in the form of an ordered cluster surrounded by a less ordered structure. The heterogeneity in the membrane can contribute to an increase in the number of defects in it, which can lead to an increase in conductance. The mechanism under consideration is similar to the processes taking place at the phase transitions of lipids. The phase transition from the liquid state to the gel phase is accompanied by the appearance of discrete pores (see, for example, [42]).

On the other hand, the protein can be immersed into the membrane. This is easier to do at a higher pH when the charge of the cyt c molecule decreases. In [5,6], the immersion of cyt c into membranes of dioleoylphosphatidylglycerol at a low ionic strength was considered. In [6], it was stated that cyt c insertion changes the mechanical properties of the bilayer significantly. Note that anchoring was not considered in [5,6]. The degree of immersion of the cyt c molecule in a bilayer containing cardiolipin was studied in [1]. It has been shown that the estimates of the heme distance from the bilayer center suggest a shallow bilayer location of cyt c at a physiological pH, whereas at a pH below 6.0, the protein tends to insert into the membrane core. These data show that the change in pH affects the immersion of the cyt c molecule in the bilayer, but they are obtained for bilayers containing cardiolipin, and immersion occurs at a pH < 6, whereas our data suggest a possible increase in the cyt c immersion in the bilayer at a pH ~8.4.

It was shown in [10,15,16] that cyt c can form pores in a membrane containing cardiolipin, both in the absence [10] and in the presence of hydrogen peroxide [15,16]. The authors consider the formation of pores to be the result of cyt c peroxidase activity. Note that neither cardiolipin nor hydrogen peroxide were used in our experiments; however, the formation of pores, which is quite intense (see Figures 2–4), has been registered. It cannot be excluded that the occurrence of pores is associated with weak cyt c peroxidase activity, but other possible causes of pore formation should not be neglected.

If we assume that cyt c molecules are significantly immersed in the bilayer, then the increase in conductance may not be related to the specifics of cyt c. In this case, the globules disrupt the structure of the bilayer, which can lead to the formation of pores in the membrane. This interaction of cyt c with the bilayer is similar to the interaction of hydrophobic nanoparticles with the membrane. It was shown in [43] that cobalt ferrite nanoparticles form the through pores in asolectin and diphytanoylphosphatidylcholine membranes.

The nonlinearity of the conductance versus concentration at pH 6.4 and 7.4 indicates that the increase in conductance is determined by several concentration-dependent processes. Perhaps these are the ones described above: the adsorption of cyt c on the membrane surface, which increases with the increasing concentration of cyt c (see Figure 6, dotted lines), and the penetration of the protein into the membrane, which is greater, when the concentration of cyt c is greater. Each effect independently increases membrane conductance, but they can act together. In this case, the increase in conductance with an increasing concentration is quadratic, which we can see in Figure 5 at pH 6.4 and 7.4. Comparing the data presented in Figures 5 and 6 at pH 8.4, it can be concluded that the adsorption of cyt c on the membrane surface does not affect the increase in conductance. Consequently, at pH 8.4, only the penetration of the protein into the membrane sets the linear increase in conductance.

Let us consider the pores registered in the experiment. Pores appeared after the cyt c concentration reached 5, 12, and 14 μM at pH 6.4, 7.4, and 8.4, respectively. At higher concentrations the duty cycle τf increases with an increase in the cyt c concentration. This value, τf, determines the average number of pores in the membrane at a given moment. An increase in the lifetime leads to an increase in the integral membrane conductance. Thus, only the duration of the open state of the pore increases significantly with an increase in cyt c concentration. Other parameters of the pore (the size and the frequency of occurrence) are not changed significantly.

Due to the decrease in hydrogen bonds between the polar heads of acid lipids, the packing density of lipid heads decreases with increasing pH. This simplifies the penetration of cyt c molecules into the membrane. The lateral interaction of discrete charges of cyt c can also reduce the packing density. The presence of cyt c globules in the membrane leads to the occurrence of mechanical stresses, increased fluctuations, and the formation of defects, which contributes to the formation of through pores. At high concentrations of cyt c, mechanical stresses in the membrane increase, and the presence of the through pore is energetically more favorable than its absence. This is confirmed by the occurrence of long-lived pores.

Rectangular current pulses imply the presence of two local minima of energy associated with the open and closed states of the pore. The times of the open τ and closed $\Theta = \frac{1}{f} - \tau$ states of the pore are related to its energy characteristics by these ratios (see, for example, [44]):

$$\tau = \frac{1}{\nu V_{pore}} \exp\left(\frac{E_{max} - E_{open}}{kT}\right), \quad (7)$$

$$\Theta = \frac{1}{\nu V} \exp\left(\frac{E_{max}}{kT}\right), \quad (8)$$

where ν is the attempt rate density of membrane lipids, V_{pore} and V are the membrane volumes whose molecule fluctuations can lead to the closure and opening of the pore, respectively, E_{max} is the value of the energy barrier between the open and closed states of the pore, E_{open} is the energy of the open pore (the energy of the closed pore is $E_{close} = 0$), k is Boltzmann's constant, and T is the temperature (see Figure S4 in Supplementary Materials). Suppose that in Figures 2c and 3, except for multilevel fluctuations, the same pore closes and opens. In this case, in Equations (7) and (8), the volumes are equal, $V_{pore} = V$, and by dividing the equations, we can find E_{open}: $\frac{\tau}{\Theta} = exp\left(-\frac{E_{open}}{kT}\right)$. For the cases shown in Figures 3a and 4a, the difference in energy between the open and closed states of the pore is in units of kT: for short-lived pores, $E_{open} = 1.4kT$; for fluctuations in Figure 3c, $E_{open} = 0$; and for long-lived pores, $E_{open} = -2.2kT$. The slight differences in the energies of the open and closed states of the pore show that the formation of a pore does not lead to a significant change in the mechanical stresses in the bilayer.

To calculate E_{max}, the literature data on the attempt rate density should be used. However, estimates of this value differ by orders of magnitude, and we did not consider it possible to present the results of calculations in the text of this article. They can be found in the Supplementary Materials (Table S1).

It is worth noting how the pores close by analyzing the intervals A and B in Figure 3c. In both cases, the conductance level decreases. However, in the case of A, fluctuations with a frequency of ~8 Hz do not stop. It follows from this that the pore changing its state with a frequency of 8 Hz has not closed, but the other conductance structure in the membrane has closed. In case B, the fluctuations stop. This means that the observed pore was closed for ~0.5 s. In Figure 2c, for ~4 s, three negative rectangular pulses of different amplitudes are visible. Perhaps it was a closure of three pores of different sizes. However, it is more likely that the same pore changed its size. Moreover, in all three cases, the pore is in a metastable state. Note that a similar result was obtained in [45], where the metastable pores were formed in the diphytanoylphosphatidylcholine membrane, which interacted with cobalt ferrite nanoparticles.

5. Conclusions

The cyt c-induced conductance of asolectin membranes depends on its concentration and significantly depends on the pH at a medium range from 6.4 to 8.4. For the same concentration of cyt c, the conductance is greater at a higher pH. Note that in the mitochondria and in the cytosol, the pH values are different and changed in the initial stage of apoptosis.

Thus, by changing the pH, it is possible to change the conductance of the membrane with which cyt c interacts.

The unilateral addition of cyt c leads to the appearance of a difference in surface potentials. The measurement of this value at various concentrations and pHs shows an increase in cyt c binding at a low pH. A comparison of the experimental data with the calculation by the Gouy–Chapman formula shows that in order to agree with the experiment at low pH values, the binding constant must be three orders of magnitude higher compared to pH 8.4. This shows that at a low pH, cyt c binding occurs mainly through hydrogen bonding, a hydrophobic interaction, or other mechanisms.

At all three pH values, long-lived pores occur—a state in which a high level of conductance (state of an open pore) is maintained most of the time. The concentration value at which such a state is reached is minimal for a high pH. This suggests that the structure of the lipid bilayer changes in this region due to lateral electrostatic interactions of deprotonated acid lipids and cyt c molecules associated with the bilayer, and due to possible perturbations of the hydrophobic part of the lipid bilayer, which leads to an increase in fluctuations in the bilayer, as well as the appearance of structural defects and mechanical stresses that contribute to the appearance of pores. A possible reason for the increase in conductance at high pH values can also be a decrease in the packing density of lipids in the bilayer and an increase in fluidity due to the breaking of hydrogen bonds between deprotonated acid lipids.

In the presence of cyt c, through pores with a diameter of ~0.8 nm (which is less than the size of the cyt c globule ~3 nm) appear in the membrane. Note that previously, cyt c-induced pores in membranes were registered only in the presence of cardiolipin, which was not used in our study. At low concentrations of cyt c, short-lived pores appear in the membrane, in which the time spent in the open state is less than that spend in the closed state. With an increase in the concentration of cyt c, long-lived pores appear, the lifetime of which is longer in the open state than in the closed one. Thus, the increase in conductance is due, among other things, to an increase in the lifetime of pores, and not to an increase in the frequency of their appearance.

Supplementary Materials: The following supporting information can be downloaded at: https://www.mdpi.com/article/10.3390/membranes13030268/s1, Figure S1: Experimental (markers) and approximating (dotted lines) dependences of conductivity increment on cyt c concentration at different pHs; Figure S2: Distributions of the durations of the open and closed states of pores presented in Figure 2e; Figure S3: The half-difference of membrane currents in control and after one-sided addition of 5 μM cyt c; Figure S4: Energy diagram for the long-lived pore; Table S1: Energy barrier between the open and closed states of a pore. Refs. [46,47] have been cited in supplementary materials.

Author Contributions: Conceptualization, A.A. and A.O.; methodology, E.S.; software, E.B.; validation, A.A., E.B. and E.K.; formal analysis, E.B.; investigation, E.K.; resources, A.O.; data curation, A.A.; writing—original draft preparation, A.A. and E.S.; writing—review and editing, E.S.; visualization, A.A.; supervision, A.A.; project administration, A.O.; funding acquisition, A.A. All authors have read and agreed to the published version of the manuscript.

Funding: This study was carried out within the state task of the Kotelnikov Institute of Radioengineering and Electronics of RAS (State Registration no. AAAA-A19-119041590070-01).

Institutional Review Board Statement: Not applicable.

Data Availability Statement: Not applicable.

Conflicts of Interest: The authors declare no conflict of interest. The funders had no role in the design of the study; in the collection, analyses, or interpretation of data; in the writing of the manuscript; or in the decision to publish the results.

References

1. Gorbenko, G.P.; Molotkovsky, J.G.; Kinnunen, P.K.J. Cytochrome c interaction with cardiolipin/phosphatidylcholine model membranes: Effect of cardiolipin protonation. *Biophys. J.* **2006**, *90*, 4093–4103. [CrossRef] [PubMed]
2. Sinibaldi, F.; Howes, B.D.; Piro, M.C.; Polticelli, F.; Bombelli, C.; Ferri, T.; Coletta, M.; Smulevich, G.; Santucci, R. Extended cardiolipin anchorage to cytochrome c: A model for protein–mitochondrial membrane binding. *JBIC J. Biol. Inorg. Chem.* **2010**, *15*, 689–700. [CrossRef] [PubMed]
3. Rytömaa, M.; Kinnunen, P.K. Reversibility of the Binding of Cytochrome c to Liposomes: Implications for Lipid-Protein Interactions (∗). *J. Biol. Chem.* **1995**, *270*, 3197–3202. [CrossRef]
4. Milorey, B.; Schweitzer-Stenner, R.; Kurbaj, R.; Malyshka, D. pH-induced switch between different modes of cytochrome c binding to cardiolipin-containing liposomes. *ACS Omega* **2019**, *4*, 1386–1400. [CrossRef]
5. Zuckermann, M.J.; Heimburg, T. Insertion and pore formation driven by adsorption of proteins onto lipid bilayer membrane–water interfaces. *Biophys. J.* **2001**, *81*, 2458–2472. [CrossRef]
6. Choi, E.J.; Dimitriadis, E.K. Cytochrome c adsorption to supported, anionic lipid bilayers studied via atomic force microscopy. *Biophys. J.* **2004**, *87*, 3234–3241. [CrossRef] [PubMed]
7. Schweitzer-Stenner, R. Relating the multi-functionality of cytochrome c to membrane binding and structural conversion. *Biophys. Rev.* **2018**, *10*, 1151–1185. [CrossRef]
8. Tuominen, E.K.J.; Wallace, C.J.A.; Kinnunen, P.K.J. Phospholipid-cytochrome c interaction: Evidence for the extended lipid anchorage. *J. Biol. Chem.* **2002**, *277*, 8822–8826. [CrossRef]
9. Bernabeu, A.; Contreras, L.M.; Villalaín, J. Two-dimensional infrared correlation spectroscopy study of the interaction of oxidized and reduced cytochrome c with phospholipid model membranes. *Biochim. Biophys. Acta (BBA) Biomembr.* **2007**, *1768*, 2409–2420. [CrossRef]
10. Bergstrom, C.L.; Beales, P.A.; Lv, Y.; Vanderlick, T.K.; Groves, J.T. Cytochrome c causes pore formation in cardiolipin-containing membranes. *Proc. Natl. Acad. Sci.* **2013**, *110*, 6269–6274. [CrossRef]
11. Kitt, J.P.; Bryce, D.A.; Minteer, S.D.; Harris, J.M. Raman spectroscopy reveals selective interactions of cytochrome c with cardiolipin that correlate with membrane permeability. *J. Am. Chem. Soc.* **2017**, *139*, 3851–3860. [CrossRef] [PubMed]
12. Belikova, N.A.; Vladimirov, Y.A.; Osipov, A.N.; Kapralov, A.A.; Tyurin, V.A.; Potapovich, M.V.; Basova, L.V.; Peterson, J.; Kurnikov, I.V.; Kagan, V.E. Peroxidase activity and structural transitions of cytochrome c bound to cardiolipin-containing membranes. *Biochemistry* **2006**, *45*, 4998–5009. [CrossRef] [PubMed]
13. Kagan, V.E.; Tyurin, V.A.; Jiang, J.; Tyurina, Y.Y.; Ritov, V.B.; Amoscato, A.A.; Osipov, A.N.; Belikova, N.A.; Kapralov, A.A.; Kini, V.; et al. Cytochrome c acts as a cardiolipin oxygenase required for release of proapoptotic factors. *Nat. Chem. Biol.* **2005**, *1*, 223–232. [CrossRef]
14. Firsov, A.M.; Kotova, E.A.; Korepanova, E.A.; Osipov, A.N.; Antonenko, Y.N. Peroxidative permeabilization of liposomes induced by cytochrome c/cardiolipin complex. *Biochim. Biophys. Acta (BBA) Biomembr.* **2015**, *1848*, 767–774. [CrossRef] [PubMed]
15. Puchkov, M.N.; Vassarais, R.A.; Korepanova, E.A.; Osipov, A.N. Cytochrome c produces pores in cardiolipin-containing planar bilayer lipid membranes in the presence of hydrogen peroxide. *Biochim. Biophys. Acta (BBA) Biomembr.* **2013**, *1828*, 208–212. [CrossRef] [PubMed]
16. Antonov, V.F.; Puchkov, M.N.; Korepanova, E.A.; Nemchenko, O.Y.; Borodulin, V. Soft perforation of cardiolipin-containing planar lipid bilayer membrane by cytochrome c and H_2O_2. *Eur. Biophys. J.* **2014**, *43*, 469–476. [CrossRef]
17. Kagan, V.E.; Borisenko, G.G.; Tyurina, Y.Y.; Tyurin, V.A.; Jiang, J.; Potapovich, A.I.; Fujii, Y. Oxidative lipidomics of apoptosis: Redox catalytic interactions of cytochrome c with cardiolipin and phosphatidylserine. *Free. Radic. Biol. Med.* **2004**, *37*, 1963–1985. [CrossRef]
18. Zhang, C.; Li, X.; Wang, Z.; Huang, X.; Ge, Z.; Hu, B. Influence of structured water layers on protein adsorption process: A case study of cytochrome c and carbon nanotube interactions and its implications. *J. Phys. Chem. B* **2019**, *124*, 684–694. [CrossRef]
19. Zhang, W.; Wang, C.; Li, Z.; Lu, Z.; Li, Y.; Yin, J.J.; Zhou, Y.T.; Gao, X.; Fang, Y.; Nie, G.; et al. Unraveling stress-induced toxicity properties of graphene oxide and the underlying mechanism. *Adv. Mater.* **2012**, *24*, 5391–5397. [CrossRef]
20. Xu, J.; Vanderlick, T.K.; Beales, P.A. Lytic and non-lytic permeabilization of cardiolipin-containing lipid bilayers induced by cytochrome c. *PLoS ONE* **2013**, *8*, e69492. [CrossRef]
21. Horvath, S.E.; Daum, G. Lipids of mitochondria. *Prog. Lipid Res.* **2013**, *52*, 590–614. [CrossRef] [PubMed]
22. Matsuyama, S.; Llopis, J.; Deveraux, Q.L.; Tsien, R.Y.; Reed, J.C. Changes in intramitochondrial and cytosolic pH: Early events that modulate caspase activation during apoptosis. *Nat. Cell Biol.* **2000**, *2*, 318–325. [CrossRef]
23. Mueller, P.; Rudin, D.O.; Ti Tien, H.; Wescott, W.C. Reconstitution of cell membrane structure in vitro and its transformation into an excitable system. *Nature* **1962**, *194*, 979–980. [CrossRef] [PubMed]
24. DeBlois, R.W.; Bean, C.P.; Wesley, R.K. Electrokinetic measurements with submicron particles and pores by the resistive pulse technique. *J. Colloid Interface Sci.* **1977**, *61*, 323–335. [CrossRef]
25. Glaser, R.W.; Leikin, S.L.; Chernomordik, L.V.; Pastushenko, V.F.; Sokirko, A.I. Reversible electrical breakdown of lipid bilayers: Formation and evolution of pores. *Biochim. Biophys. Acta (BBA) Biomembr.* **1988**, *940*, 275–287. [CrossRef]
26. McLaughlin, S. The electrostatic properties of membranes. *Annu. Rev. Biophys. Biophys. Chem.* **1989**, *18*, 113–136. [CrossRef]

27. Konstantinova, A.N.; Sokolov, V.S.; Jimenez-Munguia, I.; Finogenova, O.A.; Ermakov, Y.A.; Gorbunova, Y.G. Adsorption and photodynamic efficiency of meso-tetrakis (p-sulfonatophenyl) porphyrin on the surface of bilayer lipid membranes. *J. Photochem. Photobiol. B Biol.* **2018**, *189*, 74–80. [CrossRef]
28. Alvarez, O.; Latorre, R. Voltage-dependent capacitance in lipid bilayers made from monolayers. *Biophys. J.* **1978**, *21*, 1–17. [CrossRef]
29. Smirnova, E.Y.; Anosov, A.A. Bilayer Lipid Membrane as Memcapacitance: Capacitance–Voltage Pinched Hysteresis and Negative Insertion Conductance. *Membranes* **2023**, *13*, 97. [CrossRef]
30. Cherny, V.V.; Sokolov, V.S.; Abidor, I.G. 330-Determination of surface charge of bilayer lipid membranes. *Bioelectrochem. Bioenerg.* **1980**, *7*, 413–420. [CrossRef]
31. Anosov, A.A.; Smirnova, E.Y.; Korepanova, E.A.; Shogenov, I.M. The effects of SDS at subsolubilizing concentrations on the planar lipid bilayer permeability: Two kinds of current fluctuations. *Chem. Phys. Lipids* **2019**, *218*, 10–15. [CrossRef] [PubMed]
32. Jacobson, K.; Papahadjopoulos, D. Phase transitions and phase separations in phospholipid membranes induced by changes in temperature, pH, and concentration of bivalent cations. *Biochemistry* **1975**, *14*, 152–161. [CrossRef] [PubMed]
33. Nagle, J.F.; Tristram-Nagle, S. Structure of lipid bilayers. *Biochim. Biophys. Acta (BBA)-Rev. Biomembr.* **2000**, *1469*, 159–195. [CrossRef]
34. Gordon, J.C.; Myers, J.B.; Folta, T.; Shoja, V.; Heath, L.S.; Onufriev, A. H++: A server for estimating p K as and adding missing hydrogens to macromolecules. *Nucleic Acids Res.* **2005**, *33* (Suppl. S2), W368–W371. [CrossRef] [PubMed]
35. Anandakrishnan, R.; Aguilar, B.; Onufriev, A.V. H++ 3.0: Automating p K prediction and the preparation of biomolecular structures for atomistic molecular modeling and simulations. *Nucleic Acids Res.* **2012**, *40*, W537–W541. [CrossRef] [PubMed]
36. Smirnova, E.Y.; Kozhomkulov, E.T.; Vasserman, A.N.; Vosnesensky, S.A.; Shevchenko, E.V.; Morozov, Y.V.; Antonov, V.F. Permeability of bilayer lipid membranes in phase transition. The significance of intermolecular phosphate-phosphate hydrogen bonding. *Chem. Phys. Lipids* **1986**, *41*, 173–180. [CrossRef]
37. Verkleij, A.J.; De Kruyff, B.; Ververgaert, P.T.; Tocanne, J.F.; Van Deenen, L.L.M. The influence of pH, Ca^{2+} and protein on the thermotropic behaviour of the negatively charged phospholipid, phosphatidylglycerol. *Biochim. Biophys. Acta (BBA) Biomembr.* **1974**, *339*, 432–437. [CrossRef]
38. Papahadjopoulos, D. Surface properties of acidic phospholipids: Interaction of monolayers and hydrated liquid crystals with uni- and bi-valent metal ions. *Biochim. Biophys. Acta (BBA) Biomembr.* **1968**, *163*, 240–254. [CrossRef]
39. Teissie, J. Interaction of cytochrome c with phospholipid monolayers. Orientation and penetration of protein as functions of the packing density of film, nature of the phospholipids, and ionic content of the aqueous phase. *Biochemistry* **1981**, *20*, 1554–1560. [CrossRef]
40. Peschke, J.; Möhwald, H. Cytochrome c interaction with phospholipid monolayers and vesicles. *Colloids Surf.* **1987**, *27*, 305–323. [CrossRef]
41. Li, J.; Sun, R.; Hao, C.; He, G.; Zhang, L.; Wang, J. The behavior of the adsorption of cytochrome c on lipid monolayers: A study by the Langmuir–Blodgett technique and theoretical analysis. *Biophys. Chem.* **2015**, *205*, 33–40. [CrossRef] [PubMed]
42. Antonov, V.F.; Anosov, A.A.; Norik, V.P.; Smirnova, E.Y. Soft perforation of planar bilayer lipid membranes of dipalmitoylphosphatidylcholine at the temperature of the phase transition from the liquid crystalline to the gel state. *Eur. Biophys. J.* **2005**, *34*, 155–162. [CrossRef] [PubMed]
43. Anosov, A.; Astanina, P.; Proskuryakov, I.; Koplak, O.; Morgunov, R. Surface and Structure of Phosphatidylcholine Membranes Reconstructed with CoFe2O4 Nanoparticles. *Langmuir* **2022**, *38*, 14517–14526. [CrossRef] [PubMed]
44. Freeman, S.A.; Wang, M.A.; Weaver, J.C. Theory of electroporation of planar bilayer membranes: Predictions of the aqueous area, change in capacitance, and pore-pore separation. *Biophys. J.* **1994**, *67*, 42–56. [CrossRef]
45. Anosov, A.; Koplak, O.; Smirnova, E.; Borisova, E.; Korepanova, E.; Derunets, A. Effect of Cobalt Ferrite Nanoparticles in a Hydrophilic Shell on the Conductance of Bilayer Lipid Membrane. *Membranes* **2022**, *12*, 1106. [CrossRef]
46. Powell, K.T.; Derrick, E.G.; Weaver, J.C. A quantitative theory of reversible electrical breakdown in bilayer membranes. *Bioelectrochem. Bioenerg.* **1986**, *15*, 243–255. [CrossRef]
47. Anosov, A.A.; Smirnova, E.Y.; Ryleeva, E.D.; Gligonov, I.A.; Korepanova, E.A.; Sharakshane, A.A. Estimation of the parameters of the Smoluchowski equation describing the occurrence of pores in a bilayer lipid membrane under soft poration. *Eur. Phys. J. E* **2020**, *43*, 1–9. [CrossRef]

membranes MDPI

Review

The Role of Outer Membrane Proteins in UPEC Antimicrobial Resistance: A Systematic Review

Inês C. Rodrigues [1,*], Sílvia C. Rodrigues [2], Filipe V. Duarte [3], Paula M. da Costa [4] and Paulo M. da Costa [1,5]

1 Laboratório de Microbiologia e Tecnologia Alimentar, Departamento de Produção Aquática, Instituto de Ciências Biomédicas Abel Salazar (ICBAS), Rua de Jorge Viterbo Ferreira, 228, 4050-313 Porto, Portugal
2 Pharmaissues, Consultoria, Lda, Rua da Esperança n° 101, Ribeira de Frades, 3045-420 Coimbra, Portugal
3 Centro de Neurociências e Biologia Celular (CNC), Faculdade de Medicina, Pólo 1, Universidade de Coimbra, Rua Larga, 3004-504 Coimbra, Portugal
4 Microbiology Department, Centro Hospitalar Universitário do Porto, Largo do Prof. Abel Salazar, 4099-001 Porto, Portugal
5 Interdisciplinary Centre of Marine and Environmental Research (CIIMAR), Terminal de Cruzeiros do Porto, de Lexões, Av. General Norton de Matos s/n, 4450-208 Matosinhos, Portugal
* Correspondence: icrodrigues@icbas.up.pt; Tel.: +351-917045825

Abstract: Uropathogenic *Escherichia coli* (UPEC) are one of the most common agents of urinary tract infection. In the last decade, several UPEC strains have acquired antibiotic resistance mechanisms and some have become resistant to all classes of antibiotics. UPEC outer membrane proteins (OMPs) seem to have a decisive role not only in the processes of invasion and colonization of the bladder mucosa, but also in mechanisms of drug resistance, by which bacteria avoid killing by antimicrobial molecules. This systematic review was performed according to the PRISMA guidelines, aiming to characterize UPEC OMPs and identify their potential role in antimicrobial resistance. The search was limited to studies in English published during the last decade. Twenty-nine studies were included for revision and, among the 76 proteins identified, seven were associated with antibiotic resistance. Indeed, OmpC was associated with β-lactams resistance and OmpF with β-lactams and fluoroquinolone resistance. In turn, TolC, OmpX, YddB, TosA and murein lipoprotein (Lpp) were associated with fluoroquinolones, enrofloxacin, novobiocin, β-lactams and globomycin resistances, respectively. The clinical implications of UPEC resistance to antimicrobial agents in both veterinary and human medicine must propel the implementation of new strategies of administration of antimicrobial agents, while also promoting the development of improved antimicrobials, protective vaccines and specific inhibitors of virulence and resistance factors.

Keywords: antimicrobial resistance; bacterial proteins; outer membrane proteins; uropathogenic *Escherichia coli*

Citation: Rodrigues, I.C.; Rodrigues, S.C.; Duarte, F.V.; Costa, P.M.d.; Costa, P.M.d. The Role of Outer Membrane Proteins in UPEC Antimicrobial Resistance: A Systematic Review. *Membranes* **2022**, *12*, 981. https://doi.org/10.3390/membranes12100981

Academic Editors: Yosuke Senju and Shiro Suetsugu

Received: 28 September 2022
Accepted: 7 October 2022
Published: 10 October 2022

1. Introduction

Uropathogenic *Escherichia coli* (UPEC) are the most common agents of urinary tract infection (UTI), in both humans and pets [1,2]. It is estimated that UPEC are responsible for more than 80% of UTIs in humans and between 30 to 69% of UTIs in pets [1,3]. A successful combination of virulence factors confers to these strains an increased capacity of ascending through the urinary tract, colonizing, invading and disseminating in the bladder mucosa. From there on, they can continue to progress to the kidneys (pyelonephritis) as well as enter the bloodstream, causing bacteremia [4]. UPEC strains are mainly classified into virulent-phylogroup B2, possessing specific and diverse virulence factors responsible for colonization, invasion and dissemination; to a lesser extent, they are also included in phylogroup D [1,5].

The treatment of UTI relies on antibiotic therapy, such as trimethoprim-sulfamethoxazole, fluoroquinolones and cephalosporins, representing the first-line empirical antibiotics [1,6,7].

According to the European Center for Disease Prevention and Control (ECDC), more than half (58.3%) of the *E. coli* isolates reported in 2018 were resistant to at least one of the antimicrobial groups (i.e., aminopenicillins, fluoroquinolones, third-generation cephalosporins, aminoglycosides and carbapenems) [8]. The dramatically increasing rate of multidrug-resistant strains carries a higher risk of treatment failure, entailing increased costs in health care [2,9]. In fact, resistance to carbapenems, a class of antibiotics used to treat some of the most severe bacterial infections, and resistance to colistin, also a last-line antibiotic, have been described in UPEC isolates [10,11].

Despite a deepening awareness of antimicrobial resistance selection among companion animals, many antimicrobials paramount for human health are used in pets [12]. As people and companion animals share routines and living spaces, the circulation of multi-resistant UPEC strains between companion animals and their owners is of concern, as these strains carry with them virulence and resistance factors [12,13]. Indeed, multiple human associated extended-spectrum cephalosporin-resistant UPEC strains have been isolated from cats and dogs, suggesting clonal dissemination [1].

Whenever antimicrobials are used, bacteria inevitably develop resistance mechanisms, including the modification of proteins of the outer membrane (for example, the loss of porins) [14]. Among Gram-negative bacteria, as is the case of *E. coli*, the presence of the outer membrane is key feature, conferring a crucial and impermeable barrier to the passage of toxic chemicals, such as antibiotics [15,16]. The outer membrane is composed by phospholipids, lipopolysaccharides (LPS), and a myriad of proteins (outer membrane proteins—OMPs), from which OMPs represent approximately half of the cell wall of Gram-negative bacteria [17]. OMPs are responsible for several functions, such as antibiotic and iron transportation, host mucosal adhesion and membrane integrity [5].

OMPs of UPEC, such as flagella, fimbriae, porins, iron receptors and efflux pumps, possess distinctive features, which allow them to invade and colonize the bladder mucosa, representing a crucial tool for both UTI development and antimicrobial resistance. Therefore, since the OMPs are on the bacterial surface, they represent critical targets for the development of improved antimicrobials, protective vaccines and new therapeutic strategies [18].

Through an extensive systematic method, this review aims at an in-depth characterization of the proteins of the outer membrane of UPEC, highlighting the OMPs and their role in antimicrobial resistance.

2. Materials and Methods

This systematic review was performed according to the relevant points of the PRISMA (Preferred Reporting Items for Systematic Reviews and Meta-analyses) guidelines [19].

2.1. Selection Strategy

In September 2022, an independent researcher (I.C.R.) searched the PubMed database without language restrictions in the past ten years (since 2012). Peer-reviewed studies describing antimicrobial susceptibility patterns in terms of proteins located in the outer membrane of UPEC isolated from human or pet samples of any age and region were included in this review. Studies published in a language other than English or Portuguese, publications comprising editorials, comments, letters to the editor, guidelines, theses, books and scientific meeting abstracts, literature reviews or case reports, studies performed specifically in ESBL and/or related to other bacteria and that did not mention uropathogenic *E. coli* or UTI in the title, publications without a description of outer membrane proteins, and studies published before 2012 were excluded. The author (I.C.R.) also reviewed the reference lists from the review articles reported in the PubMed and Web of Science searches to identify possible additional articles for inclusion. A combination of the following search terms was used: Uropathogenic *Escherichia coli* AND membrane proteins. In order to evaluate a potential inclusion in this study, titles and search results were examined.

2.2. Selection Process and Data Extraction

All search results were exported to Microsoft Office™ Excel. Results from the initial search were evaluated separately by the two review authors (I.C.R and S.C.R.) according to the inclusion criteria. First, the results were screened by reading the article titles and excluding articles that were not relevant according to the inclusion criteria. Afterwards, the study abstracts were evaluated and non-relevant articles were excluded. Finally, the full-text articles selected by the two reviewers were collected and assessed for their relevance relative to the inclusion criteria. Any disagreements regarding the eligibility of studies were reconciled at the final step by discussion and consensus.

Upon the final consensus, the following data were extracted from each selected study and validated by the second author (S.C.R.) elaborating a systematic database:

Title, authors and outer(s) membrane(s) detected;
Samples origin and type;
Method used for the detection of outer membrane protein;
Group of outer membrane protein;
Description of outer membrane protein;
Function of outer membrane protein.

2.3. Quality Assessment

Risk of bias (RoB) was examined using the Newcastle–Ottawa scale (NOS) quality assessment scale for cohort studies [20]. The used scale was adapted from the NOS and the assessment was based on three criteria: sample (maximum of 3 points), comparability (maximum of 2 points) and outcome (maximum of 3 points). This was done by I.C.R. and S.C.R. and three categories were decided: low RoB (rating 7 to 8 points), moderate RoB (rating 5 to 6 points) and high RoB (rating from 0 to 4 points). No article was excluded based on this assessment.

3. Results

3.1. Description of Studies

The PRISMA flow diagram summarizes the number of records screened and included (Figure 1). The characteristics and methodological quality of the included studies are presented in Table 1. The literature search using PRISMA identified a total of 1122 studies. After removing the duplicates, 569 were screened for eligibility. After the screening of titles and abstracts, 520 studies were excluded. Full texts of the remaining 48 studies were read and 22 more studies were excluded. At the end, 29 publications were included in this systematic review.

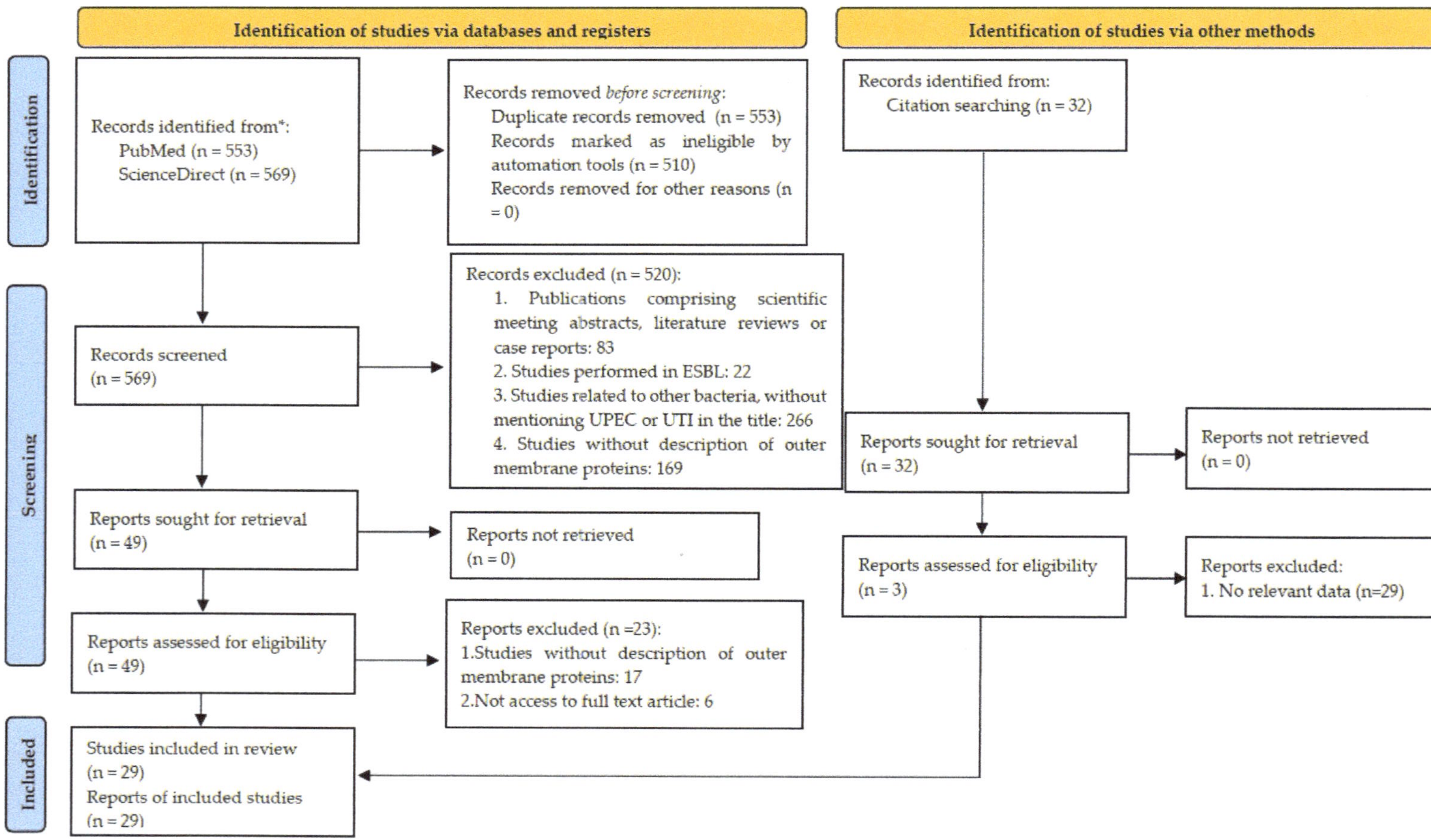

Figure 1. PRISMA 2020 flow diagram. * Consider, if feasible to do so, reporting the number of records indenified from each database or register searched (rather than the total number across all databases/registers). Adapted from [19].

Table 1. Characteristics of included studies after full assessment. Data from the final twenty-nine studies were extracted and systematized. Author; year of publication; the species of the isolates and the type of the sample collected; and the method used for protein characterization and the OMP detected were listed. The RoB assessed through the adaptation of NOS quality scale was also included for each study.

Author	Year of Publication	Origin of Isolates	Sample Type	Method	Risk of Bias	OMP Detected	Reference
Alteri et al.	2008	Human	Urine and blood	2D-DIGE, MS/MS, SDS-PAGE	Low	IutA	[18]
Beck et al.	2016	Human	Urine, blood and faecal	Adhesion assays, immunoblot analysis	Moderate	BamA OmpC OmpF	[21]
Brannon et al.	2013	Human	Urine, blood and fecal	Membrane integrity assays, proteolytic activity, SDS-PAGE, WB	Low	OmpT	[22]
Choi et al.	2020	Human	Urine and blood	Membrane integrity assays	Low	OmpA OmpC OmpF	[23]
Conover et al.	2016	Human	Urine and blood	Crystallization, structure determination and further vaccination of an infected mouse	Low	FimIH	[24]
Dehghani et al.	2016	Human	Urine	SDS-PAGE	Low	OmpA OmpC OmpF	[17]
Desloges et al.	2019	Human	Urine, blood and fecal	Proteolytic activity, SDS-PAGE, WB	Low	OmpP OmpT	[25]

Table 1. *Cont.*

Author	Year of Publication	Origin of Isolates	Sample Type	Method	Risk of Bias	OMP Detected	Reference
Diao et al.	2018	Human	Urine and blood	LC-MS/MS, SDS-PAGE, WB	Low	Bam A BamD BamE c1765 c2482 c3153 chuA FimD IreA KpsD lolB Lpp nmpC OmpA OmpC OmpF OmpX Slp SurA tolC	[26]
Geibel et al.	2013	Non-specified origin	-	Crystallization, structure determination, computational methods	Moderate	Complex FimD–FimC–FimF–FimG–FimH	[27]
Ghosh et al.	2019	Human	Urine	Polymorphism study	Low	FimH	[9]
Greene et al.	2015	Human and non-specified origin	Urine and blood	Electron microscopy, HA, immunoblot analysis	Moderate	Type 1 pili	[28]
He et al.	2015	Human and non-specified origin	Urine and blood	Adhesion and invasion assays, HA, proteolytic activity	Moderate	OmpT	[29]

Table 1. *Cont.*

Author	Year of Publication	Origin of Isolates	Sample Type	Method	Risk of Bias	OMP Detected	Reference
Hederson & Thanassi	2013	Non-specified origin	-	FPLC	Moderate	FimC–FimH chaperone–adhesin Type 1 pilus usher FimD P pilus usher PapC PapD–PapG chaperone–adhesin	[30]
Hirakawa et al.	2019	Human	Urine and blood	Adhesion and biofilm assays, HA, motility assays, SDS-PAGE	Low	TolB	[31]
Huang et al.	2020	Human	Urine	LC-MS/MS, invasion assay, macrophage and serum survival methods, SDS-PAGE, survival in SDS and low pH conditions assays, WB	Low	FliC	[4]
Kalas et al.	2018	Human	Urine and blood	Crystallization, structure determination, in silico docking, immunofluorescence microscopy	Moderate	Type 1 pili F9 pili	[10]
Nielsen et al.	2020	Human and avian	Fecal, vaginal and CSF	In silico analysis	Moderate	OmpA	[32]
Pantel et al.	2016	Human	Urine and blood	Biofilm, nematode-killing and motility assays, immunoblot analysis, SDS-PAGE	Low	AcrA-AcrB-TolC OmpC OmpF	[33]

Table 1. *Cont.*

Author	Year of Publication	Origin of Isolates	Sample Type	Method	Risk of Bias	OMP Detected	Reference
Pantua et al.	2022	Human and non-specified origin	Urine and blood	SDS-PAGE, WB	Low	Lpp	[34]
Piras et al.	2015	Dog	Urine	2D-DIGE, nLC-MS/MS	Low	OmpW	[14]
Ribeiro et al.	2020	Human	Urine	Molecular docking studies	Moderate	QepA OqxAB	[35]
Tavio et al.	2010	Human	Urine	SDS-PAGE	Low	AcrA–AcrB–TolC	[36]
Wurpel et al.	2015	Human	Urine and blood	nLC-MS/MS	Moderate	BamA BamB BamC BtuB ChuA CirA CjrC FadL FepA FhuA FhuE FitA Fiu Flu FyuA GdhA Iha IreA IroN IutA Lpp LptD LptE	[5]

Table 1. *Cont.*

						MalB NlpD NmpC OmpA OmpC OmpF OmpT OmpX Pal RcsF SmpA TraT Tsh Tsx Slp SlyB UTI89_CC1129 UTI89_C4946 VacJ Wzi YbhC YddB YeaF YgeR	
Wurpel et al.	2016	Human	Urine and blood	Adhesin and biofilm assays, nLC-MS/MS, TEM	Low	BtuB ChuA CirA CjrC FepA FhuA Fiu Flu FyuA IutA	[37]

Table 1. *Cont.*

						Iha IreAIroN NmpC OmpA OmpC OmpF OmpT OmpW OmpX UidC UTI89C2234 SlyB	
Xicohtencatl-Cortes et al.	2019	Human	Urine	Adhesion assays, SDS-PAGE, WB	Low	TosA	[38]
Yep et al.	2014	Human	Urine and blood	Bacteriophage adsorption assay, chelator assay, dose-response analysis, HTS	Moderate	RecA	[6]
Zalewska-Piatek et al.	2013	Human and non-specified origin	Urine	Biofilm assays	Moderate	Dr Fimbriae	[39]
Zalewska-Piatek et al.	2015	Human and non-specified origin	Urine	Autoaggregation assay, biofilm, adhesion, invasion assays, immunofluorescence microscopy, SDS-PAGE, WB	Moderate	DraD DraE	[40]

Table 1. *Cont.*

Author	Year of Publication	Origin of Isolates	Sample Type	Method	Risk of Bias	OMP Detected	Reference
Zude et al.	2014	Non-specified origin	-	Autoaggregation and biofilm assays, extracellular matrix protein binding method, immunofluorescence microscopy, WB	Moderate	Ag43	[41]

2D-DIGE: Two-dimensional difference gel electrophoresis; CSF: cerebrospinal fluid; HA: hemagglutination assays; FPLC: fast protein liquid chromatography; HTS: high-throughput screen; LC-MS/MS: liquid chromatography coupled to tandem mass spectrometry; MS/MS: tandem mass spectrometry; NA: not available; nLC-MS/MS: nanoscale liquid chromatography coupled to tandem mass spectrometry; SDS-PAGE: sodium dodecyl sulfate–polyacrylamide gel electrophoresis; TEM: transmission electron microscopy; WB: Western blot.

3.2. Quality Assessment

Based on the quality assessment of studies using the NOS assessment, three studies scored 8 points [14,17,25], which could be regarded as good studies. Another fourteen recorded 7 points, being classified as low Rob [4,9,18,22–24,26,31,33,34,37,38,40,42]. Lastly, twelve studies scored 5–6 points [5,6,10,21,27–30,32,35,39,41], and could be regarded as satisfactory studies.

3.3. Characteristics of Studies and Outcomes Measures

Characteristics of the included studies are presented in Table 1. In a total of 29 studies, 24 encompassed only human isolates [4–6,9,10,17,18,21–26,28,29,31,33–41], one study comprised human and avian isolates [32] and one regarded only dog isolates [14]. Among the first group, seven studies included urine isolates [4,17,35,36,38–40]; fourteen comprised urine and blood isolates [5,6,9,10,18,23,24,26,28,29,31,33,34,37]; three studies involved urine, blood and fecal isolates [21,22,25]; and one study addressed fecal, urine and vaginal samples as well as cerebrospinal liquid. Regarding animal samples, only one study evaluated dog samples [14] and a different one included avian fecal samples along with human isolates [32].

Considering the methodology used to identify directly the OMPs of UPEC strains, sodium dodecyl sulfate–polyacrylamide gel electrophoresis (SDS–PAGE) was performed in eleven studies [4,17,18,22,25,26,31,33,34,38,40], Western blot (WB) in eight [4,22,25,26,34,38,40,41], nano-liquid chromatography tandem mass spectrometry (nLC-MS/MS) in three [5,14,37], liquid chromatography coupled to tandem mass spectrometry (LC-MS/MS) in two [5,24], two-dimensional difference gel electrophoresis (2D DIGE) in two [14,18], fast protein liquid chromatography (FPLC) in one [30] and tandem mass spectrometry in one [18]. In order to evaluate the function of a specific OMP and/or their indirect presence, adhesion methods were performed in seven [21,28,29,31,37,38,40], biofilm assays were performed in six studies [31,33,37,39–41], hemagglutination assays (HA) in three [28,29,31], an invasion method in four [4,28,29,40], crystallization and structure determination in three [10,24,27], immunofluorescence imaging in three [10,40,41], proteolytic activity in three [22,25,29], autoaggregation assays in two [40,41], immunoblot analysis in two [21,33], membrane integrity assays in two [22,23], motility assays in two [31,33], bacteriophage adsorption assay in one [6], a chelator assay in one [6], dose response analysis in one [6], electron microscopy in one [28], an extracellular matrix binding method in one [41], high-throughput screen (HTS) in one [6], in silico docking in one [10], molecular docking studies in one [35], a nematode killing assay in one [33], a polymorphism study in one [9], survival studies in one [4] and transmission electron microscopy (TEM) in one [37].

3.4. Characterization of Outer Membrane Proteins of UPEC

Several proteins were identified as constitutively part of the UPEC strains outer membrane. One study described the role of the BamA protein [21], nine studies described fimbriae and their function [4,5,10,24,27,28,37,39,40] and another ten characterized the porins found in UPEC isolates [5,14,17,22,23,25,29,32,33,35]. Seven studies described the role of lipoproteins [4–6,18,26,31,34], three studies described the importance of efflux pumps [33,35,36], three studies identified iron receptors and siderophores in the outer membrane of UPEC [5,6,18], one study identified TosA protein and its role [38] and one study identified Traf and Ydef [5]. In addition, two studies described the importance of flagella in UPEC motility, encoded by flagellin (FliC) [4,31], and three studies characterized the role of the phase-variable antigen 43 autotransporter protein (Ag43) [5,40,41].

ChuA, FepA, FyuA, NmpC, OmpA, OmpC, OmpF, OmpT, OmpX and SlyB proteins were identified in more than 80% of UPEC strains [5]. According to Wurpel et al. [5], Omp A, OmpX and OmpC were present in all UPEC isolates, while OmpT, NmpC and OmpF were identified in 96%, 89% and 83% of UPEC isolates, respectively. Dehghani et al. [17] have also identified OmpA and OmpC as the most prevalent proteins in the outer membrane of the studied UPEC.

All of the identified proteins have an essential function as described in Table 2. The functional diversity of the OMPs ranges from motility and adhesion (flagella, fimbriae, adhesins) to survival within the urinary tract (iron receptors, siderophores and efflux pumps) [4,5]. Their individual or joint action in the host tissue entails a high level of virulence and pathogenicity, leading to severe UTI symptoms [33,35,36,39].

Table 2. Overview of UPEC OMPs and their association to antimicrobial resistance. Integrated view of results is described, listing the OMP found in UPEC's outer membrane as well as their function.

Group of OMP	OMP	Description	Function	Association to Antimicrobial Resistance
Autotransporter	Ag43	Phase-variable antigen 43 autotransporter protein	Autotransporter	-
BAM complex	BamA	Outer membrane protein assembly factor A	OMP assembly	-
	BamB	Outer membrane protein assembly factor B	OMP assembly	-
	BamC	Outer membrane protein assembly factor C	OMP assembly	-
	BamD	Outer membrane protein assembly factor D	OMP assembly	-
	BamE	Outer membrane protein assembly factor E	OMP assembly	-
Efflux pumps	TolC	Outer membrane protein ToC	Substance efflux	Yes
Fimbriae	DraE	Dr fimbriae subunit	Dr fimbriae	-
	DraD	Dr fimbriae subunit tip	Adhesin	-
	FimA	Type 1 fimbrial protein, A chain	Type 1 pili structure	-
	FimC	Type 1 pilus chaperone, FimC	Type 1 pili structure	-
	FimD	Outer membrane usher protein FimD	Type 1 pili structure	-
	FimF	Outer membrane protein FimF	Type 1 pili structure	-
	FimG	Outer membrane protein FimG	Type 1 pili structure	-
	FimH	Type 1 pilus adhesin, FimH	Adhesin	-
	FimIH	Pili FimH-like adhesin	Adhesin	-
	PapC	P pilus protein PapC	P pili structure	-
	PapD	P pilus chaperone PapD	P pili structure	-
	PapG	P pilus protein PapG	Adhesin	-
Flagella	FliC	Flagellin	Flagella struture	-

Table 2. *Cont.*

Group of OMP	OMP	Description	Function	Association to Antimicrobial Resistance
Iron receptor and siderophores	c2482	Putative outer membrane receptor for iron or colicin	Metal ion receptor and transport	-
	ChuA	Heme/hemoglobin receptor	Metal ion receptor	-
	CirA	Colicin I receptor and iron receptor	Metal ion receptor and transport	-
	CjrC	Putative siderophore receptor	Metal ion receptor	-
	FepA	Ferrienterobactin receptor	Metal ion receptor	-
	FhuA	Ferrichrome–iron receptor	Metal ion receptor	-
	FhuE	Ferri–rhodotorulic siderophore receptor	Metal ion receptor	-
	FitA	Ferrichrome iron transport receptor	Metal ion receptor	-
	Fiu	Catecholate siderophore recetor	Metal ion receptor	-
	FyuA	Yersiniabactin receptor	Metal ion receptor	-
	Iha	Siderophore receptor/adhesion	Metal ion receptor	-
	IreA	Putative siderophore receptor	Metal ion receptor	-
	IroN	Samochelin receptor	Metal ion receptor	-
	IutA	Ferric aerobactin receotor	Metal ion receptor	-
	TonB–ExbB–ExbD	TonB system	Metal ion receptor and transport	-
	UTI89_C1129	Putative heme/hemoglobin receptor	Metal ion receptor	-
	UTI89C2234	Putative iron compound receptor	Metal ion receptor	-
Lipoproteins	lolB	Outer-membrane lipoprotein LolB	LPS organization/synthesis	-
	NlpD	Lipoprotein	LPS organization/synthesis	-
	Lpp	Murein lipoprotein	LPS organization/synthesis	Yes
	LptD	Putative uncharacterized protein	LPS organization/synthesis	-
	LptE	LPS assembly lipoprotein	LPS organization/synthesis	-
	Slp	Outer membrane protein	Unknown	-
	SlyB	Outer membrane protein	Unknown	-
	VacJ	Lipoprotein	Unknown	-
	Wzi	Outer membrane protein	LPS organization/synthesis	-
	YbhC	Putative pectinesterase	LPS organization/synthesis	-
	YgeR	Lipoprotein	LPS organization/synthesis	-

Table 2. *Cont.*

Group of OMP	OMP	Description	Function	Association to Antimicrobial Resistance
Other	c1765	Partial Putative outer membrane channel protein	Unknown	-
	c3153	Putative outer membrane protein of prophage	Unknown	-
	Flu	Antigen 43 AT	Autoaggregatiom	-
	GdhA	NADP-specific glutamate dehydrogenase	Aminoacid metabolismo	-
	KpsD	Putative outer membrane translocon for export of group 2 capsular polysaccharides	Transport	-
	RcsF	Regulator in colanic acid synthesis	Unknown	-
	SmpA	Small protein A	OMP assembly	Yes
	TosA	Putative repeats-in-toxin protein	Nonfimbrial adhesin	-
	TraT	Conjugal transfer surface exclusion protein	Conjugation	-
	Tsh	Temperature sensitive hemagglutinin AT	Unknown	-
	SurA	Outer membrane protein chaperone	OMP assembly	-
	UTI89_C4946	Putative filamentous hemagglutinin	Unknown	-
	YeaF	Putative LPS scaffolding protein	Unknown	
Porin	MalB	Maltose-inducible porin	Porin	-
	NmpC/c2348	Outer membrane porin protein	Porin	-
	OmpA	Outer membrane protein A	Porin	-
	OmpC	Outer membrane protein C	Porin	Yes
	OmpF	Outer membrane protein F	Porin	Yes
	OmpP	Outer membrane protein P	Porin	-
	OmpT	Outer membrane protein T, protéase VII	Proteolysis	-
	OmpW	Outer membrane protein W	Unknown	Yes
	OmpX	Outer membrane protein X	Unknown	-
	Tsx	Nucleoside-specific channel forming protein	Porin	-
	UidC	Outer membrane porin protein	Porin	-
	YddB	Putative porin protein	Porin	Yes
Tol–Pal System	Pal	Peptidoglycan-associated lipoprotein	Membrane integrity	-
Transporters	BtuB	Vitamin B-12 receptor	Transporter	-
	FadL	Bifunctional long-chain fatty acid transporter	Transporter	-

(-) not reported.

3.5. Association to Antimicrobial Resistance

Association to antimicrobial resistance was found for TosA, TolC, Lpp, OmpC, OmpF, OmpW and YddB [14,17,23,32–36,38]. Xicohtencatl-Cortes et al. [38] found that TosA, a nonfimbrial adhesin, binds to host epithelial cells receptors from the upper urinary tract, contributing to the pathogenesis of UPEC. The same authors suggested that UPEC strains producing TosA exhibited a strong association and antimicrobial resistance to β-lactam and non β-lactam antibiotics, such as penicillin, β-lactamase inhibitors and inhibitors of the folate pathway [38].

Overexpression of the AcrA–AcrB–TolC efflux pump complex (QepA and OqxAB) increases antimicrobial resistance of UPEC strains towards β-lactams (mainly cefoxitin), chloramphenicol, cyclines and, notably, quinolones [33,35]. Pantel et al. [33] suggested that the increased efflux capacity of *E. coli* strains involves several efflux pumps from different families, as well as other genes encoding efflux systems less frequently described in clinical resistance (acrEF, mdfA, yhiV, acrD and tehA) that were highly overexpressed.

According to Tavio et al. [36], the AcrA–AcrB–TolC efflux pump is the major multidrug efflux transporter in *E. coli*, allowing the passage of fluoroquinolones, among other antimicrobials.

Pantua et al. [34] found that the deficient production of Lpp leads to an increased OM permeability, a leakage of periplasmic components and an increased outer membrane vesicle release. The same authors also described that complete deletion of Lpp causes globomycin resistance [34].

Among the identified porins, OmpC, OmpF, OmpW and YddB were shown to be the preponderant in antimicrobial resistance strategies of UPEC strains. OmpC is involved in both the transport of small molecular weight hydrophilic materials across the outer membrane and in antimicrobial resistance (e.g., β-lactams antibiotics) [17,32]. OmpF plays an important role in antibiotic transport, such as β-lactams and fluoroquinolones [23]. In fact, a decreased *ompF* expression was related to an increase of resistance towards β-lactam antibiotics in *E. coli* [23,35], since OmpF is the main gate for enrofloxacin entrance [14]. Lastly, YddB (a putative porin protein) seemed to be the major porin involved in the passive transport of novobiocin across the outer membrane [23].

4. Discussion

UPEC is the most frequent uropathogen worldwide, entailing elevated healthcare costs and a high risk of treatment failure in both human and veterinary medicine due to the emergence of multidrug-resistant strains and limited therapeutic options to treat infections caused by these strains.

Just as a One Health unifying approach has the merit of reframing our understanding about the capacity of resistant bacteria to spread between different biomes [42,43], a wide overview of resistance mechanisms used by bacteria to evade antimicrobial treatment will be crucial for overcoming the problem and to find both new drug targets and new structural classes of antibiotics. Herein, among 76 OMPs described in the 29 studies included in this systematic review, seven were responsible for resistance to several antimicrobial drugs. Among them, OmpC is associated with β-lactams resistance and OmpF with β-lactams and fluoroquinolones resistance. With a prevalence of less than 75%, TolC, YddB, TosA and Lpp are also associated with resistance against fluoroquinolones, enrofloxacin, novobiocin, β-lactams and globomycin, respectively. Considering that fluoroquinolones and β-lactams are the antimicrobial classes more commonly used for the treatment of UTI in both human and animals, the spread of these OMPs between UPEC strains is of enormous clinical relevance. Indeed, both antimicrobial drugs are particularly affected by changes in the permeability of the outer membrane of UPEC strains, since β-lactams and fluoroquinolones often use water-filled diffusion channels (porins) to penetrate this external barrier [44]. Despite the predominant mechanism of resistance to β-lactams in UPEC being mainly attributed to the production of β-lactamases [45], OMPs seem to possess essential complementary "tools", as they control the entry of these molecules to the periplasmic space, allowing the production of β-lactamases in sufficient concentrations to achieve the destruction of the antibiotic molecules [46].

Porins are outer membrane β-barrel proteins that cross the membrane cell, responsible for uptake of nutrients, being essential for bacterial survival and growth [33]. On the other hand, they allow the "unwanted" entrance of several antimicrobial molecules; for example, OmpC, OmpF and OmpW are porins responsible for creating a size-selective channel, promoting the transport of small hydrophilic molecules, including β-lactams and fluoroquinolones [47]. Thus, the production of OmpW and OmpC was reduced in tetracycline and nalidixic acid resistant strains, indicating an antibiotic-specific pattern of porin expression [47,48]; and a decreased *ompF* expression was related to an increase of resistance towards β-lactam and fluoroquinolones antibiotics [23,35], since OmpF is the main gate for enrofloxacin entrance [14]. In addition, OmpF and OmpC production seemed to be influenced by the environment of UPEC during a UTI. Urine osmolarity appears to affect *ompF* and *ompC* expression: *ompF* is upregulated under a low osmolarity condition, while *ompC* is upregulated under high osmolarity [49]. Although it is a lesser reported porin, YddB is associated with passive transport of novobiocin across the outer membrane. Presently, novobiocin resistance is not an emerging problem; however, the mechanism by which YddB confers novobiocin resistance has not yet been characterized.

Beyond the alteration of membrane permeability, bacteria is also able to enhance its resistance by the overproduction of efflux pumps (actively expelling the antibiotic out of the cells) [50]. Efflux pumps may be the most rapid acting and effective resistance mechanism in the bacterial defense [50]. Regarding the AcrA–AcrB–TolC efflux pump, Chowdhury et al. [51] hypothesized that one of the intrinsic mechanisms of multidrug-resistant UPEC strains is the overexpression of the AcrAB–TolC efflux pump, probably caused by the mutation of the acrR gene.

Among the OMPs related to antimicrobial resistance, the TosA protein, a nonfimbrial adhesin, is a virulence marker in UPEC strains, since deletion of the tosA gene affected their capacity to colonize the bladder and kidney in a murine model [52,53]. Moreover, multidrug-resistant UPEC isolates expressing the tosA gene maintained a high correlation with virulence genes presence [53]. However, additional research is required to fully understand the role of TosA during UTI.

Another OMP associated to antimicrobial resistance is Lpp, which is the most abundant lipoprotein of *E. coli* (more than 5×10^5 molecules per bacterial cell), being responsible for maintaining the stability and integrity of the outer membrane [54]. Although Pantua et al. [34] described that the complete deletion of Lpp led to globomycin resistance, other authors stated that the deletion of this lipoprotein increased the susceptibility for vancomycin, erythromycin and rifampicin antibiotics [54,55]. In fact, the reduction of Lpp stimulated vesicle formation, decreased membrane integrity and reduced cell invasiveness by turning down biofilm formation [54–56].

In addition to the role of the OMPs, other proteins are also involved in the antimicrobial defence of bacteria. Qnr proteins protect DNA gyrase and type IV topoisomerase IV, being responsible for quinolones resistance in UPEC isolates [35]. These proteins supplement resistance to quinolones by efflux pump activation, altered quinolone target enzymes or deficiencies in OMPs [57]. RecA protein also participates in DNA reparation and activation of the bacterial SOS system [35]. At last, Spr (peptidoglycan endopeptidase) proteins assists peptidoglycan biogenesis in UPEC and flagella expression, which are correlated to a fitness advantage during bladder colonization [4].

Given the potential zoonotic of UPEC, the origin of the isolates of each study was also assessed. Although most of the isolates included in this review were from human origin, UPEC strains are being able to circulate among human and pets populations [13]. Belanger et al. [12] corroborated the phylogenetic proximity between human isolates and those obtained from cohabitant cats and dogs. Thus, we questioned whether the OMPs of UPEC isolated from pets would be different from those isolated from human origin. Piras et al. [14] suggested that if UPEC isolated from humans are exposed to the same concentration of antibiotic that animals are exposed to, then the mechanisms involved in resistance will be similar. Moreover, it requires further characterization of OMPs in UPEC isolated from both human and pet samples, as well as the elucidation of the mechanisms involved in antimicrobial resistance.

UPEC strains have cellular membrane machinery that allows them to be very effective when infecting the urinary system. Combining the fact that *E. coli* possess both a great ability of adaptation to different environments and present highly conserved proteins and organelles, a rapid acquisition of the tools necessary for antimicrobial resistance is promoted. Since available antimicrobial options have become ineffective for treatment, the access to antimicrobial molecules should be rethought and empirical therapy should be avoided. Therefore, the emergence of antimicrobial resistances is an important public health problem. Without implementing more stringent measures, the emergence of "more" multi-resistant strains will occur, leading us to a point of no return.

New therapeutic alternatives need to be discovered to combat the emergence of multidrug UPEC as well as other pathogens [58,59]. Several antimicrobial molecules that interact with OMPs have been already studied, such as darobactin, Polyphor peptide 7 and MRL-494, which affect the integrity of the bacterial outer membrane by inhibition of BamA function [60]. Pilicides and curlicides compounds have the ability to inhibit the pili of UPEC [61,62]. Pre-existing molecules used for other applications may also have antibacterial potential against UPEC, such as nitazoxanide, which is used for intestinal parasitic diseases and also inhibits the function of type 1 and Pap pili [63]. Therefore, the OMPs described in this review seem to be crucial targets in the fight against UPEC and they may also be key targets for sparing other multidrug-resistant pathogens.

This systematic review has several limitations that should be considered. Firstly, the enormous variety of methodology used and often not directly detecting the protein, but rather its effect, may create some bias. Secondly, six published studies were not included as full text articles were not accessible online; thus, some important data on OMPs and antimicrobial resistance may be missing. Thirdly, the isolates included in this review were mostly from human origin; therefore, a comparison of prevalence, function and role in antimicrobial resistance of OMPs isolated from humans and pets was not possible, skewing the research towards humans irremediably. Despite limitations on reporting, the strengths of this review are the low prevalence of high RoB studies and the broader understanding of the potential antibiotic resistance for different outer membrane proteins.

This systematic review highlights the need for further investigation about the role of OMPs in antimicrobial resistance among UPEC, as well as the prudent use of antimicrobial agents in both veterinary and human medicine.

5. Conclusions

In conclusion, our findings showed that several OMPs are related to antimicrobial resistance. ChuA, FepA, FyuA, NmpC, OmpA, OmpC, OmpF, OmpT, OmpX and SlyB were identified in more than 80% of UPEC. However, only OmpC, OmpF, TolC, OmpX, YddB, TosA and Lpp are related with antimicrobial resistance. The classes of antibiotics most affected by antimicrobial resistance conferred through OMPs are fluoroquinolones and β-lactams. These results reflect the urgency of the implementation of new strategies of administration of antimicrobial agents in both veterinary and human medicines, in order to subvert the emergence of multidrug-resistant UPEC strains.

Author Contributions: Conceptualization: I.C.R. and S.C.R.; methodology and formal analysis: I.C.R. and S.C.R.; writing—original draft preparation: I.C.R.; writing—review & editing: S.C.R., F.V.D., P.M.d.C. (Paula M. da Costa) and P.M.d.C. (Paulo M. da Costa). All authors have read and agreed to the published version of the manuscript.

Funding: This research received no external funding.

Institutional Review Board Statement: Not applicable.

Data Availability Statement: Not applicable.

Acknowledgments: The authors Inês C. Rodrigues and Paulo M. da Costa acknowledge the support and the valuable contributions added by Diana Resende in the scope of "Siderophore efflux pump inhibitors (SEPIs) conjugates: A new concept for environmental problems" project, with the reference EXPL/CTA-AMB/0810/2021.

Conflicts of Interest: The authors declare no conflict of interest.

References

1. Nittayasut, N.; Yindee, J.; Boonkham, P.; Yata, T.; Suanpairintr, N.; Chanchaithong, P. Multiple and High-Risk Clones of Extended-Spectrum Cephalosporin-Resistant and $bla_{\text{NDM-5}}$-Harbouring Uropathogenic *Escherichia coli* from Cats and Dogs in Thailand. *Antibiotics* **2021**, *10*, 1374. [CrossRef]
2. Terlizzi, M.E.; Gribaudo, G.; Maffei, M.E. UroPathogenic *Escherichia coli* (UPEC) Infections: Virulence Factors, Bladder Responses, Antibiotic, and Non-antibiotic Antimicrobial Strategies. *Front. Microbiol.* **2017**, *8*, 1566. [CrossRef]
3. Mysorekar, I.U.; Hultgren, S.J. Mechanisms of uropathogenic *Escherichia coli* persistence and eradication from the urinary tract. *Proc. Natl. Acad. Sci. USA* **2006**, *103*, 14170–14175. [CrossRef]
4. Huang, W.-C.; Hashimoto, M.; Shih, Y.-L.; Wu, C.-C.; Lee, M.-F.; Chen, Y.-L.; Wu, J.-J.; Wang, M.-C.; Lin, W.-H.; Hong, M.-Y.; et al. Peptidoglycan Endopeptidase Spr of Uropathogenic *Escherichia coli* Contributes to Kidney Infections and Competitive Fitness During Bladder Colonization. *Front. Microbiol.* **2020**, *11*, 586214. [CrossRef]
5. Wurpel, D.J.; Moriel, D.G.; Totsika, M.; Easton, D.M.; Schembri, M.A. Comparative analysis of the uropathogenic *Escherichia coli* surface proteome by tandem mass-spectrometry of artificially induced outer membrane vesicles. *J. Proteom.* **2015**, *115*, 93–106. [CrossRef]
6. Yep, A.; McQuade, T.; Kirchhoff, P.; Larsen, M.; Mobley, H.L.T. Inhibitors of TonB Function Identified by a High-Throughput Screen for Inhibitors of Iron Acquisition in Uropathogenic *Escherichia coli* CFT073. *mBio* **2014**, *5*, e01089-13. [CrossRef]
7. Zogg, A.L.; Zurfluh, K.; Schmitt, S.; Nüesch-Inderbinen, M.; Stephan, R. Antimicrobial resistance, multilocus sequence types and virulence profiles of ESBL producing and non-ESBL producing uropathogenic *Escherichia coli* isolated from cats and dogs in Switzerland. *Veter- Microbiol.* **2018**, *216*, 79–84. [CrossRef]
8. ECDC. Surveillance of antimicrobial resistance in Europe. *Br. Med. J.* **2018**, *317*, 614–615.
9. Ghosh, A.; Mukherjee, M. Incidence of multidrug resistance, pathogenicity island markers, and pathoadaptive FimH mutations in uropathogenic *Escherichia coli* isolated from asymptomatic hospitalized patients. *Folia Microbiol.* **2019**, *64*, 587–600. [CrossRef]
10. Kalas, V.; Hibbing, M.E.; Maddirala, A.R.; Chugani, R.; Pinkner, J.S.; Mydock-McGrane, L.K.; Conover, M.S.; Janetka, J.W.; Hultgren, S.J. Structure-Based Discovery of Glycomimetic FmlH Ligands As Inhibitors of Bacterial Adhesion During Urinary Tract Infection. *Proc. Natl. Acad. Sci. USA* **2018**, *115*, E2819–E2828. [CrossRef]
11. Matin, F.Z.; Rezatofighi, S.E.; Ardakani, M.R.; Akhoond, M.R.; Mahmoodi, F. Virulence characterization and clonal analysis of uropathogenic *Escherichia coli* metallo-beta-lactamase-producing isolates. *Ann. Clin. Microbiol. Antimicrob.* **2021**, *20*, 50. [CrossRef]
12. Bélanger, L.; Garenaux, A.; Harel, J.; Boulianne, M.; Nadeau, E.; Dozois, C.M. *Escherichia coli* from animal reservoirs as a potential source of human extraintestinal pathogenic *E. coli*. *FEMS Immunol. Med. Microbiol.* **2011**, *62*, 1–10. [CrossRef]
13. Salgado-Caxito, M.; Benavides, J.A.; Adell, A.D.; Paes, A.C.; Moreno-Switt, A.I. Global prevalence and molecular characterization of extended-spectrum β-lactamase producing-*Escherichia coli* in dogs and cats—A scoping review and meta-analysis. *One Health* **2021**, *12*, 100236. [CrossRef]
14. Piras, C.; Soggiu, A.; Greco, V.; Martino, P.A.; Del Chierico, F.; Putignani, L.; Urbani, A.; Nally, J.E.; Bonizzi, L.; Roncada, P. Mechanisms of antibiotic resistance to enrofloxacin in uropathogenic *Escherichia coli* in dog. *J. Proteom.* **2015**, *127*, 365–376. [CrossRef]
15. Matamoros-Recio, A.; Franco-Gonzalez, J.F.; Forgione, R.E.; Torres-Mozas, A.; Silipo, A.; Martín-Santamaría, S. Understanding the Antibacterial Resistance: Computational Explorations in Bacterial Membranes. *ACS Omega* **2021**, *6*, 6041–6054. [CrossRef]
16. Silhavy, T.J.; Kahne, D.; Walker, S. The Bacterial Cell Envelope. *Cold Spring Harb. Perspect. Biol.* **2010**, *2*, 1–16. Available online: https://www.ncbi.nlm.nih.gov/pmc/articles/PMC2857177/pdf/cshperspect-PRK-a000414.pdf (accessed on 4 September 2022). [CrossRef]
17. Dehghani, B.; Mottamedifar, M.; Khoshkharam-Roodmajani, H.; Hassanzadeh, A.; Zomorrodian, K.; Rahimi, A. SDS-PAGE Analysis of the Outer Membrane Proteins of Uropathogenic *Escherichia coli* Isolated from Patients in Different Wards of Nemazee Hospital, Shiraz, Iran. *Iran. J. Med. Sci.* **2016**, *41*, 399–405.
18. Alteri, C.; Mobley, H.L.T. Quantitative Profile of the Uropathogenic *Escherichia coli* Outer Membrane Proteome during Growth in Human Urine. *Infect. Immun.* **2007**, *75*, 2679–2688. [CrossRef]
19. Page, M.J.; McKenzie, J.E.; Bossuyt, P.M.; Boutron, I.; Hoffmann, T.C.; Mulrow, C.D.; Shamseer, L.; Tetzlaff, J.M.; Akl, E.A.; Brennan, S.E.; et al. The PRISMA 2020 statement: An updated guideline for reporting systematic reviews. *BMJ* **2021**, *372*, n71. [CrossRef]
20. Wells, G.; Shea, B.; O'Connel, D.; Peterson, J.; Welch, V.; Losos, M.; Tugwell, P. The Newcastle-Ottawa Scale (NOS) for assessing the quality of nonrandomised studies in meta-analyses. 2021. Available online: https://www.ohri.ca//programs/clinical_epidemiology/oxford.asp (accessed on 25 August 2022).

21. Beck, C.M.; Willett, J.L.E.; Cunningham, D.A.; Kim, J.J.; Low, D.A.; Hayes, C.S. CdiA Effectors from Uropathogenic *Escherichia coli* Use Heterotrimeric Osmoporins as Receptors to Recognize Target Bacteria. *PLoS Pathog.* **2016**, *12*, e1005925. [CrossRef]
22. Brannon, J.R.; Thomassin, J.L.; Desloges, I.; Gruenheid, S.; Le Moual, H. Role of uropathogenic *Escherichia coli* OmpT in the resistance against human cathelicidin LL-37. *FEMS Microbiol. Lett.* **2013**, *345*, 64–71. [CrossRef]
23. Choi, U.; Lee, C.-R. Distinct Roles of Outer Membrane Porins in Antibiotic Resistance and Membrane Integrity in *Escherichia coli*. *Front. Microbiol.* **2019**, *10*, 953. [CrossRef]
24. Conover, M.S.; Ruer, S.; Taganna, J.; Kalas, V.; De Greve, H.; Pinkner, J.; Dodson, K.W.; Remaut, H.; Hultgren, S.J. Inflammation-Induced Adhesin-Receptor Interaction Provides a Fitness Advantage to Uropathogenic *E. coli* during Chronic Infection. *Cell Host. Microbe.* **2016**, *176*, 498–503. [CrossRef]
25. Desloges, I.; Taylor, J.A.; Leclerc, J.M.; Brannon, J.R.; Portt, A.; Spencer, J.D.; Dewar, K.; Marczynski, G.T.; Manges, A.; Gruenheid, S.; et al. Identification and characterization of OmpT-like proteases in uropathogenic *Escherichia coli* clinical isolates. *Microbiologyopen* **2019**, *8*, e915. [CrossRef]
26. Diao, J.; Bouwman, C.; Yan, D.; Kang, J.; Katakam, A.; Liu, P.; Pantua, H.; Xu, M.; Kapadia, S. Peptidoglycan Association of Murein Lipoprotein is required for KpsD-Dependent Group 2 Capsular Polysaccharide Expression and Serum Resistance in a Uropathogenic *Escherichia coli* Isolate. *Am. Soc. Fir. Microbiol.* **2018**, *200*, 500–502. [CrossRef]
27. Geibel, S.; Procko, E.; Hultgren, S.J.; Baker, D.; Waksman, G. Structural and energetic basis of folded-protein transport by the FimD usher. *Nature* **2013**, *496*, 243–246. [CrossRef]
28. Greene, S.E.; Hibbing, M.E.; Janetka, J.; Chen, S.; Hultgren, S.J. Human Urine Decreases Function and Expression of Type 1 Pili in Uropathogenic *Escherichia coli*. *mBio* **2015**, *6*, e00820-15. [CrossRef]
29. He, X.L.; Wang, Q.; Peng, L.; Qu, Y.-R.; Puthiyakunnon, S.; Liu, X.-L.; Hui, C.Y.; Boddu, S.; Cao, H.; Huang, S.-H. Role of uropathogenic *Escherichia coli* outer membrane protein T in pathogenesis of urinary tract infection. *Pathog. Dis.* **2015**, *73*, ftv006. [CrossRef]
30. Henderson, N.S.; Thanassi, D.G. Purification of the Outer Membrane Usher Protein and Periplasmic Chaperone-Subunit Complexes from the P and Type 1 Pilus Systems. *Methods Mol. Biol.* **2013**, *966*, 37–52. [CrossRef]
31. Hirakawa, H.; Suzue, K.; Kurabayashi, K.; Tomita, H. The Tol-Pal System of Uropathogenic *Escherichia coli* Is Responsible for Optimal Internalization Into and Aggregation Within Bladder Epithelial Cells, Colonization of the Urinary Tract of Mice, and Bacterial Motility. *Front. Microbiol.* **2019**, *10*, 1827. [CrossRef]
32. Nielsen, D.W.; Ricker, N.; Barbieri, N.L.; Allen, H.K.; Nolan, L.K.; Logue, C.M. Outer membrane protein A (OmpA) of extraintestinal pathogenic *Escherichia coli*. *BMC Res. Notes* **2020**, *13*, 51. [CrossRef]
33. Pantel, A.; Dunyach-Remy, C.; Essebe, C.N.; Mesureur, J.; Sotto, A.; Pagès, J.-M.; Nicolas-Chanoine, M.-H.; Lavigne, J.-P. Modulation of Membrane Influx and Efflux in *Escherichia coli* Sequence Type 131 Has an Impact on Bacterial Motility, Biofilm Formation, and Virulence in a Caenorhabditis elegans Model. *Antimicrob. Agents Chemother.* **2016**, *60*, 2901–2911. [CrossRef]
34. Pantua, H.; Skippington, E.; Braun, M.-G.; Noland, C.L.; Diao, J.; Peng, Y.; Gloor, S.L.; Yan, D.; Kang, J.; Katakam, A.K.; et al. Unstable Mechanisms of Resistance to Inhibitors of *Escherichia coli* Lipoprotein Signal Peptidase. *mBio* **2020**, *11*, e02018-20. [CrossRef]
35. Ribeiro, C.D.S.; Martins, W.M.B.D.S.; da Silva, A.A.; Gales, A.C.; Rando, D.G.G.; Minarini, L.A.D.R. Exposure to sub-inhibitory ciprofloxacin and nitrofurantoin concentrations increases recA gene expression in uropathogenic *Escherichia coli*: The role of RecA protein as a drug target. *Eur. J. Pharm. Sci.* **2020**, *146*, 105268. [CrossRef]
36. Tavio, M.M.; Aquili, V.D.; Poveda, J.B.; Antunes, N.T.; Sanchez-Cespedes, J.; Vila, J. Quorum-sensing regulator sdiA and marA overexpression is involved in in vitro-selected multidrug resistance of *Escherichia coli*. *J. Antimicrob. Chemother.* **2010**, *65*, 1178–1186. [CrossRef]
37. Wurpel, D.J.; Totsika, M.; Allsopp, L.; Webb, R.I.; Moriel, D.G.; Schembri, M.A. Comparative proteomics of uropathogenic *Escherichia coli* during growth in human urine identify UCA-like (UCL) fimbriae as an adherence factor involved in biofilm formation and binding to uroepithelial cells. *J. Proteom.* **2016**, *131*, 177–189. [CrossRef]
38. Xicohtencatl-Cortes, J.; Cruz-Córdova, A.; Cázares-Domínguez, V.; Escalona-Venegas, G.; Zavala-Vega, S.; Arellano-Galindo, J.; Romo-Castillo, M.; Hernández-Castro, R.; Ochoa, S.A.; Luna-Pineda, V.M. Uropathogenic *Escherichia coli* strains harboring tosA gene were associated to high virulence genes and a multidrug-resistant profile. *Microb. Pathog.* **2019**, *134*, 103593. [CrossRef]
39. Zalewska-Piątek, B.; Wilkanowicz, S.; Bruździak, P.; Piątek, R.; Kur, J. Biochemical characteristic of biofilm of uropathogenic *Escherichia coli* Dr + strains. *Microbiol. Res.* **2013**, *168*, 367–378. [CrossRef]
40. Zalewska-Piątek, B.M.; Piatek, R.; Olszewski, M.; Kur, J. Identification of antigen Ag43 in uropathogenic *Escherichia coli* Dr+ strains and defining its role in the pathogenesis of urinary tract infections. *Microbiology* **2015**, *161*, 1034–1049. [CrossRef]
41. Zude, I.; Leimbach, A.; Dobrindt, U. Prevalence of autotransporters in *Escherichia coli*: What is the impact of phylogeny and pathotype? *Int. J. Med. Microbiol.* **2014**, *304*, 243–256. [CrossRef]
42. Martins da Costa, P.; Loureiro, L.; Matos, A.J.F. Transfer of Multidrug-Resistant Bacteria Between Intermingled Ecological Niches: The Interface Between Humans, Animals and the Environment. *Int. J. Environ. Res. Public Health* **2013**, *10*, 278–294. [CrossRef]
43. Collignon, P.J.; McEwen, S.A. One Health—Its Importance in Helping to Better Control Antimicrobial Resistance. *Trop. Med. Infect. Dis.* **2019**, *4*, 22. [CrossRef]
44. Pagès, J.-M.; James, C.E.; Winterhalter, M. The porin and the permeating antibiotic: A selective diffusion barrier in Gram-negative bacteria. *Nat. Rev. Microbiol.* **2008**, *6*, 893–903. [CrossRef]

45. Worthington, R.J.; Melander, C. Overcoming Resistance to β-Lactam Antibiotics. *J. Org. Chem.* **2014**, *78*, 4207–4213. [CrossRef]
46. Munita, J.; Arias, C. Mechanisms of Antibiotic Resistance. *Microbiol. Spectr.* **2016**, *17*, 119–127.
47. Cardoso, M.H.; Almeida, K.C.D.; Cândido, E.D.S.; Murad, A.M. Comparative NanoUPLC-MS E analysis between magainin I-susceptible and -resistant *Escherichia coli* strains. *Sci. Rep.* **2017**, *7*, 4197. [CrossRef]
48. Liu, Y.-F.; Yan, J.-J.; Lei, H.-Y.; Teng, C.-H.; Wang, M.-C.; Tseng, C.-C.; Wu, J.-J. Loss of Outer Membrane Protein C in *Escherichia coli* Contributes to Both Antibiotic Resistance and Escaping Antibody-Dependent Bactericidal Activity. *Infect. Immun.* **2012**, *80*, 1815–1822. [CrossRef]
49. Kishii, R.; Takei, M. Relationship between the expression of ompF and quinolone resistance in *Escherichia coli*. *J. Infect. Chemother.* **2009**, *15*, 361–366. [CrossRef]
50. Fernández, L.; Hancock, R.E.W. Adaptive and Mutational Resistance: Role of Porins and Efflux Pumps in Drug Resistance. *Clin. Microbiol. Rev.* **2012**, *25*, 661–681. [CrossRef]
51. Chowdhury, N.; Suhani, S.; Purkaystha, A.; Begum, M.K.; Raihan, T.; Alam, J. Identification of AcrAB-TolC Efflux Pump Genes and Detection of Mutation in Efflux Repressor AcrR from Omeprazole Responsive Multidrug-Resistant *Escherichia coli* Isolates Causing Urinary Tract Infections. *Microbiol. Insights* **2019**, *12*, 1178636119889629. [CrossRef]
52. Ochoa, S.A.; Cruz-Córdova, A.; Luna-Pineda, V.M.; Reyes, J.P.; Cázares-Domínguez, V.; Escalona, G.; Sepúlveda-González, M.E.; López-Montiel, F.; Arellano-Galindo, J.; López-Martínez, B.; et al. Multidrug- and Extensively Drug-Resistant Uropathogenic *Escherichia coli* Clinical Strains: Phylogenetic Groups Widely Associated with Integrons Maintain High Genetic Diversity. *Front. Microbiol.* **2016**, *7*, 2042. [CrossRef]
53. Contreras-Alvarado, L.M.; Zavala-Vega, S.; Cruz-Córdova, A.; Reyes-Grajeda, J.P.; Escalona-Venegas, G.; Flores, V.; Alcázar-López, V.; Arellano-Galindo, J. Molecular Epidemiology of Multidrug-Resistant Uropathogenic *Escherichia coli* O25b Strains Associated with Complicated Urinary Tract Infection in Children. *Microorganisms* **2021**, *9*, 2299. [CrossRef]
54. Ebbensgaard, A.E.; Mordhorst, H.; Aarestrup, F.; Hansen, E.B. The Role of Outer Membrane Proteins and Lipopolysaccharides for the Sensitivity of *Escherichia coli* to Antimicrobial Peptides. *Front. Microbiol.* **2018**, *9*, 2153. [CrossRef]
55. Mathelié-Guinlet, M.; Asmar, A.T.; Collet, J.-F.; Dufrêne, Y.F. Lipoprotein Lpp regulates the mechanical properties of the E. coli cell envelope. *Nat. Commun.* **2020**, *11*, 1789. [CrossRef]
56. Asmar, A.T. Lpp, the Braun lipoprotein, turns 50—Major achievements and remaining issues. *FEMS Microbiol. Lett.* **2018**, *365*, fny199. [CrossRef]
57. Poirel, L.; Cattoir, V.; Nordmann, P. Plasmid-Mediated Quinolone Resistance; Interactions between Human, Animal, and Environmental Ecologies. *Front. Microbiol.* **2012**, *3*, 24. [CrossRef]
58. Iacovelli, R.; Sokolova, N.; Haslinger, K. *Striving for Sustainable Biosynthesis: Discovery, Diversification, and Production of Antimicrobial Drugs in Escherichia coli*; Portland Press Ltd.: London, UK, 2022; p. 6.
59. Kot, B. Antibiotic Resistance Among Uropathogenic *Escherichia coli*. *Pol. J. Microbiol.* **2019**, *68*, 403–415. [CrossRef]
60. Peterson, J.H.; Doyle, M.T.; Bernstein, H.D. Small Molecule Antibiotics Inhibit Distinct Stages of Bacterial Outer Membrane Protein Assembly. *mBio* **2022**, e02286-22. [CrossRef]
61. Karam, M.R.A.; Habibi, M.; Bouzari, S. Urinary tract infection: Pathogenicity, antibiotic resistance and development of effective vaccines against Uropathogenic *Escherichia coli*. *Mol. Immunol.* **2019**, *108*, 56–67. [CrossRef]
62. Piatek, R.; Zalewska-Piatek, B.; Dzierzbicka, K.; Makowiec, S.; Pilipczuk, J.; Szemiako, K.; Cyranka-Czaja, A.; Wojciechowski, M. Pilicides inhibit the FGL chaperone/usher assisted biogenesis of the Dr fimbrial polyadhesin from uropathogenic *Escherichia coli*. *BMC Microbiol.* **2013**, *13*, 131. [CrossRef]
63. Chahales, P.; Hoffman, P.S.; Thanassi, D.G. Nitazoxanide Inhibits Pilus Biogenesis by Interfering with Folding of the Usher Protein in the Outer Membrane. *Antimicrob. Agents Chemother.* **2016**, *60*, 2028–2038. [CrossRef]

membranes

MDPI

Article

Mechanism of Action and Structure–Activity Relationships of Tetracyclic Small Molecules Acting as Universal Positive Allosteric Modulators of the Cholecystokinin Receptor

Daniela G. Dengler [1,*,†], Kaleeckal G. Harikumar [2,†], Alice Yen [1], Eduard A. Sergienko [1,*,†] and Laurence J. Miller [2,*,†]

1 Conrad Prebys Center for Chemical Genomics, Sanford Burnham Prebys Medical Discovery Institute, La Jolla, CA 92037, USA
2 Department of Molecular Pharmacology and Experimental Therapeutics, Mayo Clinic, Scottsdale, AZ 85259, USA
* Correspondence: ddengler@sbpdiscovery.org (D.G.D.); esergien@sbpdiscovery.org (E.A.S.); miller@mayo.edu (L.J.M.)
† These authors have contributed equally to this work.

Abstract: As part of an ongoing effort to develop a drug targeting the type 1 cholecystokinin receptor (CCK1R) to help prevent and/or treat obesity, we recently performed a high throughput screening effort of small molecules seeking candidates that enhanced the action of the natural agonist, CCK, thus acting as positive allosteric modulators without exhibiting intrinsic agonist action. Such probes would be expected to act in a temporally finite way to enhance CCK action to induce satiety during and after a meal and potentially even modulate activity at the CCK1R in a high cholesterol environment present in some obese patients. The current work focuses on the best scaffold, representing tetracyclic molecules identified through high throughput screening we previously reported. Extensive characterization of the two top "hits" from the previous effort demonstrated them to fulfill the desired pharmacologic profile. We undertook analog-by-catalog expansion of this scaffold using 65 commercially available analogs. In this effort, we were able to eliminate an off-target effect observed for this scaffold while retaining its activity as a positive allosteric modulator of CCK1R in both normal and high cholesterol membrane environments. These insights should be useful in the rational medicinal chemical enhancement of this scaffold and in the future development of candidates to advance to pre-clinical proof-of-concept and to clinical trials.

Keywords: cholecystokinin; cholecystokinin receptor; G protein-coupled receptor; positive allosteric modulator; obesity

Citation: Dengler, D.G.; Harikumar, K.G.; Yen, A.; Sergienko, E.A.; Miller, L.J. Mechanism of Action and Structure–Activity Relationships of Tetracyclic Small Molecules Acting as Universal Positive Allosteric Modulators of the Cholecystokinin Receptor. *Membranes* **2023**, *13*, 150. https://doi.org/10.3390/membranes13020150

Academic Editors: Shiro Suetsugu and Yosuke Senju

Received: 20 December 2022
Revised: 17 January 2023
Accepted: 20 January 2023
Published: 24 January 2023

1. Introduction

The type 1 cholecystokinin receptor (CCK1R) has been recognized as a key physiologic regulator of appetite and a potential target for anti-obesity therapy [1]. However, while multiple full agonists of CK1R have been developed, these molecules have failed to achieve the primary endpoint in clinical obesity trials of being superior to short-term dieting to induce weight loss [2–4]. The enhancement of potency and/or duration of action of such candidate molecules has not been pursued due to concerns about on-target side effects and potential toxicity [3,5]. We recently reported a high throughput screening effort to identify potential molecules with a distinct pharmacologic profile representing positive allosteric modulators (PAMs) of CCK action at this receptor that also possess minimal intrinsic agonist activity [6], a strategy we proposed to increase the safety and effectiveness of such drugs [3,5]. We were also particularly interested in identifying such compounds that were active both in a normal and high cholesterol membrane environment representing universal PAMs. Of the three chemical classes of molecules identified in that effort [6], we focused on those molecules from the initial high throughput screening effort that incorporate a tetracyclic scaffold ("hit 1" and "hit 6") as being the most promising.

In the present project, we explore the molecular basis of action of this class of CCK1R PAM ligands and expand our understanding of structure–activity relationships of molecules with this tetracyclic scaffold, exploring structural modifications present in 65 commercially available analogs. Since our preliminary work with the original "hits" having this scaffold demonstrated an effect on another class A G protein-coupled receptor, the vasopressin 2 receptor, a parallel goal was to establish the feasibility of teasing apart on-target and off-target activity as an indication of the general selectivity of this scaffold among this family of receptors. Such insights will be useful to guide further rational enhancement of the pharmacologic activity profile and development of drug candidates to advance toward future preclinical and potential clinical trials.

2. Materials and Methods

2.1. Materials

CCK peptide analogues were custom synthesized in our laboratory, purified to homogeneity, and verified by mass spectrometry [7]. These include natural CCK-26-33 (CCK-8); a partial agonist analogue, D-Tyr-Gly-[(Nle^{28,31})CCK-26-32]-phenethyl ester (CCK-OPE); and a fluorescent analogue of this hormone, alexa488-D-Tyr-Gly-[(Nle^{28,31})CCK-26-33] (alexa488-CCK) [8]. The sulfated CCK octapeptide (CCK-8) (#4033010) used in the structure–activity relationship studies and arginine vasopressin (#4012215) were purchased from Bachem AG (Bubendorf, Switzerland). CCK-33 was purchased from Peptides International (Louisville, KY, USA). Clonal receptor-bearing cell lines were prepared from non-CCK receptor-bearing CHO-K1 cells or HEK-293 cells (American Type Culture Collection, ATCC), as previously described [9]. In select experiments, the cholesterol composition of cell lines was enhanced by treatment with methyl-β-cyclodextrin-cholesterol inclusion complex, as we previously described [10].

2.2. Methods

Biological activity was quantified in the CHO cell lines described above using intracellular calcium assays that were performed, as previously described, using Fura-8-AM (AAT Bioquest, Pleasonton, CA, USA) [11]. Full agonist dose–response curves were performed along with determination of maximal responses to 0.1 mM ATP, targeting an endogenous receptor present on the parental cells. Concentration–response curves of peak intracellular calcium responses were analyzed and plotted as percentages of maximal responses using non-linear regression analysis in Prism 9.1 (GraphPad).

For structure–activity relationship studies, we performed myo-inositol-1-phosphate (IP-One) accumulation assays with TR-FRET technology applied to the HEK-293 cell line overexpressing CCK1Rs, as we had described in our high throughput screening effort [6]. In brief, thawed cryopreserved cell stocks were re-suspended in IP-One assay media consisting of phenol-red free DMEM (Corning, Corning, NY, USA, #17-205-CV) with 10% FBS, 1% penicillin (10,000 units)/streptomycin (10 mg) (Pen/Strep, Thermo Fisher Scientific, Waltham, MA, USA, Gibco #15140122), and 1% L-glutamine (200 mM) (Gibco #25030081) and diluted to required cell densities. Then, the cell suspension was dispensed into a 1536-well tissue culture microplate (Corning, #3727), and the plate was incubated overnight at 37 °C and 5% CO^2. After 16–20 h, DMSO or compounds were added with an Echo liquid handler (Labcyte, San Jose, CA, USA), resulting in a top final compound concentration of 50 μM diluted 2-fold for 16-point dose–responses. For the positive allosteric modulator (PAM) format, orthosteric stimulator (CCK) dilutions for control wells were prepared in assay media containing 150 mM lithium chloride (LiCl, 50 mM final, Sigma-Aldrich, St Louis, MO, USA, #L7026). After a 30 min incubation of test compounds at 37 °C and 5% CO_2, orthosteric stimulator dilutions (for PAM format) or equivalent dilutions of DMSO (for agonist format) were added to designated wells using BioRaptr (Beckman Coulter, Brea, CA, USA). In addition, CCK control ligand dilutions were included on each test plate. After that, the plate was incubated for 1 h at 37 °C and 5% CO^2 and equilibrated to room temperature. Detection reagents from the IP-One Gq detection kit from Cisbio (Cisbio US Inc., Bedford, MA, USA, #62IPAPEJ) were added, and after 1 h at RT, IP-One content was measured with a PHERAstar FSX microplate reader (BMG Labtech, Ortenberg, Germany). Control ligand dose–responses were analyzed using GraphPad Prism 9.3.1 (San Diego, CA, USA) to validate adequate control and test well concentrations for positive allosteric modulation (PAM) and agonist formats. Dose–response curves were performed as 16-point 2-fold dilutions in duplicates in at least three independent experiments. Large data sets were uploaded and analyzed with CBIS (Chemical and Biology Information System software, ChemInnovation Software, Inc., San Diego, CA, USA). Further data analysis for detailed SAR studies was conducted using the D360 software (Certara).

Possible off-target biological effects. In our previous report [6], the lead compounds were tested for possible off-target effects at the purinergic receptor (P2YR), representing another Gq-coupled class A GPCR, where they were found to have no off-target activity. Here, we tested them at the vasopressin-2 receptor (AVP2R), representing a class A GPCR structurally related to CCK1R, which is coupled to Gs. The activity at AVP2R was tested using AVP2R-overexpressing CHO-K1 (PAM format) or HEK-293 (agonist format) cells, stimulating the cells with vasopressin. Original hits were tested in both agonist and PAM formats. Since all recognized compound responses were present in both formats, we continued to screen analogs in only the PAM format (also able to capture agonist effects) in CHO-K1 AVP2R cells for potential enhanced sensitivity. The dose–response data sets for original hits were combined, as the PAM and agonist effects were indistinguishable. In brief, arginine vasopressin (AVP) orthosteric stimulator dilutions and ligand and compound dose–response titrations in DMSO were transferred onto a 1536-well plate (Corning #3725) using an Echo liquid handler (Labcyte). Thawed cryopreserved cell stocks were diluted in stimulation buffer consisting of HBSS (Hank's Balanced Salt Solution with Ca^{2+} and Mg^{2+}, Gibco #24020117), 5 mM HEPES (hydroxyethyl piperazineethanesulfonic acid), 0.5 mM IBMX (3-isobutyl-1-methylxanthine, Sigma-Aldrich), and 0.075% BSA (7.5% DTPA-purified bovine serum albumin, PerkinElmer). The resulting cell suspension was added to the microplate and incubated with test compounds at room temperature for 30 min. Then, the detection reagents from the Cisbio cAMP Gs Dynamic HTRF detection kit were added. The plate was kept at room temperature for 30 min and read on a PHERAstar FSX microplate reader. Dose–response curves were performed as 16-point 2-fold dilutions in duplicate in at least two independent experiments.

For CCK1R-induced Gs signaling studies, we performed cAMP accumulation assays as described above and as previously reported [6], utilizing the HEK-293 cells overexpressing CCK1Rs and the EC_{20} concentration of CCK for screening in PAM format.

Regarding CCK binding kinetics, fluorescence polarization assays for binding and dissociation of the fluorescent CCK probe, alexa488-CCK, were performed as we described using a PHERAstar FSX instrument (BMG Labtech, Cary, NC, USA) [12].

Receptor internalization assays were performed with the fluorescent alexa488-CCK ligand used to visualize cell surface receptors after treatment and fixation with 2% paraformaldehyde (Electron Microscopy Sciences, Cat# 15710), as we described [12]. Images were acquired with a Zeiss Axiovert 200M inverted epifluorescence microscope.

For structure–activity analysis, 65 commercially available compounds containing the tetracyclic scaffold were included in this analysis to gain insights into structure–activity relationships (structures shown in SAR tables).

The potency and efficacy data for these compounds were used to calculate their activity scores using the following formula: Activity Score = normalized $E_{max} \times pEC_{50}$. Normalized E_{max} represents the efficacy of the compound as a fractional response relative to that of the CCK control ligand. pEC_{50} represents the positive logarithmic value of compound EC_{50} concentrations in (M).

Because this was not a systematic prospective synthetic series, compounds often modified more than one position around the tetracyclic scaffold at a time. We elected to number the entire list of analogues in order from highest to lowest activity score (described below) in the primary screen, the PAM format CCK-stimulated IP-One assay (CMPs 1-65). In total, 37 of these compounds had a measurable activity score in this assay. We grouped the compounds in six SAR tables to focus on groups of related structures, and we ordered each group of structures from most to least active in the primary assay when possible. By comparing the position in each table and the compound number, it was clear when multiple modifications may have played a role in the series. These compounds were also characterized with the off-target AVP2R assay described above, with results listed in the same tables.

Regarding statistical analysis, all assays were performed in duplicate and repeated in at least three independent assays (number of such assays, "*n*" provided). Differences between experimental groups were evaluated using one-way ANOVA or the Mann–Whitney test, with $p < 0.05$ considered as significant.

3. Results

The two hits initially identified in our high throughput screen for positive allosteric modulators (PAMs) of CCK action at the CCK1R possessing minimal intrinsic agonist activity [6] were further characterized to gain insights into their mechanism of action. Both of these compounds possess the same tetracyclic scaffold (structures shown in Figure 1). Figure 1 also shows the ability of these compounds to shift the CCK-8 concentration–response curve for stimulating intracellular calcium in CCK1R to the left, reflecting their PAM activity (Figure 1a and Table 1). This is unique to its action at CCK1R, with no analogous impact on CCK activity at CCK2R (Figure 1b and Table 1). Neither compound had any demonstrable, statistically significant intrinsic agonist action in concentrations as high as 20 μM at the CCK1R over-expressing CHO cell line ($p = 0.1$) (Figure 1c), and neither compound exhibited any demonstrable agonist activity at the parental CHO cells, even at 20 μM concentrations (Figure 1d). A longer molecular form of CCK, CCK-33, also had its biological effect augmented significantly by both of these compounds (Table 1).

Table 1. Agonist-induced intracellular calcium responses in CHO cell lines in the absence and presence of PAM compounds.

Receptor-Ligands	pEC_{50}	*n*, *p* Values
CCK1R		
CCK-8	10.6 ± 0.1	5
CCK-8 + "hit 1"	11.1 ± 0.1 *	5, 0.048
CCK-8 + "hit 6"	11.5 ± 0.2 **	5, 0.002
CCK1R		
CCK-33	9.6 ± 0.2	5
CCK-33 + "hit 1"	10.7 ± 0.2 *	5, 0.016
CCK-33 + "hit 6"	10.7 ± 0.2 **	5, 0.008
CCK2R		
CCK-8	9.3 ± 0.1	6
CCK-8 + "hit 1"	9.5 ± 0.1	6, 0.20
CCK-8 + "hit 6"	9.5 ± 0.1	6, 0.19
CCK1R(Y140A)		
CCK-8	9.0 ± 0.1	5
CCK-8 + "hit 1"	9.7 ± 0.1 ***	5, 0.0008
CCK-8 + "hit 6"	10 ± 0.1 ***	5, <0.0001
CCK1R		
CCK-OPE	7.3 ± 0.2	5
CCK-OPE + "hit 1"	8.0 ± 0.2	5, 0.08
CCK-OPE + "hit 6"	8.1 ± 0.2 *	6, 0.04
CCK1R + excess cholesterol		
CCK-8	9.4 ± 0.2	5
CCK-8 + "hit 1"	10.2 ± 0.2 **	5, 0.005
CCK-8 + "hit 6"	10.2 ± 0.2 **	6, 0.005

Values are expressed as means ± SEM of "*n*" independent experiments performed in duplicate. "Hits" 1 and 6 were utilized in 10 μM concentrations. Differences between control and the presence of "hit 1" and "hit 6" were determined using one-way ANOVA. * $p < 0.05$; ** $p < 0.01$; *** $p < 0.001$.

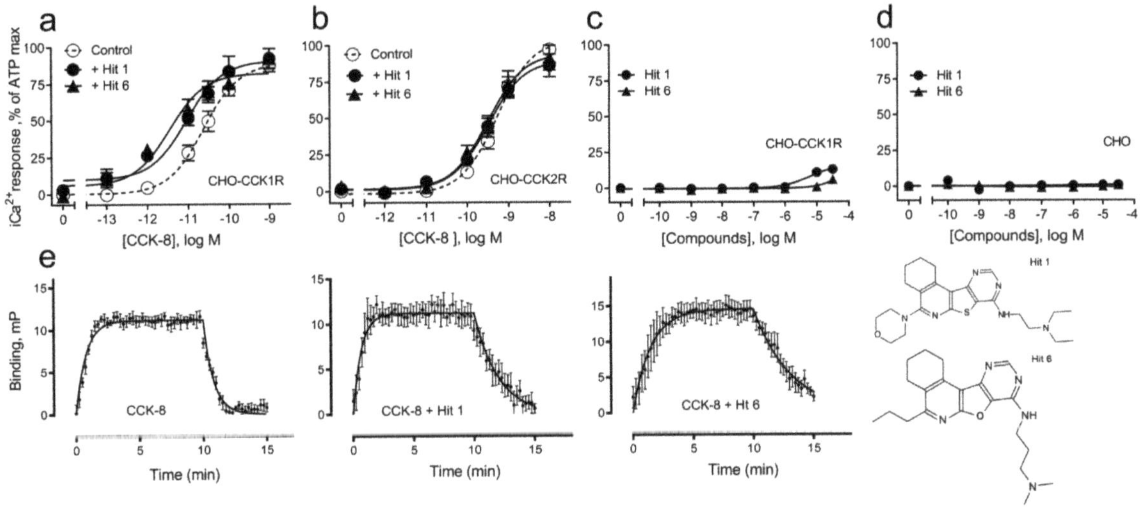

Figure 1. Pharmacologic profiles of "hits" 1 and 6. Shown are the structures of these two ligands, as well as evidence for their positive allosteric modulation of CCK action to stimulate intracellular calcium using Fura-8-AM at the CCK1R expressed on the CHO-CCK1R cell line (**a**), with no effect on CCK2R in the analogous CHO-CCK2R cell line (**b**) and no significant endogenous agonist activity at CCK1R in the CHO-CCK1R cell line (**c**) or at parental CHO cells (**d**). "Hits" 1 and 6 were utilized in 10 μM concentrations. Maximal intracellular calcium responses were determined using 0.1 mM ATP targeting an endogenous CHO cell receptor. Values represent means ± SEM of data from a minimum of 5 independent experiments performed in duplicate and are analyzed in Table 1. Both ligands exerted their positive allosteric modulatory effect on CCK action at CCK1R by prolonging the peptide off rate (**e**). Kinetic data represent 5 independent experiments, with results analyzed in Table 2.

Table 2. Kinetic parameters for CCK binding to CCK1R-expressing membranes from CHO-CCK1R cells in the absence or presence of tetracyclic compounds.

	CCK-8	*n*	CCK-8 + "Hit 1"	*n*, *p* Values	CCK-8 + "Hit 6"	*n*, *p* Values
K_{on} rate, × 10^6 M^{-1} min^{-1}	0.7 ± 0.1	5	2.6 ± 0.6 *	3, 0.04	0.7 ± 0.4	5, 0.31
K_{off} rate, min^{-1}	1.1 ± 0.1	5	0.4 ± 0.1 *	3, 0.04	0.5 ± 0.1 **	5, 0.01
pK_i	7.8 ± 0.1	5	8.8 ± 0.1 *	3, 0.04	8.1 ± 0.2	5, 0.06

Values are expressed as means ± SEM from "*n*" independent experiments performed in triplicate. "Hits" 1 and 6 were utilized in 10 μM concentrations. Levels of significance for differences relative to CCK-8 controls were calculated using the Mann-Whitney test. * $p < 0.05$, ** $p < 0.01$.

The PAM activity of these compounds was explained by their slower dissociation rate for CCK, shown in Figure 1e, resulting in prolonged receptor occupation times. The kinetic parameters are shown in Table 2.

The allosteric constants for these compounds were calculated based on their concentration–response curves for enhancing CCK-stimulated intracellular calcium responses in the CCK1R-expressing cells (Figure 2). Using the operational model for allosterism (Prism 9.1, GraphPad), we determined log(αβ) values of 1.5 and 1.0 for hit 1 and hit 6, respectively (Table 3), which confirmed their positive cooperativity with natural CCK peptide at CCK1Rs.

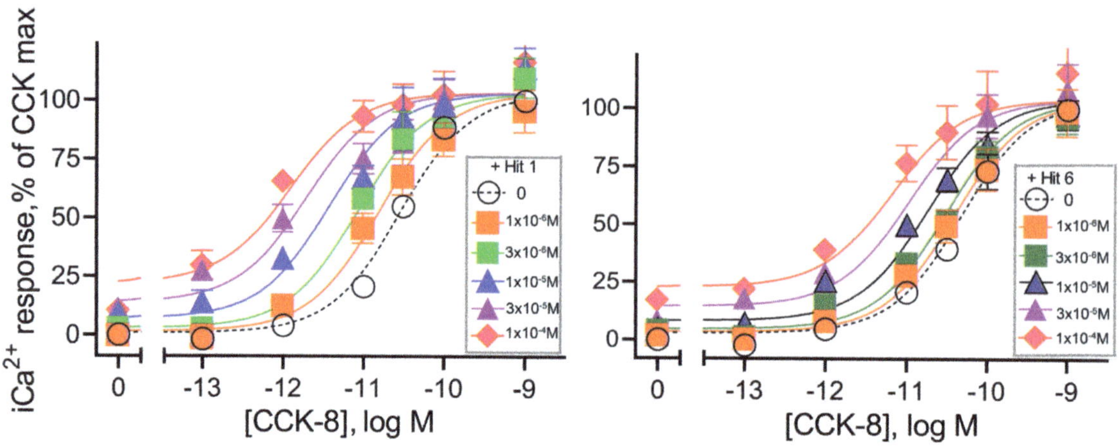

Figure 2. Positive allosteric modulation of CCK action at CCK1R expressed on CHO-CCK1R cells. Shown are the abilities of "hits" 1 and 6 utilized in 10 μM concentrations to shift the CCK concentration–response curves to the left in a dose-dependent manner. Values were plotted as percentages of maximal intracellular calcium responses to CCK in each replicate curve. There were no statistically significant differences in the maximal responses achieved under any condition. Data plotted represent means ± SEM and are analyzed in Table 3.

Table 3. Cooperativity analysis of the impact of tetracyclic compounds on CCK-stimulated intracellular calcium responses in CCK1R-expressing cells.

	CCK-8 + "Hit 1"	**CCK-8 + "Hit 6"**
pK_b	4.5 ± 0.4	4.6 ± 0.2
Tau K_b	0.5 ± 0.1	0.4 ± 0.1
Log$\alpha\beta$	1.5 ± 0.3	1.0 ± 0.2
n	5	5

Values are expressed as means ± SEM from five experiments performed in duplicate. "Hits" 1 and 6 were utilized in 10 μM concentrations.

Part of the rationale for developing these compounds relates to the possibility of correcting the aberrant stimulus–activity coupling observed at the CCK1R in a membrane environment with high cholesterol, as sometimes seen in obesity. Figure 3 shows the ability of both compounds to exhibit PAM action of a partial agonist acting at CCK1R and CCK-OPE (Figure 3a), as well as at the CCK1R(Y140A) receptor construct known to mimic CCK1R in high cholesterol [11] (Figure 3b), and at wild type CCK1R in the setting of elevated cholesterol (Figure 3c). The rationale for using this partial agonist was to attempt to amplify the PAM activity even though this is not a physiologic ligand. Indeed, using CCK-OPE, we observed not only a left shift in the concentration–response curves, but also increases in maximal responses. This continues to be quite encouraging and to further fulfill the pharmacologic profile of interest.

For a small molecule to modulate the action of endogenous CCK, it is important that it occupies the receptor while on the cell surface and does not stimulate its internalization. The receptor is therefore primed for its enhanced response to endogenous CCK when released after a meal. Figure 4 shows that these compounds when used in concentrations as high as 10 μM did not stimulate CCK1R internalization.

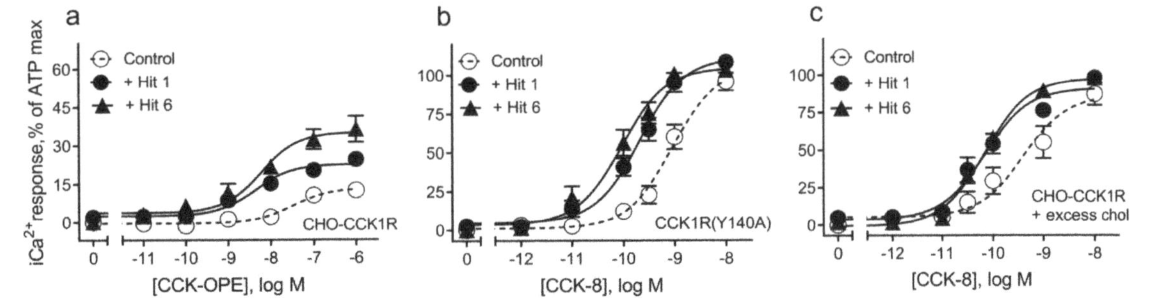

Figure 3. Activity of "hits" 1 and 6 on partial agonist activity and CCK activity in CHO cell lines expressing the noted receptor constructs. "Hits" 1 and 6 were utilized in 10 μM concentrations. Shown is the ability of these ligands to affect the intracellular calcium activity measured with Fura-8-AM of a partial agonist of CCK1R and CCK-OPE (**a**), and their ability to act as positive allosteric modulators of CCK action at a mimic of the high cholesterol state of CCK1R and CCK1R(Y140A) (**b**) and at CCK1R in a high cholesterol environment (**c**). Values represent means ± SEM of data from 5 independent experiments performed in duplicate.

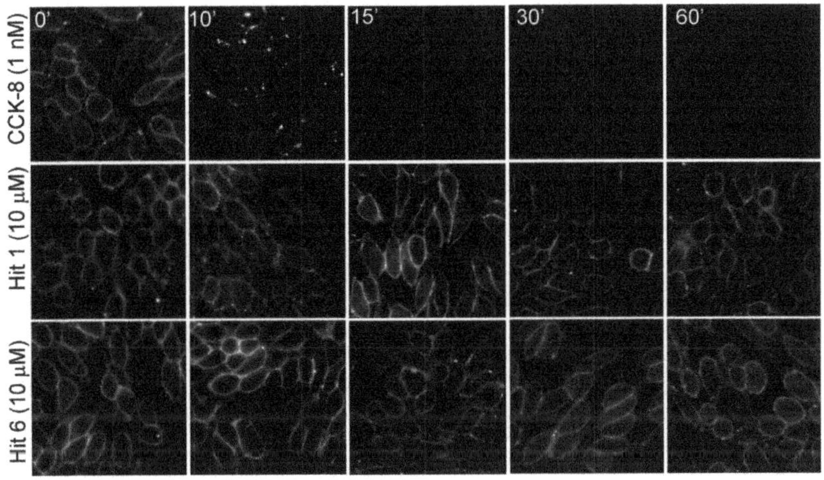

Figure 4. Impact of "hits" 1 and 6 on CCK1R internalization. "Hits" 1 and 6 were utilized in 10 μM concentrations. Alexa488-CCK was utilized to label the cell surface CCK receptor. Shown are time-dependent fluorescence images of cell surface CCK1R after exposure to CCK and the two "hits". CCK stimulated prompt receptor internalization, while neither "hit" cleared the receptor from the cell surface. Images shown are representative of data from three independent experiments.

To determine potential off-target effects, we screened the original hits, "hit 1" (CMP-1) and "hit 6" (CMP-28), against other GPCRs, including P2YRs [6] and vasopressin-2 receptors (AVP2Rs). Neither compound exhibited any activity at the P2YR. We found that CMP-1 (EC_{50} 12 μM, E_{max} 21%) and particularly CMP-28 (EC_{50} 15 μM, E_{max} 73%) showed significant activity in AVP2R cAMP assays in both agonist and PAM formats (combined data). Since the PAM format captures both the agonist and PAM responses with comparable sensitivity, we included this assay as part of our effort to characterize structure–activity relationships (SARs) of the tetracyclic scaffold to identify structural determinants that could eliminate AVP2R activity while maintaining or increasing PAM effects at CCK1R.

In exploring SARs of the tetracyclic scaffold, seeking PAMs, the CCK1R IP-One assay performed in PAM mode provided EC_{50} and E_{max} values, reflecting both potency and efficacy. Cooperative effects on either of these parameters could contribute to the desired PAM impact. We therefore utilized a composite activity score (multiplying pEC_{50} and E_{max} normalized to the maximal response to the control compound, CCK) to help prioritize the data. This score represents an approximation of the area under the curve (AUC) of a compound dose–response curve using normalized data respective the screening format (Figure 5). We found that combining EC_{50} and E_{max} into one score facilitated the comparison of PAMs. We determined an activity score of 2.9 for "hit 1" and a score of 1.7 for "hit 6" and found these scores to be more representative of the allosteric activities (log(αβ) 1.5 (hit 1) and 1.0 (hit 6)) than focusing separately on potencies or efficacies. For comparison, the activity score of CCK in the CCK1R IP-One assay was calculated at 10.1. Hence, the AUC of "hit 1" was approximately 29% of the AUC of the endogenous agonist CCK. In general, we found that activity scores below 1.5 represented non-significant activation in the tested format. Shown in Tables 4–9 are the data from the CCK1R IP-One PAM screening for the 65 compounds representing analogs built on the tetracyclic scaffold of interest. These have been numbered based on descending order of PAM activity scores.

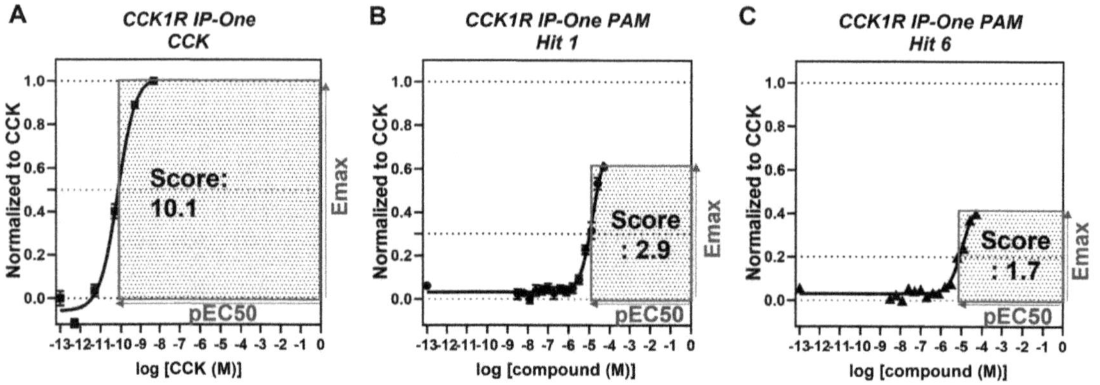

Figure 5. Activity score represents an approximation of the AUC of compound dose–response curves. IP-One dose–response of (**A**) CCK-8, (**B**) "hit 1", and (**C**) "hit 6" in HEK-293 CCK1R cells; TR-FRET ratios resulting from IP-One accumulation normalized to CCK (E_{max} = 1) with (**A**) DMSO or (**B**,**C**) EC_{20} CCK as negative control (E_{max} = 0), representing (**A**) agonist and (**B**,**C**) PAM screening format, respectively; activity score was calculated by multiplying pEC_{50} with normalized E_{max}, and the corresponding area is highlighted in grey; graphs plotted using GraphPad Prism; experiments performed in duplicate in at least three independent experiments; data points shown as mean ± SEM.

An examination of the structures of "hit 1" (CMP-1) and "hit 6" (CMP-28) identified five sites amenable for chemical modifications, as depicted in Figure 6. We recognized that both hits contained a basic tertiary amine attached to the tetracyclic core through a diamino alkyl linker at the position marked as R2 with predicted pKas values of 9.2 and 9.7, respectively. In our previous report [6], we hypothesized that this basic amine might play a crucial role in the PAM activity of our hits. To further explore this hypothesis, we examined 25 analogs with non-basic substituents at R2 attached to the pyrimidine of the tetracyclic core (Tables 4 and 5).

Table 4. Structure–activity studies of 15 morpholine analogs with non-basic amine R2 substitutions.

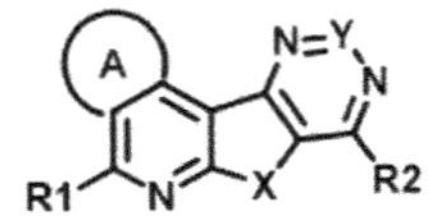

ID	Structure					CCK1R PAM (IP-One)			CCK1R Ago (IP-One)			AVP2R PAM (cAMP)		
	R1	R2	A	X	Y	EC_{50} (μM)	E_{max} (%)	Score	EC_{50} (μM)	E_{max} (%)	Score	EC_{50} (μM)	E_{max} (%)	Score
CMP-40	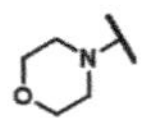	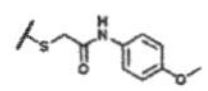	cyclohexyl	S	C	0	0	0	0	0	0	2.9 ± 0.1	96 ± 9	5.33
CMP-41	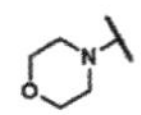	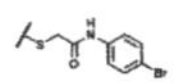	cyclohexyl	S	C	0	0	0	0	0	0	0	0	0
CMP-65	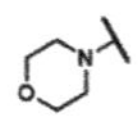	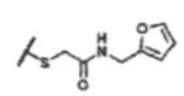	cyclohexyl	S	C	0	0	0	0	0	0	10.0 ± 1.7	96 ± 5	4.78
CMP-42	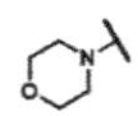	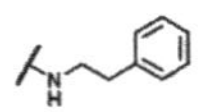	cyclohexyl	S	C	0	0	0	0	0	0	0	0	0
CMP-44	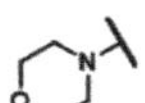	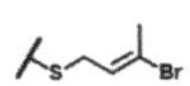	cyclohexyl	S	C	0	0	0	0	0	0	0	0	0
CMP-45	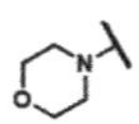	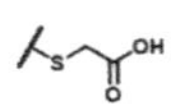	cyclohexyl	S	C	0	0	0	0	0	0	37.9 ± 7.0	33 ± 13	1.46
CMP-64	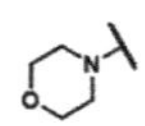	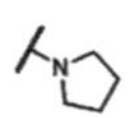	cyclohexyl	S	C	N.D.	N.D.	N.D.	0	0	0	0	0	0
CMP-46	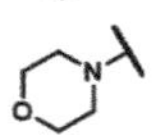	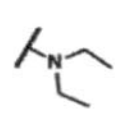	cyclohexyl	S	C	0	0	0	0	0	0	0	0	0
CMP-47	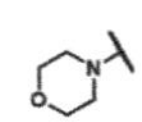	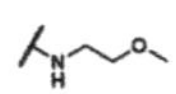	cyclohexyl	S	C	0	0	0	0	0	0	0	0	0
CMP-48	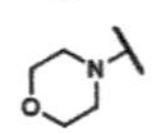	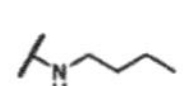	cyclohexyl	S	C	0	0	0	0	0	0	0	0	0

Table 4. *Cont.*

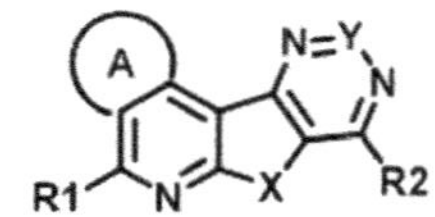

ID	Structure					CCK1R PAM (IP-One)			CCK1R Ago (IP-One)			AVP2R PAM (cAMP)		
	R1	R2	A	X	Y	EC_{50} (μM)	E_{max} (%)	Score	EC_{50} (μM)	E_{max} (%)	Score	EC_{50} (μM)	E_{max} (%)	Score
CMP-49	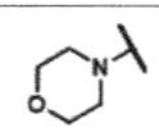	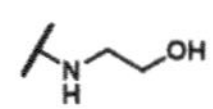	cyclohexyl	S	C	0	0	0	0	0	0	0	0	0
CMP-50	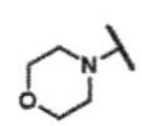	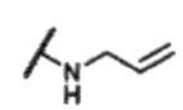	cyclohexyl	S	C	0	0	0	0	0	0	0	0	0
CMP-51	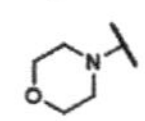	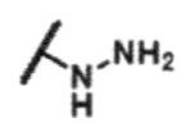	cyclohexyl	S	C	0	0	0	0	0	0	0	0	0
CMP-52	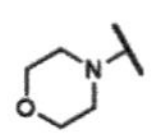		cyclohexyl	S	N	0	0	0	0	0	0	N.D.	N.D.	N.D.
CMP-59	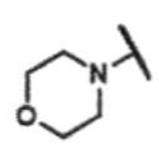	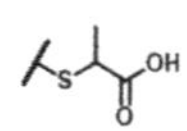	cyclohexyl	S	C	N.D.	N.D.	N.D.	0	0	0	22.3 ± 7.2	20 ± 0	0.93

Average potencies (EC_{50} (μM) ± standard deviation (SD)), average compound efficacy (E_{max} (%) ± SD), and calculated activity score ($E_{max} \times pEC_{50}$) of compound dose–response studies in screening mode. Compounds were tested in Cisbio IP-One and cAMP accumulation assays; N.D. = not determinable (response less than mean ± 3 × SD of negative control).

Table 5. Structure–activity studies of 10 alkyl analogs with non-basic amine R2 substitutions.

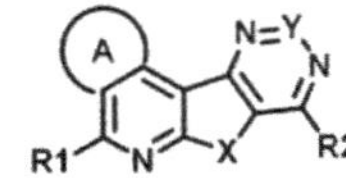

ID	Structure					CCK1R PAM (IP-One)			CCK1R Ago (IP-One)			AVP2R PAM (cAMP)		
	R1	R2	A	X	Y	EC_{50} (μM)	E_{max} (%)	Score	EC_{50} (μM)	E_{max} (%)	Score	EC_{50} (μM)	E_{max} (%)	Score
CMP-56	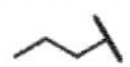	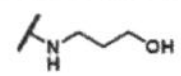	cyclopentyl	S	C	N.D.	N.D.	N.D.	0	0	0	0	0	0
CMP-57	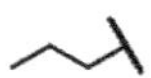	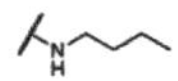	cyclohexyl	S	C	N.D.	N.D.	N.D.	0	0	0	0	0	0
CMP-58	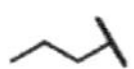	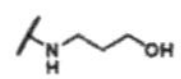	cyclohexyl	O	C	N.D.	N.D.	N.D.	0	0	0	0	0	0
CMP-60	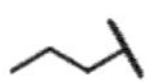	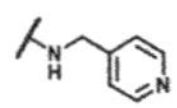	cyclohexyl	O	C	N.D.	N.D.	N.D.	0	0	0	0	0	0
CMP-38	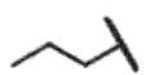		cyclohexyl	S	C	7.6 ± 3.3	20 ± 7.6	1.01	0	0	0	0	0	0
CMP-62		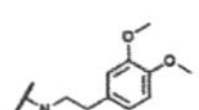	cyclopentyl	O	C	N.D.	N.D.	N.D.	0	0	0	0	0	0
CMP-39	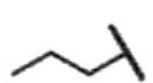	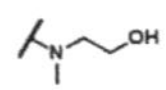	cyclohexyl	O	C	N.D.	N.D.	N.D.	0	0	0	0	0	0
CMP-63		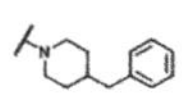	cyclopentyl	O	C	N.D.	N.D.	N.D.	0	0	0	0	0	0
CMP-55	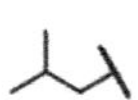	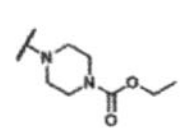	cyclohexyl	O	C	0	0	0	0	0	0	0	0	0
CMP-53	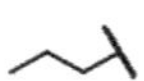	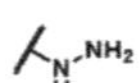	cyclohexyl	O	C	0	0	0	0	0	0	15.3 ± 9.8	61 ± 12	2.95

Average potencies (EC_{50} (μM) ± standard deviation (SD)), average compound efficacy (E_{max} (%) ± SD), and calculated activity score ($E_{max} \times pEC_{50}$) of compound dose–response studies in screening mode. Compounds were tested in Cisbio IP-One and cAMP accumulation assays; N.D. = not determinable (response less than mean ± 3 × SD of negative control).

Table 6. Structure–activity studies of 10 *n*-alkyl analogs.

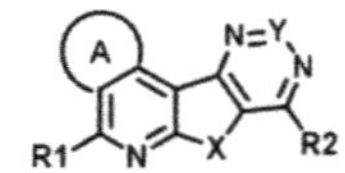

ID	Structure					CCK1R PAM (IP-One)			CCK1R ago (IP-One)			AVP2R PAM (cAMP)		
	R1	R2	A	X	Y	EC_{50} (μM)	E_{max} (%)	Score	EC_{50} (μM)	E_{max} (%)	Score	EC_{50} (μM)	E_{max} (%)	Score
CMP-5	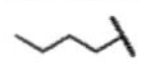	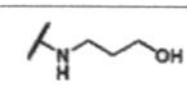	cyclopentyl	S	C	7.5 ± 2.9	53 ± 6	2.66	N.D.	N.D.	N.D.	8.1 ± 8.5	32 ± 11	1.61
CMP-7	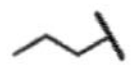	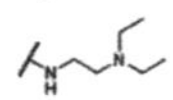	cyclohexyl	S	C	8.0 ± 2.1	48 ± 11	2.39	0	0	0	6.5 ± 4.2	31 ± 2	1.61
CMP-8	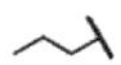	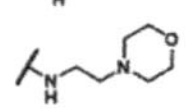	cyclohexyl	S	C	37.8 ± 3.9	53 ± 1	2.34	0	0	0	15.7 ± 8.4	48 ± 12	2.31
CMP-12	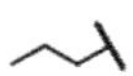	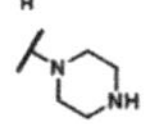	cyclohexyl	O	C	14.1 ± 4.3	46 ± 7	2.23	N.D.	N.D.	N.D.	11.4 ± 3.6	66 ± 3	3.28
CMP-18	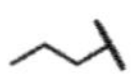	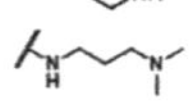	cyclohexyl	S	C	12.7 ± 5.5	43 ± 21	1.96	0	0	0	11.8 ± 14.0	40 ± 17	1.99
CMP-28 ("hit 6")	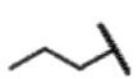	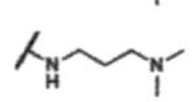	cyclohexyl	O	C	9.5 ± 3.5	35 ± 13	1.66	N.D.	N.D.	N.D.	15.0 ± 13.6	73 ± 16	3.53
CMP-30	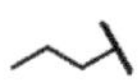	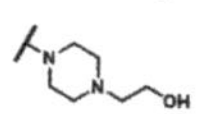	cyclohexyl	O	C	27.1 ± 10.7	36 ± 16	1.55	0	0	0	5.9 ± 3.8	29 ± 0	1.52
CMP-32	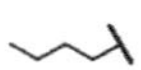	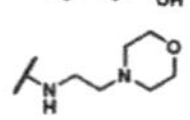	cyclohexyl	S	C	23.9 ± 25.8	41 ± 34	1.52	0	0	0	9.3 ± 11.1	33 ± 26	1.67
CMP-36	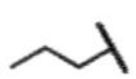	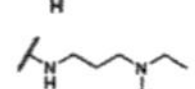	cyclohexyl	O	C	3.3 ± 5.4	25 ± 10	1.32	0	0	0	6.1 ± 2.8	27 ± 20	1.41
CMP-61		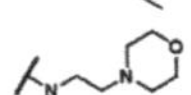	cyclohexyl	S	C	N.D.	N.D.	N.D.	0	0	0	0	0	0

Average potencies (EC_{50} (μM) ± standard deviation (SD)), average compound efficacy (E_{max} (%) ± SD), and calculated activity score ($E_{max} \times pEC_{50}$) of compound dose–response studies in screening mode. Compounds were tested in Cisbio IP-One and cAMP accumulation assays; N.D. = not determinable (response less than mean ± 3 × SD of negative control).

Table 7. Structure–activity studies of 10 branched alkyl analogs.

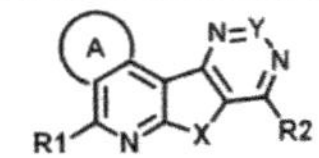

ID	Structure					CCK1R PAM (IP-One)			CCK1R ago (IP-One)			AVP2R PAM (cAMP)		
	R1	R2	A	X	Y	EC_{50} (μM)	E_{max} (%)	Score	EC_{50} (μM)	E_{max} (%)	Score	EC_{50} (μM)	E_{max} (%)	Score
CMP-10	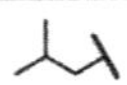	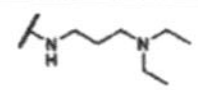	cyclopentyl	C	C	22.5 ± 9.0	51 ± 16	2.28	N.D.	N.D.	N.D.	21.7 ± 11.3	46 ± 9	2.15
CMP-13	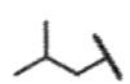	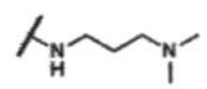	cyclopentyl	C	C	11.2 ± 4.2	52 ± 36	2.23	0	0	0	12.6 ± 7.9	28 ± 14	1.35
CMP-15	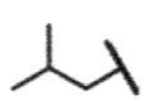	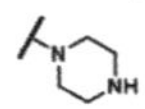	cyclohexyl	C	C	17.8 ± 6.2	43 ± 3	2.04	0	0	0	7.3 ± 3.3	58 ± 15	2.97
CMP-20		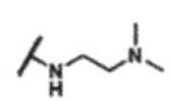	cyclopentyl	C	C	23.6 ± 1.0	42 ± 12	1.90	0	0	0	21.2 ± 19.7	25 ± 1	1.17
CMP-21	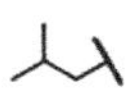	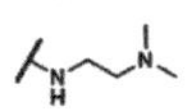	cyclopentyl	C	C	23.8 ± 10.5	45 ± 23	1.90	0	0	0	22.8 ± 3.8	37 ± 4	1.74
CMP-24		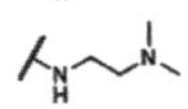	cyclopentyl	C	C	12.7 ± 15.5	48 ± 45	1.81	0	0	0	19.6 ± 5.1	34 ± 23	1.62
CMP-25	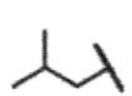	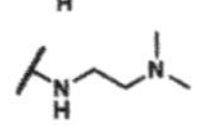	cyclohexyl	C	C	14.5 ± 5.3	38 ± 11	1.79	N.D.	N.D.	N.D.	8.6 ± 5.4	74 ± 12	3.75
CMP-33	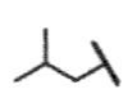	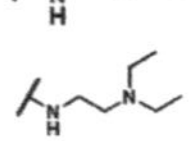	cyclopentyl	C	C	18.8 ± 7.7	35 ± 19	1.51	0	0	0	21.0 ± 6.5	34 ± 1	1.60
CMP-34	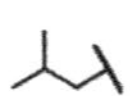	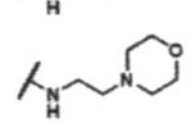	cyclohexyl	S	C	29.0 ± 13.1	34 ± 13	1.50	0	0	0	7.8 ± 4.0	42 ± 17	2.13
CMP-54	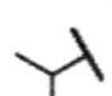	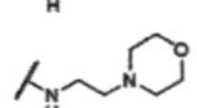	cyclohexyl	S	C	N.D.	N.D.	N.D.	0	0	0	26.2 ± 2.4	43 ± 7	1.96

Average potencies (EC_{50} (μM) ± standard deviation (SD)), average compound efficacy (E_{max} (%) ± SD), and calculated activity score ($E_{max} \times pEC_{50}$) of compound dose–response studies in screening mode. Compounds were tested in Cisbio IP-One and cAMP accumulation assays; N.D. = not determinable (response less than mean ± 3 × SD of negative control).

Table 8. Structure–activity studies of seven piperazine and four phenyl analogs.

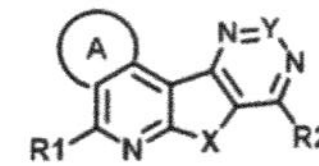

ID	Structure					CCK1R PAM (IP-One)			CCK1R ago (IP-One)			AVP2R PAM (cAMP)		
	R1	R2	A	X	Y	EC_{50} (μM)	E_{max} (%)	Score	EC_{50} (μM)	E_{max} (%)	Score	EC_{50} (μM)	E_{max} (%)	Score
CMP-6	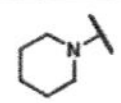	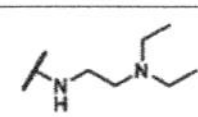	cyclohexyl	S	C	13.8 ± 3.2	57 ± 27	2.58	0	0	0	9.9 ± 0.5	33 ± 1	1.67
CMP-11	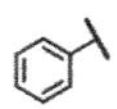	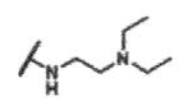	cyclohexyl	S	C	9.1 ± 4.8	45 ± 6	2.27	27.1 ± 15.3	21 ± 2	0.96	3.2 ± 1.1	44 ± 8	2.44
CMP-14	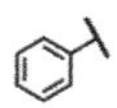	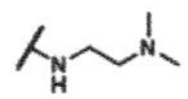	cyclohexyl	S	C	8.9 ± 1.7	44 ± 10	2.17	29.7 ± 8.0	17 ± 5	0.72	4.7 ± 4.1	56 ± 33	2.97
CMP-22	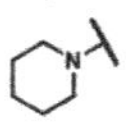	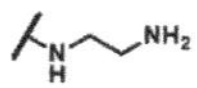	cyclohexyl	S	C	12.2 ± 4.6	39 ± 11	1.87	26.9 ± 1.1	60 ± 4.5	2.76	2.7 ± 0.7	52 ± 4	2.92
CMP-23	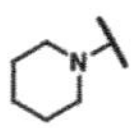	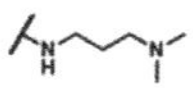	cyclohexyl	S	N	9.6 ± 3.6	38 ± 13	1.86	30.0 ± 4.9	41 ± 1	1.85	3.1 ± 1.3	53 ± 3	2.93
CMP-26	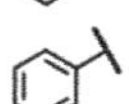	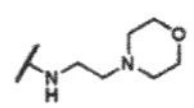	cyclohexyl	S	C	30.2 ± 11.2	41 ± 15	1.76	N.D.	N.D.	N.D.	3.7 ± 0.5	16 ± 4	0.87
CMP-27	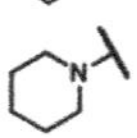	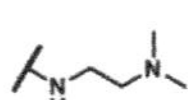	cyclohexyl	S	N	18.4 ± 8.1	38 ± 14	1.70	0	0	0	9.1 ± 1.0	70 ± 4	3.51
CMP-29	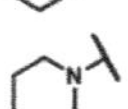	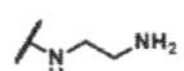	4-methyl	S	C	15.1 ± 1.8	33 ± 7	1.59	52.9 ± 39	39 ± 12	1.67	19.2 ± 0.8	35 ± 0	1.66
CMP-31	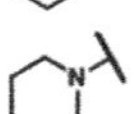	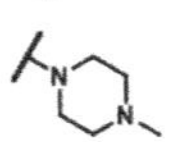	cyclohexyl	S	C	29.6 ± 1.8	35 ± 14	1.54	0	0	0	5.4 ± 1.7	24 ± 3	1.26
CMP-35	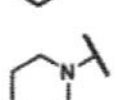	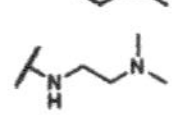	cyclohexyl	S	C	31.3 ± 14.9	36 ± 20	1.49	0	0	0	7.9 ± 2.5	31 ± 0	1.59
CMP-37	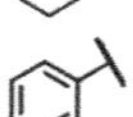	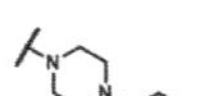	cyclopentyl	S	C	31.7 ± 1.2	28 ± 9	1.26	0	0	0	0	0	0

Average potencies (EC_{50} (μM) ± standard deviation (SD)), average compound efficacy (E_{max} (%) ± SD), and calculated activity score ($E_{max} \times pEC_{50}$) of compound dose–response studies in screening mode. Compounds were tested in Cisbio IP-One and cAMP accumulation assays; N.D. = not determinable (response less than mean ± 3 × SD of negative control).

Table 9. Structure–activity studies of eight morpholine and one N-diethyl analogs.

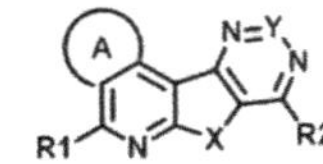

ID	Structure					CCK1R PAM (IP-One)			CCK1R Ago (IP-One)			AVP2R PAM (cAMP)		
	R1	R2	A	X	Y	EC_{50} (μM)	E_{max} (%)	Score	EC_{50} (μM)	E_{max} (%)	Score	EC_{50} (μM)	E_{max} (%)	Score
CMP-1 ("hit 1")	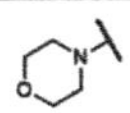	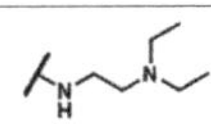	cyclohexyl	S	C	12.8 ± 4.6	63 ± 18	2.94	N.D.	N.D.	N.D.	12.0 ± 7.1	20.7 ± 9.8	1.02
CMP-2	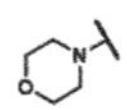	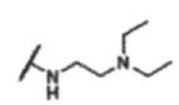	dimethyl thiacyclohe-xyl	S	C	9.2 ± 2.0	57 ± 12	2.82	31.3 ± 3.1	31 ± 15	1.31	N.D.	N.D.	N.D.
CMP-3	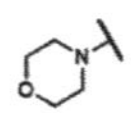	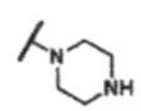	cyclohexyl	S	C	16.6 ± 15.9	60 ± 23	2.77	N.D.	N.D.	N.D.	14.8 ± 10.3	34.0 ± 5.3	1.64
CMP-4	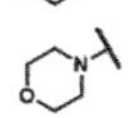	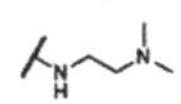	cyclohexyl	S	C	14.4 ± 1.5	62 ± 33	2.71	0	0	0	12.8 ± 14.0	24.5 ± 7.1	1.20
CMP-9	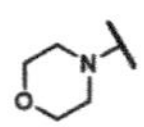	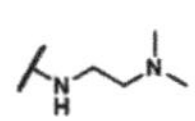	dimethyl thiacyclohe-xyl	S	C	9.3 ± 1.1	46 ± 3	2.31	N.D.	N.D.	N.D.	N.D.	N.D.	N.D.
CMP-16	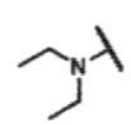	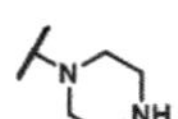	methyl	S	C	32.6 ± 12.5	45 ± 7	2.02	0	0	0	7.6 ± 8.2	17 ± 4	0.87
CMP-17	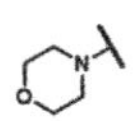	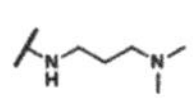	cyclohexyl	S	N	23.0 ± 17.9	50 ± 32	1.99	N.D.	N.D.	N.D.	22.4 ± 9.2	26 ± 1	1.20
CMP-19	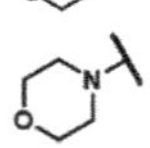	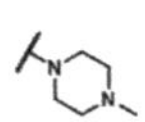	cyclohexyl	S	C	18.2 ± 8.6	51 ± 43	1.94	0	0	0	N.D.	N.D.	N.D.
CMP-43	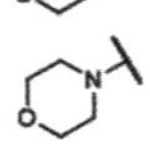	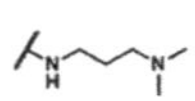	cyclopentyl	S	C	N.D.	N.D.	N.D.	0	0	0	N.D.	N.D.	N.D.

Average potencies (EC_{50} (μM) ± standard deviation (SD)), average compound efficacy (E_{max} (%) ± SD), and calculated activity score ($E_{max} \times pEC_{50}$) of compound dose–response studies in screening mode. Compounds were tested in Cisbio IP-One and cAMP accumulation assays; N.D. = not determinable (response less than mean ± 3 × SD of negative control).

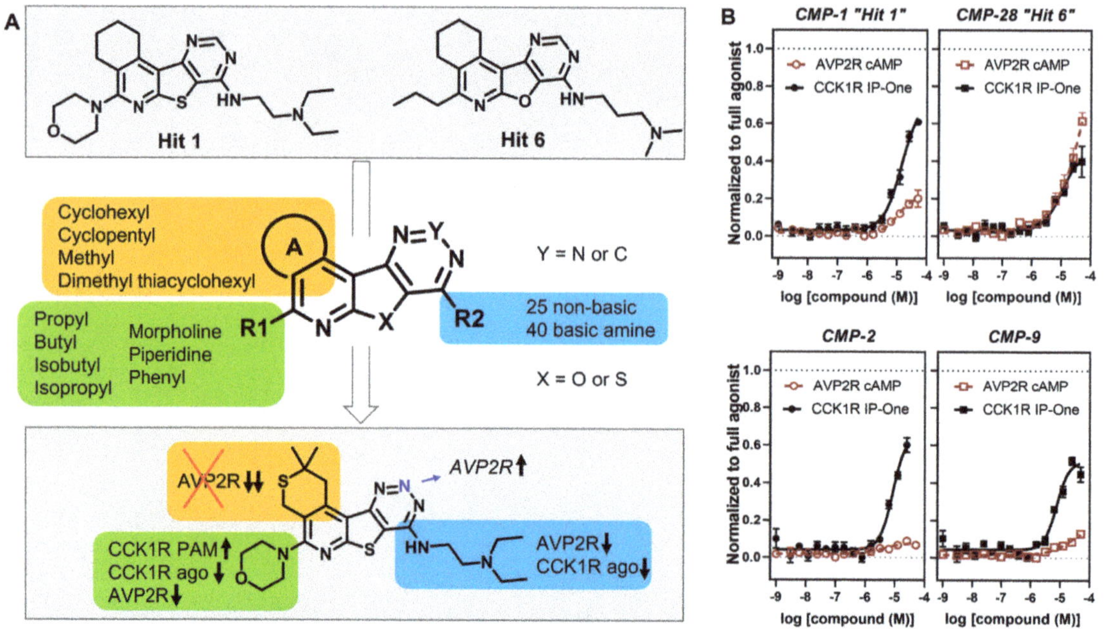

Figure 6. Elucidation of structural determinants impacting the pharmacological profile of tetracyclic analogs (**A**). "Hits" 1 and 6 were subjected to a comprehensive SAR campaign elucidating structural features with distinct effects on CCK1R PAM, CCK1R agonist and AVP2R off-target activity. (**B**) Dose–response curves of "hit 1" and "hit 6" (top) and optimized dimethyl thiacyclohexyl analogs CMP-2 and CMP-9 (bottom) in CCK1R IP-One and AVP2R cAMP assays, both conducted in PAM format. Data are shown in Tables 4–9.

With the aim to develop PAMs with no or minimal intrinsic agonist activity at CCK1Rs and no off-target effects, the compounds were profiled in a 16-point dose–response format in CCK1R IP-One PAM and agonist formats, as well as in AVP2R cAMP assays. We tested 15 direct analogs of "hit 1" with sulfur in the tetracyclic core, a cyclohexyl group as saturated ring A, and morpholine as residue R1. The results are listed in Table 4.

To explore whether the basic amine is part of an ionic interaction or rather acts via hydrogen bonds with the receptor, we investigated analogs with a methoxy (CMP-47) or hydroxy (CMP-49) function instead of the diethyl amine group, as well as analogs with a butyl (CMP-48), allyl (CMP-50), or phenyl ethyl (CMP-42) attached to an aromatic amine at the R2 position. None of these analogs showed any CCK1R activity. In addition, hydrazine (CMP-51), pyrrolidine CMP-64, or diethyl amine CMP-46, directly attached to the pyrimidine of the core, had significantly diminished responses as well. Furthermore, we evaluated derivatives connecting distinct side chains to the core via a sulfide bridge. The acetic acid analogs CMP-45 and CMP-59, the bromo butenyl (CMP-44), or the thiol (CMP-52) derivatives also displayed no effects. We also tested three compounds with N-substituted sulfanyl acetamides. Interestingly, the methoxy phenyl (CMP-40) and furanyl methyl (CMP-65) analogs, but not the bromo phenyl variant (CMP-41), showed strong activation of AVP2R-mediated cAMP accumulation; however, none of the sulfanyl acetamides showed any responses in the CCK1R IP-One PAM and agonist assays. Overall, we concluded that the CCK1R PAM activity likely relies on an ionic interaction of a negatively charged receptor residue with a basic amine attached to the tetracyclic core via a short alkyl linker important for correct positioning and orientation. These findings were further corroborated with 10 molecules containing a propyl or other alkyl side chain at the R1 position and non-basic functional groups as R2 substituents (Table 5), since none of these derivatives showed any significant activity in CCK1R IP-One assays.

Hence, we proceeded to focus on analogs of the tetracyclic scaffold that incorporated a basic amino group at the R2 side chain.

We were interested to see how much influence the oxygen or sulfur in the tetracyclic core has on the pharmacological profile (Table 6). We tested CMP-18, the sulfur analog of "hit 6", with a propyl side chain at R1, a cyclohexyl ring A, and a dimethyl amine connected via a propyl linker to the aminopyrimidine of the core (R2). In CCK1R IP-One assays we found slightly elevated responses in the PAM format (score 2.0) compared to "hit 6", while maintaining negligible activity in agonist mode. However, even though the response was attenuated, the sulfur derivative also maintained a significant off-target effect at AVP2Rs (score 2.0). We tested three further derivatives of "hit 6" incorporating oxygen in the core. The extension of the dimethyl amine to diethyl amine (CMP-36) appeared to attenuate agonist and PAM effects. Removing the hydroxyethyl to obtain the piperazine analog CMP-12, we were able to increase CCK1R IP-One PAM activity but also effects on AVP2R with activity scores of 2.2 and 3.3, respectively. In addition, we evaluated the propyl analog of "hit 1", CMP-7, which contained a *N*-diethyl diamino ethyl at R2. We found improved CCK1R PAM activities (score 2.4) and attenuated off-target effects (score 1.6 AVP2R) compared to the *N*-dimethyl diamino propyl analog CMP-18. Exchange of the diethyl amino group into a morpholine (CMP-8) led to increased AVP2R activity (score 2.3). Truncation of the propyl at R1 to methyl (CMP-61) or extension to a butyl side chain (CMP-32) attenuated agonist and PAM activities. However, keeping the butyl side chain at R1, condensing ring A to a cyclopentane, and attaching a diamino propyl with a primary amino group at R2, we obtained analog CMP-5 with improved CCK1R PAM activities (score 2.7 (IP-One), non-significant intrinsic agonist activity, and minimal AVP2R off-target effects (score 1.6).

Moreover, we studied 10 analogs with branched alkyl side chains at R1 (Table 7). The isopropyl analog of "hit 6", CMP-24, with *N*-dimethyl diamino propyl at R2 and a cyclopentyl ring A, showed minimal activity in the IP-One PAM assay (score 1.8) and minimal off-target effects in AVP2Rs (score 1.6). Truncation to *N*-dimethyl diamino ethyl resulted in CMP-20, which showed a very similar profile with slightly increased PAM activity in CCK1R IP-One (score 1.9). Exchanging the dimethyl amine to a morpholine, oxygen to sulfur, and cyclopentyl to cyclohexyl (CMP-54), we found that the compound had an inverted selectivity profile, with no significant activity in CCK1R PAM format, while maintaining activity at AVP2Rs (score 2.0). Additionally, the direct analog of this compound but with an isobutyl sidechain at R1 (CMP-34) displayed the same AVP2R-selective profile. We investigated other isobutyl analogs with oxygen in the tetracyclic core, four of which contained a cyclopentyl ring A. The *N*-dimethyl diamino ethyl analog (CMP-21) showed similar activity in CCK1R IP-One PAM and AVP2R counter-screen assays (score 1.9 (CCK1R), 1.7 (AVP2R)). Substitution of the dimethyl amine to diethyl amine (CMP-33) resulted in attenuated responses. Elongation of the linker to diamino propyl (CMP-10) resulted in enhanced responses in both CCK1R IP-One PAM (score 2.3) and AVP2R cAMP (score 2.2) assays. Interestingly, the *N*-dimethyl diamino propyl analog (CMP-13) maintained CCK1R PAM activity in IP-One assays (score 2.2) and showed non-significant effects in the AVP2R counter-screen (score 1.4). In contrast, the cyclohexyl analogs CMP-25 and CMP-15, with a *N*-dimethyl diamino ethyl and piperazine group at R2, respectively, appeared to be AVP2R-selective (scores 3.0–3.8) with moderate CCK1R PAM activities (scores 1.8–2.0).

Then, we evaluated seven analogs with a piperazine side chain at R1 (Table 8). CMP-6, the direct analog of "hit 1", demonstrated robust CCK1R IP-One PAM activity (score 2.6) with minimal activity at AVP2Rs (score 1.7). Removal of the two ethyl groups from the basic amine resulted in the primary amine analog CMP-22, which showed a very different activity profile. CMP-22 maintained moderate CCK1R IP-One PAM activity (score 1.9). However, it demonstrated strong off-target activity (score 2.9 (AVP2R)) and substantial

CCK1R intrinsic agonist activity (score 2.5 (IP-One). Interestingly, the dimethyl amino analog CMP-35 showed attenuated activities in CCK1R IP-One PAM (score 1.5) and AVP2R cAMP (score 1.6) assays. Moreover, changing the pyrimidine ring to a triazine within the tetracyclic core but keeping the *N*-dimethyl diamino ethyl group (CMP-27), we found the desired CCK1R PAM profile with an activity score of 1.7, and no intrinsic agonist activity. However, CMP-27 induced strong responses in AVP2R cAMP assays (score 3.5). Additionally, the triazine analog with a slightly elongated *N*-dimethyl diamino propyl moiety at R2 (CMP-23) demonstrated a similar profile, but substantial off-target activity (score 2.9 (AVP2R)). We also restrained the tertiary amine with an N-methyl piperazine analog containing the pyrimidine ring at the core (CMP-31), which resulted in weak PAM activities at CCK1Rs (scores 1.5 (CCK1R IP-One). As the last analog in this subset, we tested a diamino ethyl primary amine analog containing a methyl instead of a cyclohexyl ring (CMP-29). This modification led to increased agonist activity (score 2.3) over CCK1R PAM responses (score 1.6).

Next, we characterized four analogs incorporating a phenyl side chain at R1. The direct analog of "hit 1" (CMP-11) with *N*-diethyl diamino ethyl function at R2 displayed an overall promising profile with good CCK1R PAM activity (scores 2.3), and no to minimal intrinsic agonist activities (scores 1.0). However, CMP-11 was not selective towards CCK1R with an AVP2R activity score of 2.4. Truncation of the ethyl groups to a dimethyl amine led to a compound (CMP-14) with a very similar CCK1R activity profile but that appeared to enhance the preference for AVP2Rs (score 3.0). Interestingly, incorporating a morpholine with reduced basicity as a tertiary amino function (CMP-26), we found a molecule with moderate CCK1R PAM activities (scores 1.8), no to minimal intrinsic agonist activity, and, moreover, negligible off-target effects (score 0.9 (AVP2R)). For structural variations, we also tested an analog with cyclopentyl ring A and an N-hydroxyethyl piperazine at R2 (CMP-37). This combination of residues resulted in a loss of activities across all formats.

The final subset of analogs we investigated in more detail were "hit 1" derivatives containing a morpholine as side chain R1 (Table 9). "Hit 1" (CMP-1) demonstrated its highest activity in the CCK1R IP-One PAM primary screening assay (score 2.9), with no significant intrinsic agonist activity and minimal off-target effects (score 1.0 (AVP2R). Truncation of the diethyl to dimethyl amine (CMP-4) resulted in slightly higher AVP2R activity (score 1.2) with overall similar, but slightly attenuated responses in CCK1R assays. CMP-19, with a conformationally restrained N-methyl piperazine residue, R2, showed diminished signaling across all assays with modest effects in CCK1R IP-One PAM mode (score 1.9). Interestingly, removal of the methyl group from the piperazine resulted in analog CMP-3 with improved CCK1R IP-One activity (score 2.8) and minimal AVP2R off-target response (score 1.6). At this point, we also compared a piperazine analog with a diethyl amino group as R1 and a methyl instead of ring A (CMP-16). This overall truncated analog showed attenuated responses across all formats with a CCK1R IP-One PAM activity score of 2.0. CMP-17, the triazine analog with cyclohexyl ring A and a *N*-dimethyl diamino propyl side chain at R2, displayed a weak but overall promising pharmacological profile with modest CCK1R PAM activity (scores 2.0), no intrinsic agonist activity, and minimal AVP2R off-target effects (score 1.2). However, the pyrimidine analog with *N*-dimethyl diamino propyl at R2 and a cyclopentyl ring A (CMP-43) showed no agonist or PAM activity. We found analogs with a dimethyl thiacyclohexyl ring A, which we thought would expand the molecule and add bulkiness to the cyclohexyl moiety. CMP-2, the direct analog to "hit 1" with an *N*-diethyl diamino ethyl function at R2, displayed a very promising profile with robust CCK1R IP-One PAM activity (score 2.8) and a non-significant agonist response. Moreover, the introduction of the dimethyl thiacyclohexyl moiety eradicated responses in AVP2R cAMP assays. Additionally, the truncation of the diethyl to a dimethyl amino group (CMP-9), which tended to enhance AVP2R signaling for other molecules, maintained strong CCK1R-selectivity (score 2.3) by carrying the dimethyl thiacyclohexyl ring A.

To evaluate more globally how distinct modifications influence CCK1R PAM activity, we performed an R-group analysis (Figure 7). We looked at the 40 analogs containing a basic amino function and grouped them based on their R1 and R2 side chains, whereas the distinct R2 modifications were plotted on the *x*-axis and the different R1 side chains were plotted on the *y*-axis. The order of residues was determined by the CCK1R PAM activity score as a median value for each group with the same R1 or R2 residues, ranking the best modifications at the bottom-left corner. The R-group analysis confirmed that the morpholine at R1 and a *N*-diethyl diamino ethyl side chain as R2 seemed to generally enhance the activities of the tetracyclic scaffold. CMP-5, which is constituted by a butyl side chain at R1 and a diamino propyl with a primary amino group at R2, is only a single data point and is therefore difficult to compare with other residues. However, it could indicate a potentially promising direction for a future targeted medicinal chemistry effort. Next, we wanted to verify that the activity score reflects potencies and efficacies of the analogs. Therefore, we colored points, representing a distinct molecule, based on their potencies (red—white—blue, with blue being best potency), and modified the point sizes according to their efficacies (the bigger the more efficacious). We found that red-colored compounds were largely found at the right-top, whereas blue-colored molecules with larger points had tendencies towards the bottom-left corner, indicating that the activity score is a good tool to rank order compounds and residues.

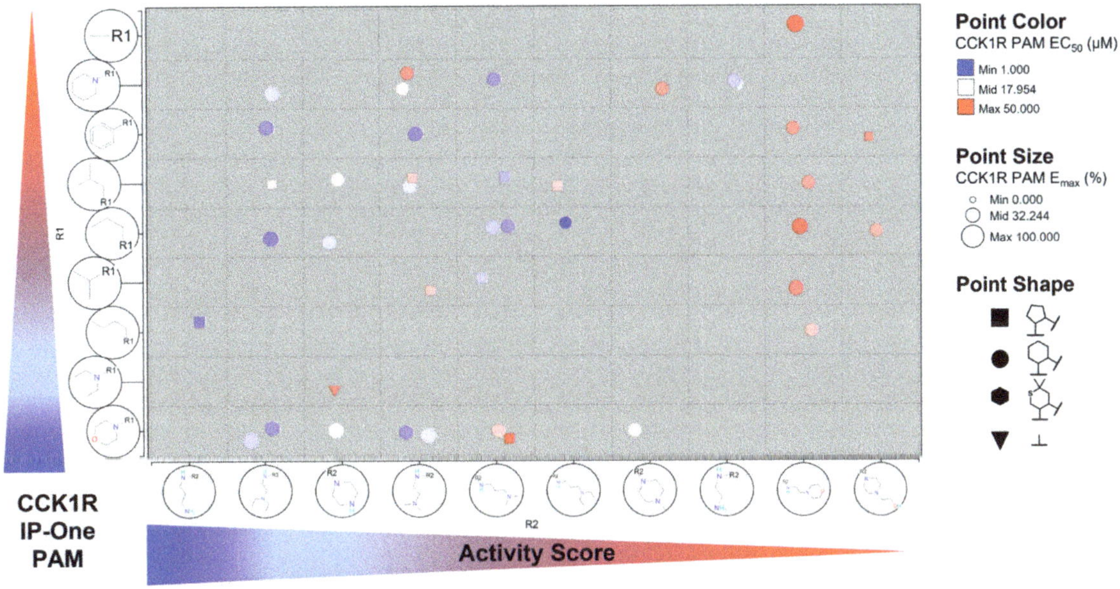

Figure 7. R-group analysis of 40 tetracyclic analogs with basic amino functions. Compounds identified by R1, R2, and point shape (ring A); point color indicates CCK1R PAM EC_{50} (µM), point size indicates CCK1R PAM E_{max} (%); R1 and R2 groups ranked via median activity score with higher scores at bottom left corner. R-group analysis was generated via the Certara D360 scientific informatics platform.

Upon CCK binding, the CCK1R preferably coupled to Gq/11 proteins; however, at higher concentrations of CCK, CCK1R-activated Gs signaling was observed as well. Therefore, we investigated our original hits and our best analogs incorporating the dimethyl thiacyclohexyl ring at R1 for their ability to enhance CCK-mediated cAMP accumulation in HEK-293 cells overexpressing CCK1R. We utilized an EC_{20} concentration of CCK for basal stimulation and found that "hit 1" (EC_{50} 14.7 ± 7.3 µM, E_{max} 48 ± 31%; n = 8) and "hit 6" (EC_{50} 15.6 ± 12.4 µM, E_{max} 56 ± 26%; n = 8) showed comparable PAM activity in cAMP as in IP-One assays. We further evaluated CCK1R-selective dimethyl thiacyclohexyl analogs CMP-2 (EC_{50} 6.0 ± 0.9 μM, E_{max} 28 ± 3%; *n* = 2) and CMP-9 (EC_{50} 5.9 ± 0.1 μM, E_{max} 30 ± 1%; *n* = 2), which demonstrated slightly improved potencies but attenuated efficacies compared to the original hits, in alignment with their activities in IP-One PAM assays. Hence, this tetracyclic CCK1R PAM scaffold seems to universally enhance CCK-mediated CCK1R signaling without discriminating between the distinct G protein pathways.

4. Discussion

Obesity is a major health problem around the world, with ongoing clinical need for new effective management strategies [13–17]. While CCK1R was long ago identified as a potential target for such therapy [1,18], full agonists of this receptor failed to meet primary end points in clinical trials, since they were no more effective than acute dieting to induce weight loss [19–22]. There was hesitance in preparing more potent and longer duration agonists, which were expected to have pronounced side effects and potential on-target toxicity [23–25]. However, we proposed a distinct strategy to utilize PAMs [26] of CCK action at CCK1R that possessed minimal or no intrinsic agonist activity [2,3,5]. Such an agent would be expected to enhance the satiety effect of endogenously released CCK during and after a meal to reduce meal size in a temporally finite manner. There was also the additional benefit that such an agent could reverse the negative impact of elevated membrane cholesterol on stimulus–activity coupling at this receptor [27,28].

We recently described our effort to identify candidates with this pharmacologic profile in a high throughput screening effort targeting small molecules [6]. Characterization and optimization of the best candidate scaffold identified in that work is the focus of the current report. This is a tetracyclic scaffold represented by "hit 1" and "hit 6" from the earlier report [6]. Here, we extensively characterize these candidates and demonstrate that they possess all the pharmacologic characteristics being sought. This includes the ability to enhance CCK action both in normal healthy cells and those with increased membrane cholesterol. They also exhibit minimal intrinsic agonist action under either condition. Under conditions in which they behave as PAMs, they do not induce CCK1R internalization, thereby priming this receptor to enhance the activity of endogenous released hormone at a temporally appropriate time.

We further conducted comprehensive structure–activity relationship studies to elucidate structural determinants for developing a PAM without intrinsic agonist activity. Therefore, we acquired 63 commercially available analogs of the original hits. We determined an activity score to facilitate the comparison and ranking of analogs that might be dependent on impact on potency and/or efficacy to exert positive allosteric effects. A previous study [29] proposed the calculation of R_{max}/R_{50} or $\Delta Log(R_{max}/R_{50})$ as a system-independent scale of relative activity, which also includes both parameters, potency and efficacy, of a compound dose–response determined in a PAM screening assay. We calculated both the activity scores and $Log(R_{max}/R_{50})$ values for our CCK1R PAMs and found that $Log(R_{max}/R_{50})$ relies excessively on the potency of a compound rather than also considering its efficacy. For example, for our most promising analogs in the CCK1R IP-One PAM assay, "hit 1" (EC_{50} 12.8 μM, E_{max} 60%) and CMP-2 (EC_{50} 9.2 μM, E_{max} 57%), we calculated $Log(R_{max}/R_{50})$ values of 4.7 and 4.8 compared to activity scores of 2.9 and 2.8, respectively. Hence, in contrast to utilizing the activity score, CMP-2 would display a slightly higher PAM activity compared to "hit 1" using $Log(R_{max}/R_{50})$. However, for an analog with minimal efficacy, such as the non-basic molecule CMP-38 (EC_{50} 7.6 μM, E_{max} 20%), we assessed a $Log(R_{max}/R_{50})$ of 4.4 and activity score of 1.0. Hence, in this case the $Log(R_{max}/R_{50})$ would have overstated the strength of CMP-38, whereas the activity score displayed it as a molecule with non-significant activity. We believe that the $Log(R_{max}/R_{50})$ might be superior for smaller sets of analogs with overall strong effects, while we found that the activity score worked better as a tool for SAR of larger sets of analogs derived from an HTS campaign that often start with moderate potency and efficacy.

One aim of this study was to further investigate the scaffolds we had identified in our HTS campaign [6] using the analog-by-catalog approach. Our target molecule would need to meet four criteria to display the desired pharmacological profile and justify expanded medicinal chemistry approaches. Firstly, the target molecule should display substantial positive modulation of CCK-mediated CCK1R Gq-signaling. Secondly, the desired compound should display no to minimal intrinsic agonist activity at CCK1Rs. As a third criterion, the molecule should maintain its CCK1R PAM profile in an environment with elevated cholesterol. As the fourth critical point, the compound should be selective toward CCK1Rs.

Initially, we tested our original "hits" having the tetracyclic scaffold and found that both displayed the desired profile of a CCK1R PAM with minimal intrinsic agonist activity [6], and moreover, both compounds maintained their PAM effects at CCK1R-Y140A cells [11] or in a system with excess cholesterol [10] (Figure 3). However, the original hits showed significant off-target effects at AVP2Rs [6]. Hence, one of our main goals for this study was to investigate structural modifications that would be able to eliminate the undesired AVP2R activity, while maintaining or enhancing clean PAM effects at CCK1R-WT.

Using the activity score as a ranking tool for PAMs, we were able to identify structural components contributing to distinct pharmacological effects. After comprehensive SAR analysis of 65 commercially available analogs, we observed the following overall trends: (i) A basic amino function at the R2 position is required for CCK1R PAM activity, but not necessarily for AVP2R signaling. (ii) Most of the tested analogs displayed no or insignificant intrinsic agonist activity in the CCK1R IP-One format, except analogs with a combination of a primary diamino ethyl group at R2 and a piperidine moiety at R1 (CMP-22, CMP-29). (iii) In general, analogs with an *N*-diethyl diamino ethyl side chain at R2 showed reduced AVP2R off-target effects compared to direct analogs with *N*-dimethyl diamino ethyl groups. This could indicate steric clashes with AVP2Rs at this position. (iv) In addition, a direct comparison of a sulfur (CMP-18) and oxygen ("hit 6") (CMP-28) analog indicated enhanced CCK1R PAM activity with reduced AVP2R signaling for CMP-18.

Overall, cyclohexyl analogs at ring A showed more promising profiles than methyl or cyclopentyl derivatives, except CMP-5, which carries a butyl sidechain at R1 and a diamino propyl function at R2. CMP-5 displayed substantial CCK1R PAM activity in IP-One (score 2.7) assays with minimal but still significant AVP2R activation (score 1.6). Thus, it is likely a combination of distinct structural components that leads to the required positioning and steric hindrance to mediate the desired selectivity.

A significant finding was that the substitution of the cyclohexyl ring to a dimethyl thiacyclohexyl moiety was able to completely eradicate AVP2R effects while maintaining CCK1R PAM activity in the IP-One format. This substantiated our hypothesis that slight structural expansion could induce steric clashes at AVP2Rs but not in the allosteric pocket of CCK1Rs.

Hence, we were able to demonstrate that structural modifications can eliminate unwanted activity at AVP2Rs but found limitations using the analog-by-catalog approach. We think that specific combinations, such as incorporating a dimethyl thiacyclohexyl ring and attaching specific residues suggested by the R-group analysis (Figures 6 and 7), would be excellent starting points for a planned medicinal chemistry effort. Therefore, we believe that the insights gained in this study will greatly improve our future hit-to-lead campaign using targeted organic synthesis of analogs or computational docking of PAMs into the recently solved cryo-EM structure of CCK1R [30].

Author Contributions: D.G.D.: data curation, formal analysis, investigation, methodology, supervision, validation, writing, review, editing; K.G.H.: data curation, formal analysis, investigation, methodology, supervision, validation, writing, review, editing; A.Y.: investigation, data curation; E.A.S.: conceptualization, formal analysis, project administration, writing, review, editing; L.J.M.: funding acquisition, conceptualization, formal analysis, project administration, writing, review, editing. All authors have read and agreed to the published version of the manuscript.

Funding: This work was supported by grants from the National Institutes of Health, including the National Institute of Diabetes and Digestive and Kidney Diseases, R01 DK115402 (LJM), and the National Institute of General Medical Sciences, R01 GM141003 (LJM).

Data Availability Statement: The data presented in this study are available upon request from the corresponding author.

Acknowledgments: We would like to thank M.L. Augustine, J. Milburn, and C. Chen at the Mayo Clinic Arizona and the personnel of the Conrad Prebys Center for Chemical Genomics at the Sanford Burnham Prebys Medical Discovery Institute for help in various parts of this work.

Conflicts of Interest: The authors declare no conflict of interest.

References

1. Smith, G.P.; Gibbs, J. The satiety effect of cholecystokinin. Recent progress and current problems. *Ann. N. Y. Acad. Sci.* **1985**, *448*, 417–423. [CrossRef]
2. Desai, A.J.; Dong, M.; Harikumar, K.G.; Miller, L.J. Cholecystokinin-induced satiety, a key gut servomechanism that is affected by the membrane microenvironment of this receptor. *Int. J. Obes. Suppl.* **2016**, *6*, S22–S27. [CrossRef]
3. Miller, L.J.; Desai, A.J. Metabolic actions of the type 1 cholecystokinin receptor: Its potential as a therapeutic target. *Trends Endocrinol. Metab.* **2016**, *27*, 609–619. [CrossRef] [PubMed]
4. Jordan, J.; Greenway, F.; Leiter, L.; Li, Z.; Jacobson, P.; Murphy, K.; Hill, J.; Kler, L.; Aftring, R. Stimulation of cholecystokinin-A receptors with GI181771X does not cause weight loss in overweight or obese patients. *Clin. Pharmacol. Ther.* **2008**, *83*, 281–287. [CrossRef] [PubMed]
5. Miller, L.; Harikumar, K.; Wootten, D.; Sexton, P. Roles of Cholecystokinin in the Nutritional Continuum. Physiology and Potential Therapeutics. *Front. Endocrinol.* **2021**, *12*, 684656. [CrossRef] [PubMed]
6. Dengler, D.; Sun, Q.; Harikumar, K.; Miller, L.; Sergienko, E. Screening for positive allosteric modulators of cholecystokinin type 1 receptor potentially useful for management of obesity. *SLAS Discov.* **2022**, *27*, 384–394. [CrossRef] [PubMed]
7. Powers, S.; Pinon, D.; Miller, L. Use of N,O-bis-Fmoc-D-Tyr-ONSu for introduction of an oxidative iodination site into cholecystokinin family peptides. *Int. J. Pept. Protein Res.* **1988**, *31*, 429–434. [CrossRef] [PubMed]
8. Harikumar, K.; Pinon, D.; Wessels, W.; Prendergast, F.; Miller, L. Environment and mobility of a series of fluorescent reporters at the amino terminus of structurally related peptide agonists and antagonists bound to the cholecystokinin receptor. *J. Biol. Chem.* **2002**, *277*, 18552–18560. [CrossRef] [PubMed]
9. Hadac, E.; Ghanekar, D.; Holicky, E.; Pinon, D.; Dougherty, R.; Miller, L. Relationship between native and recombinant cholecystokinin receptors: Role of differential glycosylation. *Pancreas* **1996**, *13*, 130–139. [CrossRef] [PubMed]
10. Potter, R.; Harikumar, K.; Wu, S.; Miller, L. Differential sensitivity of types 1 and 2 cholecystokinin receptors to membrane cholesterol. *J. Lipid Res.* **2012**, *53*, 137–148. [CrossRef]
11. Desai, A.; Harikumar, K.; Miller, L. A type 1 cholecystokinin receptor mutant that mimics the dysfunction observed for wild type receptor in a high cholesterol environment. *J. Biol. Chem.* **2014**, *289*, 18314–18326. [CrossRef]
12. Harikumar, K.; Coudrat, T.; Desai, A.; Dong, M.; Dengler, D.; Furness, S.; Christopoulos, A.; Wootten, D.; Sergienko, E.; Sexton, P.; et al. Discovery of a Positive Allosteric Modulator of Cholecystokinin Action at CCK1R in Normal and Elevated Cholesterol. *Front. Endocrinol.* **2021**, *12*, 89957. [CrossRef]
13. Kim, G.; Lin, J.; Blomain, E.; Waldman, S. Antiobesity pharmacotherapy: New drugs and emerging targets. *Clin. Pharmacol. Ther.* **2014**, *95*, 53–66. [CrossRef]
14. Rueda-Clausen, C.; Padwal, R.; Sharma, A. New pharmacological approaches for obesity management. *Nat. Rev. Endocrinol.* **2013**, *9*, 467–478. [CrossRef]
15. Pathak, V.; Flatt, P.; Irwin, N. Cholecystokinin (CCK) and related adjunct peptide therapies for the treatment of obesity and type 2 diabetes. *Peptides* **2018**, *100*, 229–235. [CrossRef]
16. Piper, N.; Whitfield, E.; Stewart, G.; Xu, X.; Furness, S. Targeting appetite and satiety in diabetes and obesity, via G protein-coupled receptors. *Biochem. Pharmacol.* **2022**, *202*, 115115. [CrossRef]
17. Muller, T.; Bluher, M.; Tschop, M.; DiMarchi, R. Anti-obesity drug discovery: Advances and challenges. *Nat. Rev. Drug. Discov.* **2022**, *21*, 201–223. [CrossRef]
18. Chandra, R.; Liddle, R. Cholecystokinin. *Curr. Opin. Endocrinol. Diabetes Obes.* **2007**, *14*, 63–67. [CrossRef]
19. Sherrill, R.; Berman, J.; Birkemo, L.; Croom, D.; Dezube, M.; Ervin, G.; Grizzle, M.; James, M.; Johnson, M.; Queen, K.; et al. 1,4-Benzodiazepine peripheral cholecystokinin (CCK-A) receptor agonists. *Bioorg. Med. Chem. Lett.* **2001**, *11*, 1145–1148. [CrossRef]
20. Berger, R.; Zhu, C.; Hansen, A.; Harper, B.; Chen, Z.; Holt, T.; Hubert, J.; Lee, S.; Pan, J.; Qian, S.; et al. 2-Substituted piperazine-derived imidazole carboxamides as potent and selective CCK1R agonists for the treatment of obesity. *Bioorg. Med. Chem. Lett.* **2008**, *18*, 4833–4837. [CrossRef]
21. Zhu, C.; Hansen, A.; Bateman, T.; Chen, Z.; Holt, T.; Hubert, J.; Karanam, B.; Lee, S.; Pan, J.; Qian, S.; et al. Discovery of imidazole carboxamides as potent and selective CCK1R agonists. *Bioorg. Med. Chem. Lett.* **2008**, *18*, 4393–4396. [CrossRef]

22. Sensfuss, U.; Kruse, T.; Skyggebjerg, R.; Uldam, H.; Vestergaard, B.; Huus, K.; Vinther, T.; Reinau, M.; Scheele, S.; Clausen, T. Structure-activity relationships and characterization of highly selective, long-acting, peptide-based cholecystokinin 1 receptor agonists. *J. Med. Chem.* **2019**, *62*, 1407–1419. [CrossRef]
23. Dawra, R.; Saluja, A.; Lerch, M.; Saluja, M.; Logsdon, C.; Steer, M. Stimulation of pancreatic growth by cholecystokinin is mediated by high affinity receptors on rat pancreatic acinar cells. *Biochem. Biophys. Res. Commun.* **1993**, *193*, 814–820. [CrossRef]
24. Hoshi, H.; Logsdon, C. Both low- and high-affinity CCK receptor states mediate trophic effects on rat pancreatic acinar cells. *Am. J. Physiol.* **1993**, *265*, G1177–G1181. [CrossRef]
25. Gitto, S.; Nakkina, S.; Beardsley, J.; Parikh, J.; Altomare, D. Induction of pancreatitis in mice with susceptibility to pancreatic cancer. *Methods Cell Biol.* **2022**, *168*, 139–159.
26. Christopoulos, A. Advances in g protein-coupled receptor allostery: From function to structure. *Mol. Pharmacol.* **2014**, *86*, 463–478. [CrossRef]
27. Paragh, G.; Kovacs, E.; Seres, I.; Keresztes, T.; Balogh, Z.; Szabo, J.; Teichmann, F.; Foris, G. Altered signal pathway in granulocytes from patients with hypercholesterolemia. *J. Lipid Res.* **1999**, *40*, 1728–1733. [CrossRef]
28. Seres, I.; Foris, G.; Varga, Z.; Kosztaczky, B.; Kassai, A.; Balogh, Z.; Fulop, P.; Paragh, G. The association between angiotensin II-induced free radical generation and membrane fluidity in neutrophils of patients with metabolic syndrome. *J. Membr. Biol.* **2006**, *214*, 91–98. [CrossRef]
29. Kenakin, T. Theoretical aspects of GPCR-ligand complex pharmacology. *Chem. Rev.* **2017**, *117*, 4–20. [CrossRef]
30. Mobbs, J.; Belousoff, M.; Harikumar, K.; Piper, S.; Xu, X.; Furness, S.; Venugopal, H.; Christopoulos, A.; Danev, R.; Wootten, D.; et al. Structures of the human cholecystokinin 1 (CCK1) receptor bound to Gs and Gq mimetic proteins provide insight into mechanisms of G protein selectivity. *PLoS Biol.* **2021**, *19*, e3001295. [CrossRef]

membranes

MDPI

Article

Cholesterol in Class C GPCRs: Role, Relevance, and Localization

Ugochi H. Isu, Shadi A Badiee, Ehsaneh Khodadadi and Mahmoud Moradi *

Department of Chemistry and Biochemistry, University of Arkansas, Fayetteville, AR 72701, USA
* Correspondence: moradi@uark.edu

Abstract: G-protein coupled receptors (GPCRs), one of the largest superfamilies of cell-surface receptors, are heptahelical integral membrane proteins that play critical roles in virtually every organ system. G-protein-coupled receptors operate in membranes rich in cholesterol, with an imbalance in cholesterol level within the vicinity of GPCR transmembrane domains affecting the structure and/or function of many GPCRs, a phenomenon that has been linked to several diseases. These effects of cholesterol could result in indirect changes by altering the mechanical properties of the lipid environment or direct changes by binding to specific sites on the protein. There are a number of studies and reviews on how cholesterol modulates class A GPCRs; however, this area of study is yet to be explored for class C GPCRs, which are characterized by a large extracellular region and often form constitutive dimers. This review highlights specific sites of interaction, functions, and structural dynamics involved in the cholesterol recognition of the class C GPCRs. We summarize recent data from some typical family members to explain the effects of membrane cholesterol on the structural features and functions of class C GPCRs and speculate on their corresponding therapeutic potential.

Keywords: G-protein-coupled-receptors; GPCRs; membrane protein; protein–lipid interactions; sterols; cholesterol; class C GPCRs

Citation: Isu, U.H.; Badiee, S.A.; Khodadadi, E.; Moradi, M. Cholesterol in Class C GPCRs: Role, Relevance, and Localization. *Membranes* **2023**, *13*, 301. https://doi.org/10.3390/membranes13030301

Academic Editors: Yosuke Senju and Shiro Suetsugu

Received: 16 January 2023
Revised: 21 February 2023
Accepted: 28 February 2023
Published: 3 March 2023

1. Introduction

Many recent studies are geared towards deciphering the structures of G-protein coupled receptors (GPCRs) through several methods, most commonly crystallography and cryogenic electron microscopy (Cryo-EM). This is simply because many medications are designed to target GPCRs due to their central role in many biological functions. All GPCRs in a lipid bilayer are not stand-alone molecules, rather they interact with other components of the bilayer e.g., lipids and sterols, most notably cholesterols [1]. Some members of GPCRs exist and function as monomers, especially within the class A GPCRs, while other GPCRs, mostly the class C GPCRs, form dimers [2,3], and oligomers with themselves or other GPCRs [4]. Most recently, several determined GPCR structures often existing as dimers, appear to indicate the presence of cholesterol. A notable example is the 2-adrenergic receptor (2AR), a class A GPCR crystallized with cholesterol molecules and a component of the dimer interface consisting of the addition of post-translational palmitate groups from each protomer [1]. GPCRs function within cholesterol-rich membranes and an insufficient or excessive amount of cholesterol within the membrane could induce conformational changes in many GPCRs which would result in various diseases [5,6]. These effects of cholesterol could result in indirect changes by altering the mechanical properties of the lipid environment or direct changes by binding to specific sites on the protein [7–9]. There are a number of studies and reviews on how cholesterol modulates class A GPCRs, but this area of study is yet to be fully explored for class C GPCRs. Recent advances in experimental and computational power have enabled researchers to investigate the role of lipids in various membranes and solvable proteins, at the atomic level using molecular dynamics simulation [10–18].

Class C GPCRs consist of about 23 receptors with a unique characterization compared to other GPCR classes and exist as obligate homo- (e.g., CaS) [19–22] or hetero- (e.g., $GABA_B$) dimers [23–26]. They are made up of three unique structural elements: a seven-transmembrane domain which is responsible for allosteric ligand recognition but is uniquely dimeric in the case of class C GPCRS [27]; an unusually large extracellular venus flytrap domain (VFT) which has a double-lobed structure with a crevice between them to serve as the orthosteric binding site; and a cysteine-rich domain (CRD) that links the VFT region to the 7TM region (Figure 1) [3]. However, some class C GPCRs, like $GABA_B$ receptors, lack the cysteine-rich domain [24,25,28–30]. Due to these distinct structural features and mandatory dimerization, the class C GPCRs have been the most complex of the GPCRs in terms of understanding their activation mechanism [31–35]. Using several methods such as crystallization [30], lipid cubic phase [36], and most commonly single particle Cryo-EM, structures of over 20 human class C GPCRs have been solved to date [37], comprising metabotropic glutamate receptors (mGluR1–5, mGluR7) [36,38–46], gamma-aminobutyric acid receptors ($GABA_B1$ and $GABA_B2$) [23,24,47], calcium-sensing receptors (CaS) [48–50], the extra-cellular domain of taste receptors (TAS1R1–TAS1R3) [51–55], and orphan receptors (GPR158, GPR179, GPR156) [56–60]. Similarly to other GPCR structures, class C GPCR structures are solved with inclusion of cholesterol or cholesteryl hemisuccinate (CHS) to the detergent mix during crystallization and recently, Cryo-EM (Table 1). However, some of the structures also have bound cholesterol or cholesteryl hemisuccinate acting as ligands to the already determined structures (Table 1). The argument for cholesterol addition varies from stabilizing the protein to aiding dimerization. Experimental analysis and, most recently, molecular dynamics simulations [16,17,61–64] have been used to decipher the possible role of cholesterol in these protein structures. In this review, we will discuss the relevance and position of cholesterol molecules in class C GPCR structures and functions.

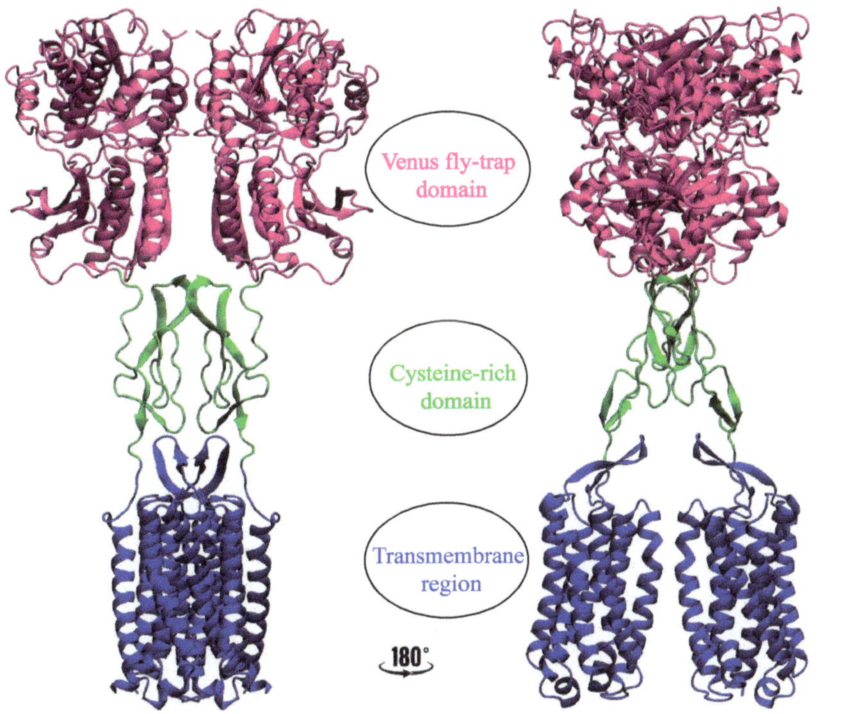

Figure 1. Representation of a class C GPCR (a full length human mGluR5) showing the different regions as: VFT (magenta), CRD (green), and 7TM region (blue). (PDB ID: 7FD8).

Table 1. Solved class C GPCR structures with cholesterol acting as ligands.

Name	PDB ID *	Number of Sterols in TM
mGluR1	4OR2 [36]	6 CLR
mGluR5	7FD8 [45]	2 CHS
GABA$_B$	6WIV [24]	10 CLR
	7CUM [65]	16 CLR
	7CA3 [65]	3 CLR
CaSR	7SIM [66]	8 CLR
	7SIL [66]	8 CLR
Orphan receptor (GPR158)	7SHF [56]	22 CLR
	7SHE [56]	22 CLR **

* Data obtained from protein data bank (PDB) database (https://www.rcsb.org, accessed on 9 February 2023). Citations to the PDB structures are included. ** CLR cholesterol; CHS cholesteryl hemisuccinate.

1.1. Cholesterol–Membrane Interactions

The plasma membrane of eukaryotic cells consists of various lipids displaying high biochemical variability in both their apolar moiety and their polar head [67,68]. Sterols are a class of lipids that are a key component of the plasma membrane and are characterized by their steroid hydrocarbon ring structure. One specific sterol, cholesterol, makes up a vital part of the plasma membrane of eukaryotic cells. Cholesterol is crucial for membrane dynamics and organization [69–71] and it is also necessary for viability and cell proliferation [72]. The structural features of cholesterol qualify it to interact with proteins and other membrane lipids in several different ways through a variety of different interaction domains [67]. Cholesterol consists of a tetracyclic fused ring skeleton with a single hydroxyl group, a double bond, and a flexible iso-octyl hydrocarbon sidechain [73] which allows it to take on a wide array of conformations [67]. The hydroxyl group is said to contribute significantly to the amphiphilic behavior of cholesterol, causing it to orient in membranes [74]. It is also essential in the hydrogen bond formation between cholesterol and water [75], as well as other lipid membranes in the cell [76]. The hydroxyl group can form two distinct types of hydrogen bonds (acceptor and donor) with a polar group belonging to either a membrane lipid or a protein. Cholesterol is able to affect the physical behavior and dynamics of the cell membrane by interacting with membranes rich in sphingolipids such as lipid "rafts" [77–79], or by being present in the liquid disordered (Ld) phase of membranes which contain a large number of glycerophospholipids such as phosphatidylcholine [80]. As a result, cholesterol can alter the properties and dynamics of proteins in the membrane [73,81–83]. When bound to cholesterol, some integral membrane proteins could become activated or inactivated [84–86]. In recent times, there has been considerable interest in cholesterol interaction sites in membrane proteins. Certain proteins function in cholesterol-rich domains, while others have direct interactions with cholesterol through their transmembrane domains, and sometimes act as ligands [11,61]. The more common cholesterol binding sites in membrane proteins include the cholesterol recognition/interaction amino acid consensus (CRAC)/CARC domain [87], the cholesterol consensus motif (CCM) [88,89] and the sterol-sensing domain (SSD) [90,91]. All of these listed are structural features in proteins that could result in preferential involvement with cholesterol. Several studies have shown that protein–cholesterol interactions are more common in proteins with sequences comprising of the CRAC motif [92], a short peptide segment at the tail of a transmembrane helix comprising of 5–13 amino acid residues. The CRAC motif consists of a well defined linear sequence of amino acids [67,92–97] identified via the following pattern: a leucine or valine residue, 1–5 non-specific amino acid residues, tyrosine, another 1–5 residues of any amino acid, and finally a lysine or arginine residue [-L/V-(X)1–5-Y-(X)1–5-R/K-, with (X)1–5 representing between one and five residues of any amino acid] [92–95]. Rhodopsin, the $\beta(2)$-adrenergic receptor, and the serotonin(1A) receptor are examples of GPCRs that have been identified with the CRAC motif recognition site [92]. The major difference between the CARC and CRAC motif is that one exhibits a preference for the outer membrane leaflet (CARC), while the mirror sequence (CRAC) is located in the inner membrane leaflet [98,99]. A double CARC-CRAC motif has been identified within the transmembrane domains of some membrane proteins [10,98]; however, the limiting factor of the CRAC/CARC sequence is that they are based on a linear (1D) sequence motif, as opposed to cholesterol-binding sites which consist of a three-dimensional (3D) structure [100]. Another common motif is the CCM, which is defined by four spatially distributed interactions with cholesterol: an aromatic Trp158, conserved in 94% of class A GPCRs; a hydrophobic Ile154 conserved in 35% of class A GPCRs (both residues in helix IV); an aromatic Tyr70 from helix II, which forms a hydrogen bond with Arg151 from helix IV [101–103]. This motif was established from the analysis of the human β2-adrenergic receptor in a complex with timolol and two molecules of cholesterol [101]. CCM can either be described as strict or less restrictive [101,104]. The strict variant is found in 21% of the class A GPCRs, while the less-restrictive variant, defined by the absence of the aromatic residue from helix II, is present in 44% of class A GPCRs [101,103,104]. Additionally, the sterol-sensing domain is another significant cholesterol recognition motif with a larger protein segment and comprises five transmembrane helices. The sterol-sensing domains usually consist of a tetrapeptide amino acid sequence— tyrosine, isoleucine, tyrosine, and phenylalanine (YIYF)— which has been found to be present in other lipid-raft associated proteins without the SSD motif [90,91,94]. Studies have shown that the presence of the YIYF amino acid sequence alone can interact with the cholesterol-rich domain [90,91,94,105,106]. Finally, START proteins have also been identified as a cholesterol binding motif, with the transport of cholesterol molecules being their primary function [107,108]. Proteins with the START domain [109] are able to transfer lipids between membranes and can interact with cholesterol [94]. While all of these are cholesterol-binding motifs in membrane proteins, the CARC-CRAC motif is the major cholesterol interaction site that has been observed in GPCRs.

1.2. GPCR–Cholesterol Interactions

G-protein-coupled receptors are a superfamily of integral membrane proteins in the human genome, constituting one of the largest classes of clinical drug targets [110–113]. Often distinguished by a characteristic seven transmembrane helices plus an eighth helix that lies underneath the surface of the layer, GPCRs depend on a relationship with the lipid membranes in their physical environment to perform their function [11,62]. As per the phylogenetic investigation, most GPCRs belong to one of four classes, i.e., A, B, C, and Frizzled. The class-C GPCR family contains metabotropic glutamate receptors (mGluR1–8), γ-aminobutyric acid receptors, a few taste-detecting receptors (TAS1R1-3), Ca^{2+}-detecting receptors (CaS), and orphan receptors [114]. One trademark highlight of the class-C GPCRs is their dimerization, either into homo- or hetero-dimers, which is requisite for their proper functioning [40]. Cholesterol assumes an essential role in the function of a significant number of GPCR structures [115]. It does this by binding to a number of GPCRs, including rhodopsin [116], oxytocin [117], μ-Opioid [118], and serotonin 1A receptors [119], at both canonical and non-canonical binding sites, consequently altering their ligand-binding activity allosterically, which could result in the activation or inactivation of the protein. For example, cholesterol is reported to influence Hedgehog (Hh) signaling as a means of activating the Smoothened orphan receptor (SMO) which belongs to GPCRs [120]. As such, it has been determined that cholesterol can influence the stability, oligomerization, and ligand-binding affinity of GPCRs [12,63,116,119,120]. Two mechanisms have been proposed by which cholesterol might influence the structure and function of GPCRs: directly, through specific interactions with the GPCRs; indirectly, by altering the physical properties of the membrane; or perhaps some combination of the two mechanisms [61,92,121,122]. Recently, several GPCR structures have been determined through X-ray diffraction and even more through Cryo-EM. A large percentage of these structures have been stabilized by site-specific cholesterol binding, although it is uncertain if these cholesterol associations are due to recurring cholesterol-binding motifs or if the experimental technique used determines the method of cholesterol binding. A comprehensive study by Taghon et al. [88] showed that cholesterol binding in both X-ray and Cryo-EM structures is much the same. They also indicate that about 92% of cholesterol molecules on GPCR surfaces are located in predictable locations that do not require cholesterol-binding motifs [88]. The importance of cholesterol in GPCR structural dynamics has been identified in some GPCR structures, especially within the class A family (e.g., the presence of CCM in the β2-adrenergic receptor [101]). In some cases, CHS has been used to substitute cholesterol in GPCRs, although the validity of this replacement has been contested [93,123–125]. The CRAC motif has been established as a characteristic feature of the serotonin (1A) receptor [126], the β2-adrenergic receptor [127], cholecystokinin [121], cannabinoid (CB1) receptor [128], etc. [92,93,129], indicating that the interaction of cholesterol with GPCRs could be specific in nature. However, another group of researchers suggested that the presence of CRAC/CARC motifs does not automatically prove that cholesterol interacts within those binding motifs [130]. On the flip side, their impact on class C GPCRs is yet to be fully explored [36]. The significance of cholesterol to GPCR structures and their functional dynamics is an ongoing question that is yet to be fully elucidated [93].

2. Significance and Interaction Sites of Cholesterol in Class C GPCRs

2.1. *Metabotropic Glutamate Receptors (mGluRs)*

Metabotropic glutamate receptors (mGluRs) are a family of G protein-coupled receptors that are significant in regulating neurotransmission [131–133]. GPCRs are membrane-bound proteins expressed in the central nervous system (CNS), and their physiological functions are dependent on their lipid environment [134]. There are three groups with eight subtypes of mGluRs that are classified based on G-protein coupling and ligand selectivity [40,135]. Group I consists of mGluR1 and mGluR5, which are linked to the activation of phospholipase C (PLC) to increase diacylglycerol (DAG) and inositol triphosphate (IP3), Group II includes mGluR2 and 3 [136], Group III is comprised of mGluR4, 6, 7, and 8. These last two groups are linked to the inhibition of adenylyl cyclases (ACs) [137,138]. In mammalian cells, cholesterol is highly concentrated in the plasma membrane but low in the intracellular membrane [139]. Cholesterol affects receptor function by affecting the membrane's fluidity or interacting with the receptor's binding site [140] and moves freely between the inner and outer leaflets [141]. In lipid rafts, plasma membranes are rich in cholesterol and sphingolipids within their lipid domains, and the cholesterol forms specific interactions with GPCRs including mGluR1 and mGluR2. Research has shown that cholesterol aids the dimerization of mGluR2 and mGluR5 through interactions with the TM4/TM5 domains and also through the TM1/TM2 domains of mGluR1 [142–144]. In mGluRs, investigations from several scientists have suggested specific interaction sites for mGluR1 and mGluR2 [40,145]. A study has revealed the presence of a CRAC motif in the transmembrane helix 5 domain of mGluR1, which is conserved for all mGluRs. The CRAC motif located in TM5 plays an important role in supporting mGluR1 recruitment to the lipid raft as a result of agonist binding [143]. It has been reported that mutations in this motif affect both signaling and the association of mGluR1 with cholesterol-rich membrane domains [143]. Another group has experimentally determined that within the transmembrane domain of mGluR1, cholesterol is localized within the helix I homodimer interface. Intriguingly, this was observed through analyses of the crystal structure of the transmembrane domain of mGluR1, bound by six cholesterol molecules mediating the dimer interface, which in this case is mainly composed of the TM1 helices from both protomers (Figure 2). It has been suggested that these cholesterol molecules stabilize the dimerization of mGluR1 (PDB:4OR2) [36]. In addition, by increasing cholesterol levels, mGluR1 signaling efficiency is enhanced upon stimulation by an agonist, while by lowering cholesterol levels, extracellular signal-regulated kinase-mitogen-activated protein kinase (ERK-MAPK) activation via mGluR1 is inhibited [143,146]. In this way, lipid rafts and membrane cholesterol act as positive allosteric modulators (PAM) of the group I mGluR signaling pathway. Therefore, it is possible to modulate abnormal group I mGluR behavior in neuropsychiatric conditions (fragile X syndrome and autism) through the use of drugs such as statins and cyclodextrins, which affect membrane cholesterol levels [143]. Furthermore, the role of cholesterol has also been considered for class II members of mGluRs. A number of neuropsychiatric conditions, including depression, Alzheimer's disease, and Parkinson's disease, as well as different types of cancer, have been treated with these same classes of drugs [147–149]. The binding of glutamate to mGluR2 dimers results in the transmission of a signal across the transmembrane domain of the receptor that prevents the activity of adenylate cyclase via the $G_{i/o}$ protein [135]. In a study, the interaction of cholesterol with mGluR2 was demonstrated across 2 to 5 sites in the transmembrane domain of mGluR2 [135], using molecular dynamics simulations [150–152], biochemical approaches, and photocrosslinking experiments. It was observed that mGluR2 is modulated by their surrounding lipid environment, particularly cholesterol, through an unknown mechanism. The CRAC/CARC motif and a cholesterol consensus motif (CCM) were suggested as cholesterol-binding motifs in GPCRs [67,153]. A central aromatic amino acid that interacts with sterols is a common characteristic of some of the motifs described [135]. A recent computational study conducted by Bruno et al. [154] found that the conformational differences observed in the helical structure of the mGluR2-TM8 domain can be used as an indicator to detect the presence of cholesterol in metabotropic glutamate receptors and GPCRs. They observed that the inclusion of higher levels of cholesterol in the membrane stabilizes the transmembrane helix 8 (TM8) of mGluR2, while a lack of cholesterol results in destabilization of the TM8 domain [154]. However, the role of cholesterol in the third group of mGluRs remains unknown.

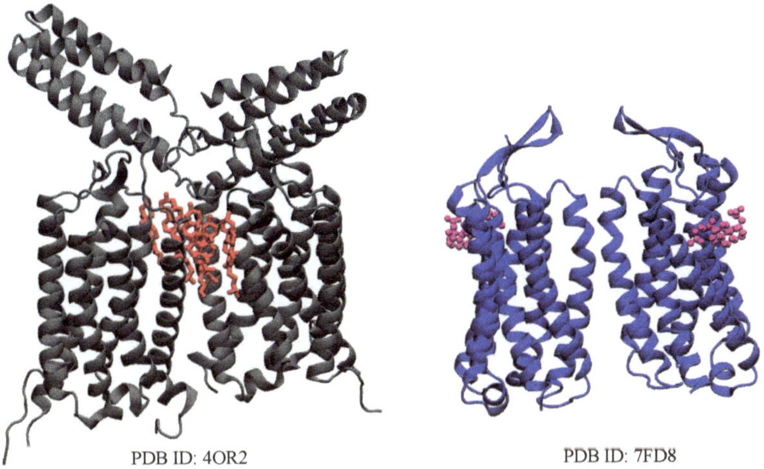

Figure 2. Crystal structure of mGluR1 (grey) in an inactive state, and a Cryo-EM structure of an intermediate-active mGluR5 (blue), determined with 6 molecules of cholesterol (red) and 2 molecules of CHS (magenta), respectively.

2.2. GABAB Receptors

In mammals, GABA (γ-Aminobutyric acid) is one of the major inhibitory neurotransmitters. In order for GABA to exert its effects, it must bind to at least two different receptor classes: $GABA_A$ and $GABA_B$. Approximately 20 to 50% of the brain's synapses contain $GABA_A$ receptors [155]. They are pentameric receptors belonging to a superfamily of ligand-gated ion channels [29]. Unlike $GABA_A$, $GABA_B$ receptors are members of class C GPCRs with the typical classification of an N-terminal VFT region: a 7TM domain, and a C-terminal intracellular domain [24,30,156]. $GABA_B$ receptors function as inhibitor receptors by opening potassium channels, reducing the activity of adenylate cyclase and calcium channels [157]. There are few solved structures of $GABA_B$ receptors containing cholesterol, deposited on the protein data bank (Figure 3), and subsequently, there is little knowledge of the effect of membrane cholesterol on the $GABA_B$ receptors. Experimental investigations have shown that cholesterol enrichment and depletion both decrease GABA potency, resulting in an up to fourfold increase in EC50 [158]. The structures of $GABA_B$ receptors with cholesterol were determined based on ligand type because the presence of a ligand can change how cholesterol interacts with the receptor. For instance, in absence of ligands, it is feasible that there is no interaction between cholesterol and the receptors. However, for systems bound to an antagonist, 10 and 16 molecules of cholesterol [24,65] were bound between the protomers of the transmembrane dimers [115] (Figure 3). Thus, It can be suggested that the ligands may have caused some conformational changes in receptors allowing greater binding to cholesterol. Moreover, three cholesterol molecules were attached to the $GABA_B$ receptor bound to a positive allosteric modulator (PAM) [65] (Figure 3). Therefore, the variation in cholesterol binding between two different ligand classes can indicate the potentially significant role of ligands in the interaction between cholesterol and the receptors.

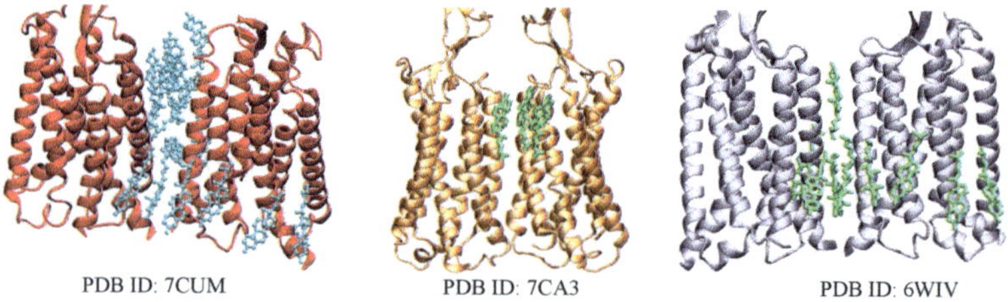

Figure 3. Visual representations of $GABA_B$ receptors in active state (orange) and inactive states (red and gray), determined by single particle Cryo-EM. They include 2 (orange), 17 (red), and 16 (gray) bound cholesterols, respectively, within the transmembrane region. The figures here show the transmembrane region only bound to cholesterol (cholesterol molecules are shown as cyan and green sticks).

2.3. *Taste Receptor*

TAS1R1 and TAS1R2 were among the first determined subfamilies of taste-related GPCRs. Prior to identifying their physiological ligands, they were originally classified as orphan receptors [159]. Subsequently, some scientists identified a member, TAS1R3, through a fusion of molecular biological and genetic approaches [160]. These three members (TAS1R1-3) code for sweet and umami tastes and are classified as class C GPCRs. The sweet taste signals are activated by TAS1R2 and TAS1R3 heterodimers, while the umami taste signals are transduced by heterodimers of TAS1R1 and TAS1R3 [161]. Therefore, the class C taste receptors consist of either TAS1R1 or TAS1R2, interacting with a common subunit TAS1R3. Similar to other class C GPCRs, they exist as obligate dimers and are characterized by a large extracellular N-terminus, which houses the orthosteric ligand-binding site, while the allosteric binding sites are present in the cysteine-rich domain and/or transmembrane region [160]. Due to these multiple binding sites, a single taste receptor is able to function for various stimuli [53]. The sweet taste receptor is able to interact with various compounds at a lower sensitivity, unlike most GPCRs, which are highly selective to specific high-affinity ligands. Cholesterol has been shown to regulate GPCR signaling in sweet taste receptors [162,163]. A study showing the presence of a CRAC motif in T2R4 (a subset of GPCRs responsible for bitter taste receptors [164]), explains that taste receptors are crucial to cholesterol sensitivity [94] and become more sensitive to cholesterol through a cellular mechanism [92]. Furthermore, they observe electrostatic interactions between the 3β-hydroxyl group of cholesterol and the positively charged residue in the cholesterol binding motif [162]. Site-directed mutagenesis and functional assays have been optimized in the study of putative cholesterol-binding motifs (CRAC and CARC) to determine the mechanism of cholesterol binding to taste receptors. A comparison of the dynamics of wild-type T2R14 receptors and mutant T2R14 receptors revealed that the amino acid residues K110, F236, and L239 are required for the receptor to function appropriately when cholesterol is present. Based on this study, it could be suggested that cholesterol influences taste receptors by directly interacting with the receptor [165].

2.4. Retinoic Acid-Inducible Orphan G Protein-Coupled Receptors (RAIGs)

Retinoic-acid inducible receptors belong to a group of class C GPCRs [166]. Although containing a characteristic secondary structure of seven transmembrane α-helical domains, these receptors have short amino-terminal extracellular domains, ranging from 30 to 50 amino acids [167]. In contrast, other family C members consist of a large N-terminal domain, comprising 500–600 amino acids [168]. Currently, there are four genes that make up the RAIG family: RAIG1, RAIG2 RAIG3, and GPCR5D [50]. RAIG1 was the first of these genes to be determined, and it was associated with a retinoic acid-responsive gene in human carcinoma cells [50,167,168]. Although classified as a class C GPCR, RAIG protein shares low sequence similarity with known members of GPCRs, and only shows 25% similarity [167,169] with the homology sequence of mGluR2 and 3, primarily in the transmembrane regions [167]. Due to the large variation between the sequence homology of RAIG proteins and most GPCRs, the endogenous ligands for RAIGs remain unknown [169] However, the ligand-binding regions are predicted to be found in the extracellular loops of the transmembrane domain, and also at the short amino-terminal regions [167,168,170]. A study that utilized fluorescence microscopy and immunocytochemical methods to study the formation and localization of synaptic vesicles in human SH-SY5Y neuroblastoma cells suggested that retinoic acid-induced proteins with cholesterol produced significant neurite extension and formation of cell-to-cell contacts, predicting it as a valuable tool for basic studies of neuronal metabolism [166].

2.5. Calcium-Sensing Receptor-Related Receptor

As a G-protein-coupled receptor, the calcium-sensing receptor (CaSR) is essential for controlling calcium homeostasis [171] in humans. CaSR is a Ca^{2+}-sensing protein found on the surface of cells [172] that exists as an obligate homodimer and belongs to class C GPCRs [48]. Each protomer has a Ca^{2+}-binding extracellular domain and a seven-transmembrane-helix domain (7TM) that activates heterotrimeric G-proteins [34]. The classical calcium-sensing receptor is known to be involved in the pathophysiology of parathyroid and renal-related diseases by sensing calcium ions in extracellular fluid [53,171]. Recent studies suggest that CaSR can be modulated through the interactions of its transmembrane (TM) domains with cholesterol [66]. This is especially observed in TM6 where cholesterol molecules found at the dimer interface influence the interactions of two residues (ILE816) from the TM6 helices of both subunits. This allows the side chains of both residues to pack against each other and make indirect dimer contacts. As a result of these observations, it could be suggested that the TM6-TM6 dimer interface is stabilized by cholesterol and as such it is essential in the receptor activation of CaSR [66].

Cholesterol depletion has been shown to negatively impact receptor function by decreasing basal activity and Ca^{2+} sensitivity [173]. Another study showed that vascular smooth muscle cells (VSMCs) [174] are expressed in CaSR and can be altered by cholesterol [175]. They further indicated that plaque stability can be affected due to CaSR [176] mediating MMP-2 (matrix metalloproteinase-2) production in the presence of cholesterol via the phosphatidylinositol 3-kinase (PI3K)/Akt signal pathway [177,178]. In addition, activation of CaSR in VSMCs increases cell proliferation and survival via the phospholipase C (PLC)-IP3 and MAPK-ERK1/2 pathways [179].

2.6. Orphan Receptor

Despite extensive reorganization efforts, there are over 140 receptors [180] within the GPCR family that have yet to be fully identified and these groups are referred to as orphan receptors [37,112,180,181]. Orphan GPCRs play important roles in physiology and diseases, yet they are poorly understood in terms of their structural organization, ligand identification, activation mechanisms, and signaling reactions [56,180,182] GPR156,

GPR158, and GPR179 make up the orphan receptor of class C GPCRs [56–60,182] and they are the least characterized members of the group [182]. They share 70% sequence similarity in both extracellular and TM domains, with a distinct feature of lacking the Venus flytrap-fold ligand-binding domain [182]. GPR158 are drug targets with significant roles in mood regulation, memory, depression [183], carcinogenesis, and cognition [56,182,184–186]. It is highly expressed in brain tissues [187] and functions by regulating ion channels and second messengers. One prominent characteristic of GPR158 is that it binds to the neuronal RGS7-Gβ5, a regulator of the G protein signaling protein complex [188], that directly deactivates G proteins [188,189]. GPR158 and RGS regulate the homeostasis of the second messenger cyclic adenosine monophosphate (cAMP), and control the neuronal activity with a marked impact on brain physiology [56]. Recently, two high-resolution Cryo-EM structures have been determined by Patil et al. [56]. The structures consist of GPR158 alone, and GPR158 bound to RGS complex (Figure 4), with both determined structures revealing the presence of an extracellular Cache domain and an unusual ligand-binding domain, that is not found in other GPCRs [56]. In both initial structures, cholesterol interacts between the protomers and the transmembrane helices to stabilize the protomers [56]. The determined structures show several cholesterol molecules surrounding the dimeric interface of GPR158, which acts as a shield for the cavity formed at the interface. It is also suggested that the interactions of cholesterol with the transmembrane helices could stabilize the interface between both protomers. Patil et al. [56] reported that the stability provided by these cholesterol molecules results in a more compact dimeric interface, which then prevents G protein activation [56].

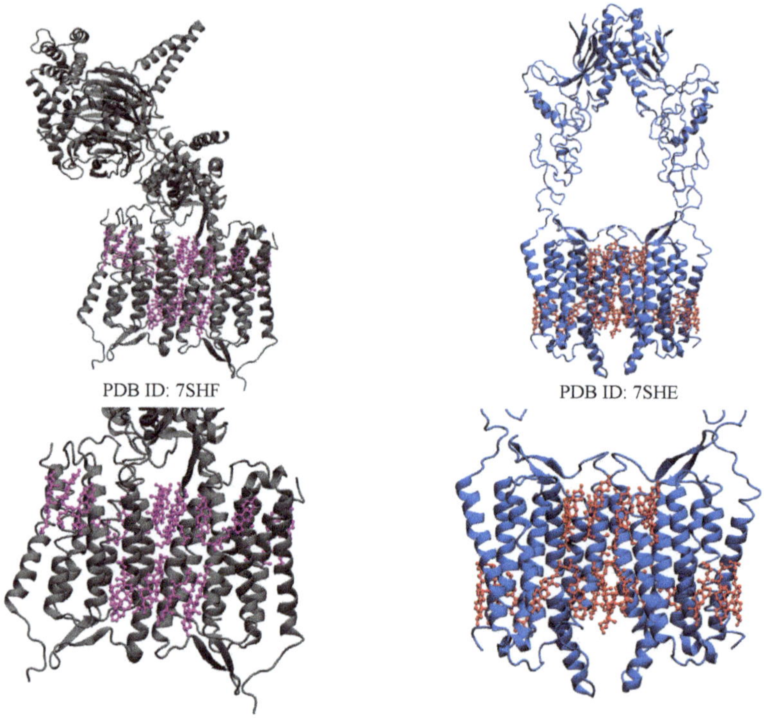

Figure 4. Visual representations of inactive states of GPR158 class C orphan receptors, showing GPR158 bound to RGS complex (gray) and GPR158 in apo form (blue). Both structures contain 22 cholesterol molecules within the transmembrane region. The upper figures show the whole protein, while the lower figures only show the transmembrane region (cholesterol molecules are shown as violet and red sticks).

3. Conclusions

Through this review and the accompanying table and figures, we have described the interaction sites of cholesterol in specific receptors of class C GPCR structures. Through the collective study of class C GPCR structures, we notice that cholesterol is mostly bound between the transmembrane dimers of the receptors and also within the surrounding groves of the transmembrane helices, which could explain why it seems to aid dimerization. Furthermore, this review highlights the significance of cholesterol within specific class C GPCRs. Consideration of several studies revealed that cholesterol is important for oligomerization, organization, function, and dynamics of class C GPCRs. In general, we see that cholesterol could affect ligand binding, G-protein coupling, and intracellular signaling of GPCRs. With the possible emergence of more cholesterol-bound GPCR structures and analyses, we picture an exciting and enlightening future in the study of cholesterol–GPCR interactions. We expect that this information will help provide insight into the molecular mechanisms of cholesterol molecules bound to particular receptors of class C GPCRs.

Author Contributions: Conceptualization, U.H.I. and M.M.; methodology, U.H.I. and M.M.; software, U.H.I. and M.M.; validation, U.H.I., S.A.B., E.K. and M.M.; formal analysis, U.H.I., S.A.B. and E.K.; investigation, U.H.I., S.A.B., E.K. and M.M.; resources, U.H.I., S.A.B., E.K. and M.M.; data curation, U.H.I., S.A.B., E.K. and M.M.; writing—original draft preparation, U.H.I., S.A.B., E.K. and M.M.; writing—review and editing, U.H.I. and M.M.; visualization, U.H.I., S.A.B., E.K. and M.M.; supervision, M.M.; project administration, U.H.I. and M.M.; funding acquisition, M.M. All authors have read and agreed to the published version of the manuscript.

Funding: This work was supported by the National Institute of General Medical Sciences of the National Institutes of Health under award number R35GM147423. This work was also supported by the Arkansas Biosciences Institute.

Institutional Review Board Statement: Not applicable.

Informed Consent Statement: Not applicable.

Data Availability Statement: No new data is reported in this review article.

Conflicts of Interest: The authors declare no conflict of interest.

Abbreviations

The following abbreviations are used in this manuscript:

GPCR	G-protein-coupled receptors
CHS	Cholesteryl hemisuccinate
CLR	Cholesterol
mGluR	Metabotropic glutamate receptors
CRAC	Cholesterol Recognition/Interaction Amino Acid Consensus (CRAC)
MD(S)	Molecular dynamics (simulation)
GABA	Gamma-Aminobutyric acid
TAS1R1	Taste 1 receptor member 1
TAS1R2	Taste 1 receptor member 2
TAS1R3	Taste 1 receptor member 3
RORs	Retinoid-related orphan receptors
CASR	calcium-sensing receptor
TM(D)	Transmembrane (domain)
NTD	N-terminal domain
VSMCs	Vascular smooth muscle cells
PLC	phospholipase C
MMP-2	Matrix metalloproteinase-2
ERK	Extracellular signal-regulated kinase
MAPK	Mitogen-activated protein kinase
AMPK	AMP-activated protein kinase

References

1. Goddard, A.D.; Watts, A. Regulation of G protein-coupled receptors by palmitoylation and cholesterol. *BMC Biol.* **2012**, *10*, 27. [CrossRef] [PubMed]
2. Liu, L.; Fan, Z.; Rovira, X.; Xue, L.; Roux, S.; Brabet, I.; Xin, M.; Pin, J.P.; Rondard, P.; Liu, J. Allosteric ligands control the activation of a class C GPCR heterodimer by acting at the transmembrane interface. *eLife* **2021**, *10*, e70188. [CrossRef] [PubMed]
3. Chun, L.; Zhang, W.H.; Liu, J.F. Structure and ligand recognition of class C GPCRs. *Acta Pharmacol. Sin.* **2012**, *33*, 312–323. [CrossRef] [PubMed]
4. Maurice, P.; Kamal, M.; Jockers, R. Asymmetry of GPCR oligomers supports their functional relevance. *Trends Pharmacol. Sci.* **2011**, *32*, 514–520. [CrossRef] [PubMed]
5. Soffientini, U.; Graham, A. Intracellular cholesterol transport proteins: roles in health and disease. *Clin. Sci.* **2016**, *130*, 1843–1859. [CrossRef]
6. Kumar, G.A.; Chattopadhyay, A. Statin-Induced Chronic Cholesterol Depletion Switches GPCR Endocytosis and Trafficking: Insights from the Serotonin 1A Receptor. *ACS Chem. Neurosci.* **2020**, *11*, 453–465. [CrossRef]
7. Lee, A.G. Interfacial Binding Sites for Cholesterol on G Protein-Coupled Receptors. *Biophys. J.* **2019**, *116*, 1586–1597. [CrossRef]
8. Zakany, F.; Kovacs, T.; Panyi, G.; Varga, Z. Direct and indirect cholesterol effects on membrane proteins with special focus on potassium channels. *Biochim. Biophys. Acta (BBA) Mol. Cell Biol. Lipids* **2020**, *1865*, 158706. [CrossRef]
9. Zocher, M.; Zhang, C.; Rasmussen, S.G.; Kobilka, B.K.; Müller, D.J. Cholesterol increases kinetic, energetic, and mechanical stability of the human β2-adrenergic receptor. *Proc. Natl. Acad. Sci. USA* **2012**, *109*, E3463–E3472. [CrossRef]
10. Sengupta, D.; Chattopadhyay, A. Molecular dynamics simulations of GPCR-cholesterol interaction: An emerging paradigm. *Biochim. Biophys. Acta (BBA) Biomembr.* **2015**, *1848*, 1775–1782. [CrossRef]
11. Sengupta, D.; Prasanna, X.; Mohole, M.; Chattopadhyay, A. exploring GPCR-lipid interactions by Molecular dynamic simulation. *J. Phys. Chem. B* **2018**, *122*, 5727–5737. [CrossRef] [PubMed]
12. Hedger, G.; Koldsø, H.; Chavent, M.; Siebold, C.; Rohatgi, R.; Sansom, M.S.P. Cholesterol Interaction Sites on the Transmembrane Domain of the Hedgehog Signal Transducer and Class F G Protein-Coupled Receptor Smoothened. *Structure* **2019**, *27*, 549–559.e2. [CrossRef] [PubMed]
13. Polasa, A.; Moradi, M. Deciphering the Inter-domain Decoupling in the Gram-negative Bacterial Membrane Insertase. *bioRxiv* **2022**. [CrossRef]
14. Govind Kumar, V.; Polasa, A.; Agrawal, S.; Kumar, T.K.S.; Moradi, M. Binding affinity estimation from restrained umbrella sampling simulations. *Nat. Comput. Sci.* **2022**, *3*, 59–70. [CrossRef]
15. Polasa, A.; Mosleh, I.; Losey, J.; Abbaspourrad, A.; Beitle, R.; Moradi, M. Developing a rational approach to designing recombinant proteins for peptide-directed nanoparticle synthesis. *Nanoscale Adv.* **2022**, *4*, 3161–3171. [CrossRef] [PubMed]
16. Govind Kumar, V.; Ogden, D.S.; Isu, U.H.; Polasa, A.; Losey, J.; Moradi, M. Prefusion spike protein conformational changes are slower in SARS-CoV-2 than in SARS-CoV-1. *J. Biol. Chem.* **2022**, *298*, 101814. [CrossRef]
17. Polasa, A.; Hettige, J.; Kalyan, I.; Moradi, M. An investigation of the YidC-mediated membrane insertion of Pf3 coat protein using molecular dynamics simulations. *Front. Mol. Biosci.* **2022**, *9*, 954262. [CrossRef]
18. Immadisetty, K.; Moradi, M. Mechanistic Picture for Chemomechanical Coupling in a Bacterial Proton-Coupled Oligopeptide Transporter from Streptococcus Thermophilus. *J. Phys. Chem. B* **2021**, *125*, 9738–9750. [CrossRef]
19. Gorvin, C.M. Calcium-sensing receptor signaling—How human disease informs biology. *Curr. Opin. Endocr. Metab. Res.* **2021**, *16*, 10–18. [CrossRef]
20. Ling, S.; Shi, P.; Liu, S.; Meng, X.; Zhou, Y.; Sun, W.; Chang, S.; Zhang, X.; Zhang, L.; Shi, C.; et al. Structural mechanism of cooperative activation of the human calcium-sensing receptor by Ca^{2+} ions and L-tryptophan. *Cell Res.* **2021**, *31*, 383–394. [CrossRef]
21. Ray, K. Chapter Six - Calcium-Sensing Receptor: Trafficking, Endocytosis, Recycling, and Importance of Interacting Proteins. In *Trafficking of GPCRs*; Wu, G., Ed.; Progress in Molecular Biology and Translational Science; Academic Press: Cambridge, MA, USA, 2015; Volume 132, pp. 127–150. [CrossRef]
22. Kniazeff, J.; Prézeau, L.; Rondard, P.; Pin, J.P.; Goudet, C. Dimers and beyond: The functional puzzles of class C GPCRs. *Pharmacol. Ther.* **2011**, *130*, 9–25. [CrossRef] [PubMed]
23. Shen, C.; Mao, C.; Xu, C.; Jin, N.; Zhang, H.; Shen, D.D.; Shen, Q.; Wang, X.; Hou, T.; Chen, Z.; et al. Structural basis of GABAB receptor—Gi protein coupling. *Nature* **2021**, *594*, 594–598. [CrossRef] [PubMed]
24. Park, J.; Fu, Z.; Frangaj, A.; Liu, J.; Mosyak, L.; Shen, T.; Slavkovich, V.N.; Ray, K.M.; Taura, J.; Cao, B.; et al. Structure of human GABAB receptor in an inactive state. *Nature* **2020**, *584*, 304–309. [CrossRef] [PubMed]
25. Terunuma, M. Diversity of structure and function of $GABA_B$ receptors: A complexity of $GABA_B$-mediated signaling. *Proc. Jpn. Acad. Ser. B* **2018**, *94*, 390–411. [CrossRef] [PubMed]
26. Shaye, H.; Ishchenko, A.; Lam, J.H.; Han, G.W.; Xue, L.; Rondard, P.; Pin, J.P.; Katritch, V.; Gati, C.; Cherezov, V. Structural basis of the activation of a metabotropic GABA receptor. *Nature* **2020**, *584*, 298–303. [CrossRef]
27. Vafabakhsh, R.; Levitz, J.; Isacoff, E.Y. Conformational dynamics of a class C G-protein-coupled receptor. *Nature* **2015**, *524*, 497–501. [CrossRef]
28. Yang, M.Y.; Kim, S.K.; Goddard, W.A. G protein coupling and activation of the metabotropic GABAB heterodimer. *Nat. Commun.* **2022**, *13*, 4612. [CrossRef]

29. Shaye, H.; Stauch, B.; Gati, C.; Cherezov, V. Molecular mechanisms of metabotropic GABAB receptor function. *Sci. Adv.* **2021**, *7*, eabg3362. [CrossRef]
30. Geng, Y.; Bush, M.; Mosyak, L.; Wang, F.; Fan, Q.R. Structural mechanism of ligand activation in human GABAB receptor. *Nature* **2013**, *504*, 254–259. [CrossRef]
31. Liu, H.; Li, Y.; Gao, Y. Chapter Four—Asymmetric activation of class C GPCRs. In *G Protein-Coupled Receptors—Part B*; Shukla, A.K., Ed.; Progress in Molecular Biology and Translational Science; Academic Press: Cambridge, MA, USA, 2023; Volume 195, pp. 77–87. [CrossRef]
32. Ellaithy, A.; Gonzalez-Maeso, J.; Logothetis, D.A.; Levitz, J. Structural and Biophysical Mechanisms of Class C G Protein-Coupled Receptor Function. *Trends Biochem. Sci.* **2020**, *45*, 1049–1064. [CrossRef]
33. McCullock, T.W.; Kammermeier, P.J. The evidence for and consequences of metabotropic glutamate receptor heterodimerization. *Neuropharmacology* **2021**, *199*, 108801. [CrossRef] [PubMed]
34. Møller, T.C.; Moreno-Delgado, D.; Pin, J.P.; Kniazeff, J. Class CG protein-coupled receptors: reviving old couples with new partners. *Biophys. Rep.* **2017**, *3*, 57–63. [CrossRef] [PubMed]
35. Salon, J.A.; Lodowski, D.T.; Palczewski, K. The Significance of G Protein-Coupled Receptor Crystallography for Drug Discovery. *Pharmacol. Rev.* **2011**, *63*, 901–937. [CrossRef] [PubMed]
36. Wu, H.; Wang, C.; Gregory, K.J.; Han, G.W.; Cho, H.P.; Xia, Y.; Niswender, C.M.; Katritch, V.; Meiler, J.; Cherezov, V.; et al. Structure of a class C GPCR metabotropic glutamate receptor 1 bound to an allosteric modulator. *Science* **2014**, *344*, 58–64. [CrossRef] [PubMed]
37. Yang, D.; Zhou, Q.; Labroska, V.; Qin, S.; Darbalaei, S.; Wu, Y.; Yuliantie, E.; Xie, L.; Tao, H.; Cheng, J.; et al. G protein-coupled receptors: Structure-and function-based drug discovery. *Signal Transduct. Target. Ther.* **2021**, *6*, 1–27. [CrossRef]
38. Gómez-Santacana, X.; Panarello, S.; Rovira, X.; Llebaria, A. Photoswitchable allosteric modulators for metabotropic glutamate receptors. *Curr. Opin. Pharmacol.* **2022**, *66*, 102266. [CrossRef]
39. Orgován, Z.; Ferenczy, G.G.; Keserű, G.M. Allosteric Molecular Switches in Metabotropic Glutamate Receptors. *ChemMedChem* **2021**, *16*, 81–93. [CrossRef]
40. Niswender, C.M.; Conn, P.J. Metabotropic glutamate receptors: physiology, pharmacology, and disease. *Annu. Rev. Pharmacol. Toxicol.* **2010**, *50*, 295. [CrossRef]
41. Doumazane, E.; Scholler, P.; Zwier, J.M.; Trinquet, E.; Rondard, P.; Pin, J.P. A new approach to analyze cell surface protein complexes reveals specific heterodimeric metabotropic glutamate receptors. *FASEB J.* **2011**, *25*, 66–77. [CrossRef]
42. Seven, A.B.; Barros-Álvarez, X.; de Lapeyrière, M.; Papasergi-Scott, M.M.; Robertson, M.J.; Zhang, C.; Nwokonko, R.M.; Gao, Y.; Meyerowitz, J.G.; Rocher, J.P.; et al. G-protein activation by a metabotropic glutamate receptor. *Nature* **2021**, *595*, 450–454. [CrossRef]
43. Fang, W.; Yang, F.; Xu, C.; Ling, S.; Lin, L.; Zhou, Y.; Sun, W.; Wang, X.; Liu, P.; Rondard, P.; et al. Structural basis of the activation of metabotropic glutamate receptor 3. *Cell Res.* **2022**, *32*, 695–698. [CrossRef] [PubMed]
44. Lin, S.; Han, S.; Cai, X.; Tan, Q.; Zhou, K.; Wang, D.; Wang, X.; Du, J.; Yi, C.; Chu, X.; et al. Structures of Gi-bound metabotropic glutamate receptors mGlu2 and mGlu4. *Nature* **2021**, *594*, 583–588. [CrossRef] [PubMed]
45. Nasrallah, C.; Cannone, G.; Briot, J.; Rottier, K.; Berizzi, A.E.; Huang, C.Y.; Quast, R.B.; Hoh, F.; Banères, J.L.; Malhaire, F.; et al. Agonists and allosteric modulators promote signaling from different metabotropic glutamate receptor 5 conformations. *Cell Rep.* **2021**, *36*, 109648. [CrossRef] [PubMed]
46. Du, J.; Wang, D.; Fan, H.; Xu, C.; Tai, L.; Lin, S.; Han, S.; Tan, Q.; Wang, X.; Xu, T.; et al. Structures of human mGlu2 and mGlu7 homo- and heterodimers. *Nature* **2021**, *594*, 589–593. [CrossRef]
47. Frangaj, A.; Fan, Q.R. Structural biology of GABAB receptor. *Neuropharmacology* **2018**, *136*, 68–79. [CrossRef]
48. Schamber, M.R.; Vafabakhsh, R. Mechanism of sensitivity modulation in the calcium-sensing receptor via electrostatic tuning. *Nat. Commun.* **2022**, *13*, 2194. [CrossRef]
49. Wen, T.; Wang, Z.; Chen, X.; Ren, Y.; Lu, X.; Xing, Y.; Lu, J.; Chang, S.; Zhang, X.; Shen, Y.; et al. Structural basis for activation and allosteric modulation of full-length calcium-sensing receptor. *Sci. Adv.* **2021**, *7*, eabg1483. [CrossRef]
50. Robbins, M.J.; Michalovich, D.; Hill, J.; Calver, A.R.; Medhurst, A.D.; Gloger, I.; Sims, M.; Middlemiss, D.N.; Pangalos, M.N. Molecular Cloning and Characterization of Two Novel Retinoic Acid-Inducible Orphan G-Protein-Coupled Receptors (GPRC5B and GPRC5C). *Genomics* **2000**, *67*, 8–18. [CrossRef]
51. Mafi, A.; Kim, S.K.; Chou, K.C.; Güthrie, B.; Goddard, W.A.I.I.I. Predicted Structure of Fully Activated Tas1R3/1R3 Homodimer Bound to G Protein and Natural Sugars: Structural Insights into G Protein Activation by a Class C Sweet Taste Homodimer with Natural Sugars. *J. Am. Chem. Soc.* **2021**, *143*, 16824–16838. [CrossRef]
52. Belloir, C.; Brulé, M.; Tornier, L.; Neiers, F.; Briand, L. Biophysical and functional characterization of the human TAS1R2 sweet taste receptor overexpressed in a HEK293S inducible cell line. *Sci. Rep.* **2021**, *11*, 22238. [CrossRef]
53. Ahmad, R.; Dalziel, J.E. G Protein-Coupled Receptors in Taste Physiology and Pharmacology. *Front. Pharmacol.* **2020**, *11*, 587664. [CrossRef] [PubMed]
54. Nuemket, N.; Yasui, N.; Kusakabe, Y.; Nomura, Y.; Atsumi, N.; Akiyama, S.; Nango, E.; Kato, Y.; Kaneko, M.K.; Takagi, J.; et al. Structural basis for perception of diverse chemical substances by T1r taste receptors. *Nat. Commun.* **2017**, *8*, 15530. [CrossRef] [PubMed]

55. Fredriksson, R.; Lagerström, M.C.; Lundin, L.G.; Schiöth, H.B. The G-protein-coupled receptors in the human genome form five main families. Phylogenetic analysis, paralogon groups, and fingerprints. *Mol. Pharmacol.* **2003**, *63*, 1256–1272. [CrossRef] [PubMed]
56. Patil, D.N.; Singh, S.; Laboute, T.; Strutzenberg, T.S.; Qiu, X.; Wu, D.; Novick, S.J.; Robinson, C.V.; Griffin, P.R.; Hunt, J.F.; et al. Cryo-EM structure of human GPR158 receptor coupled to the RGS7-Gβ5 signaling complex. *Science* **2022**, *375*, 86–91. [CrossRef]
57. Darira, S.V.; Sutton, L.P. Chapter Seven—The interaction, mechanism and function of GPR158-RGS7 cross-talk. In *G Protein-Coupled Receptors—Part A*; Shukla, A.K., Ed.; Progress in Molecular Biology and Translational Science; Academic Press: Cambridge, MA, USA, 2022; Volume 193, pp. 167–176. [CrossRef]
58. Kindt, K.S.; Akturk, A.; Jarysta, A.; Day, M.; Beirl, A.; Flonard, M.; Tarchini, B. EMX2-GPR156-Gαi reverses hair cell orientation in mechanosensory epithelia. *Nat. Commun.* **2021**, *12*, 2861. [CrossRef]
59. Jørgensen, C.V.; Bräuner-Osborne, H. Pharmacology and physiological function of the orphan GPRC6A receptor. *Basic Clin. Pharmacol. Toxicol.* **2020**, *126*, 77–87. [CrossRef]
60. Orlandi, C.; Cao, Y.; Martemyanov, K.A. Orphan Receptor GPR179 Forms Macromolecular Complexes with Components of Metabotropic Signaling Cascade in Retina ON-Bipolar Neurons. *Investig. Ophthalmol. Vis. Sci.* **2013**, *54*, 7153–7161. [CrossRef]
61. Guixà-González, R.; Albasanz, J.L.; Rodriguez-Espigares, I.; Pastor, M.; Sanz, F.; Martí-Solano, M.; Manna, M.; Martinez-Seara, H.; Hildebrand, P.W.; Martín, M.; et al. Membrane cholesterol access into a G-protein-coupled receptor. *Nat. Commun.* **2017**, *8*, 14505. [CrossRef]
62. Prasanna, X.; Mohole, M.; Chattopadhyay, A.; Sengupta, D. Role of cholesterol-mediated effects in GPCR heterodimers. *Chem. Phys. Lipids* **2020**, *227*, 104852. [CrossRef]
63. Jakubík, J.; El-Fakahany, E.E. Allosteric Modulation of GPCRs of Class A by Cholesterol. *Int. J. Mol. Sci.* **2021**, *22*, 1953. [CrossRef]
64. Harkey, T.; Govind Kumar, V.; Hettige, J.; Tabari, S.H.; Immadisetty, K.; Moradi, M. The Role of a Crystallographically Unresolved Cytoplasmic Loop in Stabilizing the Bacterial Membrane Insertase YidC2. *Sci. Rep.* **2019**, *9*, 14451. [CrossRef] [PubMed]
65. Kim, Y.; Jeong, E.; Jeong, J.H.; Kim, Y.; Cho, Y. Structural Basis for Activation of the Heterodimeric GABAB Receptor. *J. Mol. Biol.* **2020**, *432*, 5966–5984. [CrossRef]
66. Park, J.; Zuo, H.; Frangaj, A.; Fu, Z.; Yen, L.Y.; Zhang, Z.; Mosyak, L.; Slavkovich, V.N.; Liu, J.; Ray, K.M.; et al. Symmetric activation and modulation of the human calcium-sensing receptor. *Proc. Natl. Acad. Sci.* **2021**, *118*, e2115849118. [CrossRef] [PubMed]
67. Fantini, J.; Barrantes, F. How cholesterol interacts with membrane proteins: An exploration of cholesterol-binding sites including CRAC, CARC, and tilted domains. *Front. Physiol.* **2013**, *4*, 31. [CrossRef] [PubMed]
68. Dingjan, T.; Futerman, A.H. The fine-tuning of cell membrane lipid bilayers accentuates their compositional complexity. *BioEssays* **2021**, *43*, 2100021. [CrossRef]
69. Pinkwart, K.; Schneider, F.; Lukoseviciute, M.; Sauka-Spengler, T.; Lyman, E.; Eggeling, C.; Sezgin, E. Nanoscale dynamics of cholesterol in the cell membrane. *J. Biol. Chem.* **2019**, *294*, 12599–12609. [CrossRef]
70. Pucadyil, T.J.; Chattopadhyay, A. Role of cholesterol in the function and organization of G-protein coupled receptors. *Prog. Lipid Res.* **2006**, *45*, 295–333. [CrossRef]
71. Arora, A.; Raghuraman, H.; Chattopadhyay, A. Influence of cholesterol and ergosterol on membrane dynamics: a fluorescence approach. *Biochem. Biophys. Res. Commun.* **2004**, *318*, 920–926. [CrossRef]
72. Chimento, A.; Casaburi, I.; Avena, P.; Trotta, F.; De Luca, A.; Rago, V.; Pezzi, V.; Sirianni, R. Cholesterol and Its Metabolites in Tumor Growth: Therapeutic Potential of Statins in Cancer Treatment. *Front. Endocrinol.* **2019**, *9*, 807. [CrossRef]
73. Ohvo-Rekilä, H.; Ramstedt, B.; Leppimäki, P.; Peter Slotte, J. Cholesterol interactions with phospholipids in membranes. *Prog. Lipid Res.* **2002**, *41*, 66–97. [CrossRef]
74. McMullen, T.P.W.; Lewis, R.N.A.H.; McElhaney, R.N. Cholesterol–phospholipid interactions, the liquid-ordered phase and lipid rafts in model and biological membranes. *Curr. Opin. Colloid Interface Sci.* **2004**, *8*, 459–468. [CrossRef]
75. Gater, D.L.; Réat, V.; Czaplicki, G.; Saurel, O.; Milon, A.; Jolibois, F.; Cherezov, V. Hydrogen Bonding of Cholesterol in the Lipidic Cubic Phase. *Langmuir* **2013**, *29*, 8031–8038. [CrossRef] [PubMed]
76. Rowlands, L.J.; Marks, A.; Sanderson, J.M.; Law, R.V. 17O NMR spectroscopy as a tool to study hydrogen bonding of cholesterol in lipid bilayers. *Chem. Commun.* **2020**, *56*, 14499–14502. [CrossRef] [PubMed]
77. Simons, K.; Ikonen, E. Functional rafts in cell membranes. *Nature* **1997**, *387*, 569–572. [CrossRef] [PubMed]
78. Anderson, R.G.W.; Jacobson, K. A Role for Lipid Shells in Targeting Proteins to Caveolae, Rafts, and Other Lipid Domains. *Science* **2002**, *296*, 1821–1825. [CrossRef] [PubMed]
79. Fantini, J.; Yahi, N. Molecular insights into amyloid regulation by membrane cholesterol and sphingolipids: common mechanisms in neurodegenerative diseases. *Expert Rev. Mol. Med.* **2010**, *12*, e27. [CrossRef] [PubMed]
80. Fantini, J.; Garmy, N.; Mahfoud, R.; Yahi, N. Lipid rafts: structure, function and role in HIV, Alzheimer's and prion diseases. *Expert Rev. Mol. Med.* **2002**, *4*, 1–22. [CrossRef]
81. Yang, S.T.; Kreutzberger, A.J.B.; Lee, J.; Kiessling, V.; Tamm, L.K. The role of cholesterol in membrane fusion. *Chem. Phys. Lipids* **2016**, *199*, 136–143. [CrossRef]
82. Song, Y.; Kenworthy, A.K.; Sanders, C.R. Cholesterol as a co-solvent and a ligand for membrane proteins. *Protein Sci.* **2014**, *23*, 1–22. [CrossRef]

83. Najafinobar, N.; Mellander, L.J.; Kurczy, M.E.; Dunevall, J.; Angerer, T.B.; Fletcher, J.S.; Cans, A.S. Cholesterol Alters the Dynamics of Release in Protein Independent Cell Models for Exocytosis. *Sci. Rep.* **2016**, *6*, 33702. [CrossRef]
84. Yeagle, P.L. Modulation of membrane function by cholesterol. *Biochimie* **1991**, *73*, 1303–1310. [CrossRef] [PubMed]
85. Murata, M.; Peränen, J.; Schreiner, R.; Wieland, F.; Kurzchalia, T.V.; Simons, K. VIP21/caveolin is a cholesterol-binding protein. *Proc. Natl. Acad. Sci. USA* **1995**, *92*, 10339–10343. [CrossRef] [PubMed]
86. Porter, J.A.; Young, K.E.; Beachy, P.A. Cholesterol Modification of Hedgehog Signaling Proteins in Animal Development. *Science* **1996**, *274*, 255–259. [CrossRef] [PubMed]
87. Sutter, M.L.; Console, L.; Fahner, A.F.; Samodelov, S.L.; Gai, Z.; Ciarimboli, G.; Indiveri, C.; Kullak-Ublick, G.A.; Visentin, M. The role of cholesterol recognition (CARC/CRAC) mirror codes in the allosterism of the human organic cation transporter 2 (OCT2, SLC22A2). *Biochem. Pharmacol.* **2021**, *194*, 114840. [CrossRef] [PubMed]
88. Taghon, G.; Rowe, J.; Kapolka, N.; Isom, D. Predictable cholesterol binding sites in GPCRs lack consensus motifs. *Structure* **2021**, *29*, 499–506.e3. [CrossRef] [PubMed]
89. de Vries, M.; Herrmann, A.; Veit, M. A cholesterol consensus motif is required for efficient intracellular transport and raft association of a group 2 HA of influenza virus. *Biochem. J.* **2014**, *465*, 305–314. . [CrossRef]
90. Kuwabara, P.E.; Labouesse, M. The sterol-sensing domain: multiple families, a unique role? *Trends Genet.* **2002**, *18*, 193–201. [CrossRef]
91. Martín, V.; Carrillo, G.; Torroja, C.; Guerrero, I. The sterol-sensing domain of Patched protein seems to control Smoothened activity through Patched vesicular trafficking. *Curr. Biol. CB* **2001**, *11*, 601–607. [CrossRef]
92. Jafurulla, M.; Tiwari, S.; Chattopadhyay, A. Identification of cholesterol recognition amino acid consensus (CRAC) motif in G-protein coupled receptors. *Biochem. Biophys. Res. Commun.* **2011**, *404*, 569–573. [CrossRef]
93. Sarkar, P.; Chattopadhyay, A. Cholesterol in GPCR Structures: Prevalence and Relevance. *J. Membr. Biol.* **2022**, *255*, 99–106. [CrossRef]
94. Epand, R.M. Cholesterol and the interaction of proteins with membrane domains. *Prog. Lipid Res.* **2006**, *45*, 279–294. [CrossRef] [PubMed]
95. Li, H.; Yao, Z.x.; Degenhardt, B.; Teper, G.; Papadopoulos, V. Cholesterol binding at the cholesterol recognition/ interaction amino acid consensus (CRAC) of the peripheral-type benzodiazepine receptor and inhibition of steroidogenesis by an HIV TAT-CRAC peptide. *Proc. Natl. Acad. Sci. USA* **2001**, *98*, 1267–1272. [CrossRef] [PubMed]
96. Sarkar, P.; Mozumder, S.; Bej, A.; Mukherjee, S.; Sengupta, J.; Chattopadhyay, A. Structure, dynamics and lipid interactions of serotonin receptors: excitements and challenges. *Biophys. Rev.* **2021**, *13*, 101–122. [CrossRef] [PubMed]
97. Prasad, R.; Paila, Y.D.; Jafurulla, M.; Chattopadhyay, A. Membrane cholesterol depletion from live cells enhances the function of human serotonin1A receptors. *Biochem. Biophys. Res. Commun.* **2009**, *389*, 333–337. [CrossRef] [PubMed]
98. Fantini, J.; Di Scala, C.; Evans, L.S.; Williamson, P.T.F.; Barrantes, F.J. A mirror code for protein–cholesterol interactions in the two leaflets of biological membranes. *Sci. Rep.* **2016**, *6*, 21907. [CrossRef]
99. Di Scala, C.; Baier, C.; Evans, L.; Williamson, P.; Fantini, J.; Barrantes, F. Relevance of CARC and CRAC Cholesterol-Recognition Motifs in the Nicotinic Acetylcholine Receptor and Other Membrane-Bound Receptors. *Curr. Top. Membr.* **2017**, *80*, 3–23. [CrossRef]
100. Azzaz, F.; Chahinian, H.; Yahi, N.; Di Scala, C.; Baier, C.J.; Barrantes, F.J.; Fantini, J. Chapter 7—Cholesterol-recognizing amino acid consensus motifs in transmembrane proteins: Comparative analysis of in silico studies and structural data. In *Cholesterol*; Bukiya, A.N., Dopico, A.M., Eds.; Academic Press: Cambridge, MA, USA, 2022; pp. 127–145. [CrossRef]
101. Hanson, M.A.; Cherezov, V.; Griffith, M.T.; Roth, C.B.; Jaakola, V.P.; Chien, E.Y.; Velasquez, J.; Kuhn, P.; Stevens, R.C. A Specific Cholesterol Binding Site Is Established by the 2.8 Å Structure of the Human β2-Adrenergic Receptor. *Structure* **2008**, *16*, 897–905. [CrossRef]
102. McGraw, C.; Koretz, K.S.; Oseid, D.; Lyman, E.; Robinson, A.S. Cholesterol Dependent Activity of the Adenosine A2A Receptor Is Modulated via the Cholesterol Consensus Motif. *Molecules* **2022**, *27*, 3529. [CrossRef]
103. Desai, A.; Miller, L. Sensitivity of cholecystokinin receptors to membrane cholesterol content. *Front. Endocrinol.* **2012**, *3*, 123. [CrossRef]
104. Adamian, L.; Naveed, H.; Liang, J. Lipid-binding surfaces of membrane proteins: Evidence from evolutionary and structural analysis. *Biochim. Biophys. Acta (BBA) Biomembr.* **2011**, *1808*, 1092–1102. [CrossRef]
105. Yabe, D.; Xia, Z.P.; Adams, C.M.; Rawson, R.B. Three mutations in sterol-sensing domain of SCAP block interaction with insig and render SREBP cleavage insensitive to sterols. *Proc. Natl. Acad. Sci. USA* **2002**, *99*, 16672–16677. [CrossRef] [PubMed]
106. Yang, T.; Espenshade, P.J.; Wright, M.E.; Yabe, D.; Gong, Y.; Aebersold, R.; Goldstein, J.L.; Brown, M.S. Crucial Step in Cholesterol Homeostasis: Sterols Promote Binding of SCAP to INSIG-1, a Membrane Protein that Facilitates Retention of SREBPs in ER. *Cell* **2002**, *110*, 489–500. [CrossRef] [PubMed]
107. Bukiya, A.N.; Dopico, A.M. Common structural features of cholesterol binding sites in crystallized soluble proteins. *J. Lipid Res.* **2017**, *58*, 1044–1054. [CrossRef] [PubMed]
108. Rone, M.B.; Fan, J.; Papadopoulos, V. Cholesterol transport in steroid biosynthesis: Role of protein–protein interactions and implications in disease states. *Biochim. Biophys. Acta (BBA) Mol. Cell Biol. Lipids* **2009**, *1791*, 646–658. [CrossRef]
109. Naito, T.; Ercan, B.; Krshnan, L.; Triebl, A.; Koh, D.H.Z.; Wei, F.Y.; Tomizawa, K.; Torta, F.T.; Wenk, M.R.; Saheki, Y. Movement of accessible plasma membrane cholesterol by the GRAMD1 lipid transfer protein complex. *eLife* **2019**, *8*, e51401. [CrossRef]

110. Wu, Y.; Li, X.; Hua, T.; Liu, Z.J.; Liu, H.; Zhao, S. MD Simulations Revealing Special Activation Mechanism of Cannabinoid Receptor 1. *Front. Mol. Biosci.* **2022**, *9*, 860035. [CrossRef]
111. Nieto Gutierrez, A.; McDonald, P.H. GPCRs: Emerging anti-cancer drug targets. *Cell. Signal.* **2018**, *41*, 65–74. [CrossRef]
112. Hauser, A.S.; Attwood, M.M.; Rask-Andersen, M.; Schiöth, H.B.; Gloriam, D.E. Trends in GPCR drug discovery: New agents, targets and indications. *Nat. Rev. Drug Discov.* **2017**, *16*, 829–842. [CrossRef]
113. Sriram, K.; Insel, P.A. GPCRs as targets for approved drugs: How many targets and how many drugs? *Mol. Pharmacol.* **2018**, *93*, 251–258. [CrossRef]
114. Niswender, C.M.; Jones, C.K.; Lin, X.; Bubser, M.; Thompson Gray, A.; Blobaum, A.L.; Engers, D.W.; Rodriguez, A.L.; Loch, M.T.; Daniels, J.S.; et al. Development and antiparkinsonian activity of VU0418506, a selective positive allosteric modulator of metabotropic glutamate receptor 4 homomers without activity at mGlu2/4 heteromers. *ACS Chem. Neurosci.* **2016**, *7*, 1201–1211. [CrossRef]
115. Moreau, C.J.; Audic, G.; Lemel, L.; García-Fernández, M.D.; Nieścierowicz, K. Interactions of cholesterol molecules with GPCRs in different states: A comparative analysis of GPCRs' structures. *Biochim. Biophys. Acta (BBA) Biomembr.* **2023**, *1865*, 184100. [CrossRef] [PubMed]
116. Salas-Estrada, L.A.; Leioatts, N.; Romo, T.D.; Grossfield, A. Lipids Alter Rhodopsin Function via Ligand-like and Solvent-like Interactions. *Biophys. J.* **2018**, *114*, 355–367. [CrossRef] [PubMed]
117. Gimpl, G.; Fahrenholz, F. Cholesterol as stabilizer of the oxytocin receptor. *Biochim. Biophys. Acta (BBA) Biomembr.* **2002**, *1564*, 384–392. [CrossRef]
118. Marino, K.A.; Prada-Gracia, D.; Provasi, D.; Filizola, M. Impact of Lipid Composition and Receptor Conformation on the Spatio-temporal Organization of μ-Opioid Receptors in a Multi-component Plasma Membrane Model. *PLoS Comput. Biol.* **2016**, *12*, e1005240. [CrossRef]
119. Sengupta, D.; Chattopadhyay, A. Identification of Cholesterol Binding Sites in the Serotonin1A Receptor. *J. Phys. Chem. B* **2012**, *116*, 12991–12996. [CrossRef] [PubMed]
120. Luchetti, G.; Sircar, R.; Kong, J.H.; Nachtergaele, S.; Sagner, A.; Byrne, E.F.; Covey, D.F.; Siebold, C.; Rohatgi, R. Cholesterol activates the G-protein coupled receptor smoothened to promote hedgehog signaling. *eLife* **2016**, *5*, e20304. [CrossRef]
121. Geiger, J.; Sexton, R.; Al-Sahouri, Z.; Lee, M.Y.; Chun, E.; Harikumar, K.G.; Miller, L.J.; Beckstein, O.; Liu, W. Evidence that specific interactions play a role in the cholesterol sensitivity of G protein-coupled receptors. *Biochim. Biophys. Acta (BBA) Biomembr.* **2021**, *1863*, 183557. [CrossRef] [PubMed]
122. Paila, Y.D.; Chattopadhyay, A. The function of G-protein coupled receptors and membrane cholesterol: Specific or general interaction? *Glycoconj. J.* **2009**, *26*, 711–720. [CrossRef]
123. Augustyn, B.; Stepien, P.; Poojari, C.; Mobarak, E.; Polit, A.; Wisniewska-Becker, A.; Róg, T. Cholesteryl Hemisuccinate Is Not a Good Replacement for Cholesterol in Lipid Nanodiscs. *J. Phys. Chem. B* **2019**, *123*, 9839–9845. [CrossRef]
124. Kulig, W.; Tynkkynen, J.; Javanainen, M.; Manna, M.; Rog, T.; Vattulainen, I.; Jungwirth, P. How well does cholesteryl hemisuccinate mimic cholesterol in saturated phospholipid bilayers? *J. Mol. Model.* **2014**, *20*, 2121. [CrossRef]
125. Kulig, W.; Jurkiewicz, P.; Olżyńska, A.; Tynkkynen, J.; Javanainen, M.; Manna, M.; Rog, T.; Hof, M.; Vattulainen, I.; Jungwirth, P. Experimental determination and computational interpretation of biophysical properties of lipid bilayers enriched by cholesteryl hemisuccinate. *Biochim. Biophys. Acta (BBA) Biomembr.* **2015**, *1848*, 422–432. [CrossRef] [PubMed]
126. Sarkar, P.; Bhat, A.; Chattopadhyay, A. Lysine 101 in the CRAC Motif in Transmembrane Helix 2 Confers Cholesterol-Induced Thermal Stability to the Serotonin1A Receptor. *J. Membr. Biol.* **2022**, *255*, 739–746. [CrossRef] [PubMed]
127. Prasanna, X.; Chattopadhyay, A.; Sengupta, D. Cholesterol modulates the dimer interface of the β2-adrenergic receptor via cholesterol occupancy sites. *Biophys. J.* **2014**, *106*, 1290–1300. [CrossRef] [PubMed]
128. Oddi, S.; Dainese, E.; Fezza, F.; Lanuti, M.; Barcaroli, D.; De Laurenzi, V.; Centonze, D.; Maccarrone, M. Functional characterization of putative cholesterol binding sequence (CRAC) in human type-1 cannabinoid receptor. *J. Neurochem.* **2011**, *116*, 858–865. [CrossRef]
129. Oddi, S.; Dainese, E.; Sandiford, S.; Fezza, F.; Lanuti, M.; Chiurchiù, V.; Totaro, A.; Catanzaro, G.; Barcaroli, D.; De Laurenzi, V.; et al. Effects of palmitoylation of Cys 415 in helix 8 of the CB 1 cannabinoid receptor on membrane localization and signaling. *Br. J. Pharmacol.* **2012**, *165*, 2635–2651. [CrossRef]
130. Sarkar, P.; Chattopadhyay, A. Cholesterol interaction motifs in G protein-coupled receptors: Slippery hot spots? *Wiley Interdiscip. Rev. Syst. Biol. Med.* **2020**, *12*, e1481. [CrossRef]
131. Mazzitelli, M.; Presto, P.; Antenucci, N.; Meltan, S.; Neugebauer, V. Recent Advances in the Modulation of Pain by the Metabotropic Glutamate Receptors. *Cells* **2022**, *11*, 2608. [CrossRef]
132. Kroon, T.; Dawitz, J.; Kramvis, I.; Anink, J.; Obermayer, J.; Verhoog, M.B.; Wilbers, R.; Goriounova, N.A.; Idema, S.; Baayen, J.C.; et al. Group I mGluR-mediated activation of martinotti cells inhibits local cortical circuitry in human cortex. *Front. Cell. Neurosci.* **2019**, *13*, 315. [CrossRef]
133. Pin, J.P.; De Colle, C.; Bessis, A.S.; Acher, F. New perspectives for the development of selective metabotropic glutamate receptor ligands. *Eur. J. Pharmacol.* **1999**, *375*, 277–294. [CrossRef]
134. Sejdiu, B.I.; Tieleman, D.P. Lipid-protein interactions are a unique property and defining feature of G protein-coupled receptors. *Biophys. J.* **2020**, *118*, 1887–1900. [CrossRef]
135. Kurth, M.; Lolicato, F.; Sandoval, A.; Amaya-Espinosa, H.; Teslenko, A.; Sinning, I.; Beck, R.; Brügger, B.; Aponte-Santamaria, C. Cholesterol Localization Around the Metabotropic Glutamate Receptor 2. *J. Phys. Chem. B* **2020**, *124*, 9061–9078. [CrossRef] [PubMed]

136. Liauw, B.W.H.; Foroutan, A.; Schamber, M.R.; Lu, W.; Samareh Afsari, H.; Vafabakhsh, R. Conformational fingerprinting of allosteric modulators in metabotropic glutamate receptor 2. *eLife* **2022**, *11*, e78982. [CrossRef] [PubMed]
137. Bordi, F.; Ugolini, A. Group I metabotropic glutamate receptors: implications for brain diseases. *Prog. Neurobiol.* **1999**, *59*, 55–79. [CrossRef] [PubMed]
138. Liauw, B.W.H.; Afsari, H.S.; Vafabakhsh, R. Conformational rearrangement during activation of a metabotropic glutamate receptor. *Nat. Chem. Biol.* **2021**, *17*, 291–297. [CrossRef] [PubMed]
139. Yeagle, P. The biophysics and cell biology of cholesterol: an hypothesis for the essential role of cholesterol in mammalian cells. In *Cholesterol in Membrane Models*; CRC Press: Boca Raton, FL, USA, 1993; pp. 1–12.
140. Sooksawate, T.; Simmonds, M. Influence of membrane cholesterol on modulation of the GABAA receptor by neuroactive steroids and other potentiators. *Br. J. Pharmacol.* **2001**, *134*, 1303–1311. [CrossRef]
141. Schroeder, F.; Jefferson, J.R.; Kier, A.B.; Knittel, J.; Scallen, T.J.; Wood, W.G.; Hapala, I. Membrane cholesterol dynamics: cholesterol domains and kinetic pools. *Proc. Soc. Exp. Biol. Med.* **1991**, *196*, 235–252. [CrossRef]
142. Marlow, B.; Kuenze, G.; Li, B.; Sanders, C.R.; Meiler, J. Structural determinants of cholesterol recognition in helical integral membrane proteins. *Biophys. J.* **2021**, *120*, 1592–1604. [CrossRef]
143. Kumari, R.; Castillo, C.; Francesconi, A. Agonist-dependent signaling by group I metabotropic glutamate receptors is regulated by association with lipid domains. *J. Biol. Chem.* **2013**, *288*, 32004–32019. [CrossRef]
144. Lei, T.; Hu, Z.; Ding, R.; Chen, J.; Li, S.; Zhang, F.; Pu, X.; Zhao, N. Exploring the Activation Mechanism of a Metabotropic Glutamate Receptor Homodimer via Molecular Dynamics Simulation. *ACS Chem. Neurosci.* **2020**, *11*, 133–145. [CrossRef]
145. Thibado, J.K.; Tano, J.Y.; Lee, J.; Salas-Estrada, L.; Provasi, D.; Strauss, A.; Marcelo Lamim Ribeiro, J.; Xiang, G.; Broichhagen, J.; Filizola, M.; et al. Differences in interactions between transmembrane domains tune the activation of metabotropic glutamate receptors. *eLife* **2021**, *10*, e67027. [CrossRef]
146. Francesconi, A.; Kumari, R.; Zukin, R.S. Regulation of group I metabotropic glutamate receptor trafficking and signaling by the caveolar/lipid raft pathway. *J. Neurosci.* **2009**, *29*, 3590–3602. [CrossRef] [PubMed]
147. Caraci, F.; Nicoletti, F.; Copani, A. Metabotropic glutamate receptors: the potential for therapeutic applications in Alzheimer's disease. *Curr. Opin. Pharmacol.* **2018**, *38*, 1–7. [CrossRef] [PubMed]
148. Nicoletti, F.; Bruno, V.; Ngomba, R.T.; Gradini, R.; Battaglia, G. Metabotropic glutamate receptors as drug targets: what's new? *Curr. Opin. Pharmacol.* **2015**, *20*, 89–94. [CrossRef] [PubMed]
149. Lumeng, J.Y.; Wall, B.A.; Wangari-Talbot, J.; Chen, S. Metabotropic glutamate receptors in cancer. *Neuropharmacology* **2017**, *115*, 193–202.
150. Immadisetty, K.; Hettige, J.; Moradi, M. What Can and Cannot Be Learned from Molecular Dynamics Simulations of Bacterial Proton-Coupled Oligopeptide Transporter GkPOT? *J. Phys. Chem. B* **2017**, *121*, 3644–3656. [CrossRef]
151. Moradi, M.; Babin, V.; Roland, C.; Sagui, C. The Adaptively Biased Molecular Dynamics method revisited: New capabilities and an application. *J. Phys. Conf. Ser.* **2015**, *640*, 12020. [CrossRef]
152. Govind Kumar, V.; Agrawal, S.; Kumar, T.K.S.; Moradi, M. Mechanistic Picture for Monomeric Human Fibroblast Growth Factor 1 Stabilization by Heparin Binding. *J. Phys. Chem. B* **2021**, *125*, 12690–12697. [CrossRef]
153. Baier, C.J.; Fantini, J.; Barrantes, F.J. Disclosure of cholesterol recognition motifs in transmembrane domains of the human nicotinic acetylcholine receptor. *Sci. Rep.* **2011**, *1*, 69. [CrossRef]
154. Bruno, A.; Costantino, G.; de Fabritiis, G.; Pastor, M.; Selent, J. Membrane-sensitive conformational states of helix 8 in the metabotropic Glu2 receptor, a class C GPCR. *PLoS ONE* **2012**, *7*, e42023. [CrossRef]
155. Nutt, D.J.; Malizia, A.L. New insights into the role of the GABAA–benzodiazepine receptor in psychiatric disorder. *Br. J. Psychiatry* **2001**, *179*, 390–396. [CrossRef]
156. Burmakina, S.; Geng, Y.; Chen, Y.; Fan, Q.R. Heterodimeric coiled-coil interactions of human GABAB receptor. *Proc. Natl. Acad. Sci. USA* **2014**, *111*, 6958–6963. [CrossRef] [PubMed]
157. Bassetti, D. Keeping the Balance: GABAB Receptors in the Developing Brain and Beyond. *Brain Sci.* **2022**, *12*, 419. [CrossRef]
158. Sooksawate, T.; Simmonds, M. Effects of membrane cholesterol on the sensitivity of the GABAA receptor to GABA in acutely dissociated rat hippocampal neurones. *Neuropharmacology* **2001**, *40*, 178–184. [CrossRef] [PubMed]
159. Hoon, M.A.; Adler, E.; Lindemeier, J.; Battey, J.F.; Ryba, N.J.; Zuker, C.S. Putative mammalian taste receptors: A class of taste-specific GPCRs with distinct topographic selectivity. *Cell* **1999**, *96*, 541–551. [CrossRef]
160. Vigues, S.; Dotson, C.D.; Munger, S.D. The receptor basis of sweet taste in mammals. *Results Probl. Cell Differ.* **2009**, *47*, 187–202. [CrossRef] [PubMed]
161. Prasad Pydi, S.; Singh, N.; Upadhyaya, J.; Pal Bhullar, R.; Chelikani, P. The third intracellular loop plays a critical role in bitter taste receptor activation. *Biochim. Biophys. Acta (BBA) Biomembr.* **2014**, *1838*, 231–236. [CrossRef]
162. Pydi, S.P.; Jafurulla, M.; Wai, L.; Bhullar, R.P.; Chelikani, P.; Chattopadhyay, A. Cholesterol modulates bitter taste receptor function. *Biochim. Biophys. Acta (BBA) Biomembr.* **2016**, *1858*, 2081–2087. [CrossRef]
163. Ilegems, E.; Iwatsuki, K.; Kokrashvili, Z.; Benard, O.; Ninomiya, Y.; Margolskee, R.F. REEP2 Enhances Sweet Receptor Function by Recruitment to Lipid Rafts. *J. Neurosci.* **2010**, *30*, 13774–13783. [CrossRef]
164. Chandrashekar, J.; Mueller, K.L.; Hoon, M.A.; Adler, E.; Feng, L.; Guo, W.; Zuker, C.S.; Ryba, N.J. T2Rs function as bitter taste receptors. *Cell* **2000**, *100*, 703–711. [CrossRef]

165. Shaik, F.A.; Medapati, M.R.; Chelikani, P. Cholesterol modulates the signaling of chemosensory bitter taste receptor T2R14 in human airway cells. *Am. J. Physiol. Lung Cell. Mol. Physiol.* **2019**, *316*, L45–L57. [CrossRef]
166. Sarkanen, J.R.; Nykky, J.; Siikanen, J.; Selinummi, J.; Ylikomi, T.; Jalonen, T.O. Cholesterol supports the retinoic acid-induced synaptic vesicle formation in differentiating human SH-SY5Y neuroblastoma cells. *J. Neurochem.* **2007**, *102*, 1941–1952. [CrossRef]
167. Harada, Y.; Yokota, C.; Habas, R.; Slusarski, D.C.; He, X. Retinoic acid-inducible G protein-coupled receptors bind to frizzled receptors and may activate non-canonical Wnt signaling. *Biochem. Biophys. Res. Commun.* **2007**, *358*, 968–975. [CrossRef] [PubMed]
168. Cheng, Y.; Lotan, R. Molecular Cloning and Characterization of a Novel Retinoic Acid-inducible Gene That Encodes a Putative G Protein-coupled Receptor*. *J. Biol. Chem.* **1998**, *273*, 35008–35015. [CrossRef] [PubMed]
169. Bräuner-Osborne, H.; Jensen, A.A.; Sheppard, P.O.; Brodin, B.; Krogsgaard-Larsen, P.; O'Hara, P. Cloning and characterization of a human orphan family C G-protein coupled receptor GPRC5D1GenBank accession Nos. for GPRC5C: AF207989, for Gprc5d: AF218809 and for GPRC5D: AF209923.1. *Biochim. Biophys. Acta (BBA) Gene Struct. Expr.* **2001**, *1518*, 237–248. [CrossRef] [PubMed]
170. Ji, T.H.; Grossmann, M.; Ji, I. G Protein-coupled Receptors: I. Diversity of receptor-ligand interactions. *J. Biol. Chem.* **1998**, *273*, 17299–17302. [CrossRef]
171. Brown, E.M.; MacLeod, R.J. Extracellular calcium sensing and extracellular calcium signaling. *Physiol. Rev.* **2001**, *81*, 239–297. [CrossRef]
172. Hannan, F.M.; Kallay, E.; Chang, W.; Brandi, M.L.; Thakker, R.V. The calcium-sensing receptor in physiology and in calcitropic and noncalcitropic diseases. *Nat. Rev. Endocrinol.* **2019**, *15*, 33–51. [CrossRef]
173. Kifor, O.; Diaz, R.; Butters, R.; Kifor, I.; Brown, E.M. The calcium-sensing receptor is localized in caveolin-rich plasma membrane domains of bovine parathyroid cells. *J. Biol. Chem.* **1998**, *273*, 21708–21713. [CrossRef]
174. Augé, N.; Maupas-Schwalm, F.; Elbaz, M.; Thiers, J.C.; Waysbort, A.; Itohara, S.; Krell, H.W.; Salvayre, R.; Nègre-Salvayre, A. Role for matrix metalloproteinase-2 in oxidized low-density lipoprotein–induced activation of the sphingomyelin/ceramide pathway and smooth muscle cell proliferation. *Circulation* **2004**, *110*, 571–578. [CrossRef]
175. Aoyagi, M.; Yamamoto, M.; Azuma, H.; Nagashima, G.; Niimi, Y.; Tamaki, M.; Hirakawa, K.; Yamamoto, K. Immunolocalization of matrix metalloproteinases in rabbit carotid arteries after balloon denudation. *Histochem. Cell Biol.* **1998**, *109*, 97–102. [CrossRef]
176. Sundararaman, S.S.; van der Vorst, E.P. Calcium-Sensing Receptor (CaSR), Its Impact on Inflammation and the Consequences on Cardiovascular Health. *Int. J. Mol. Sci.* **2021**, *22*, 2478. [CrossRef] [PubMed]
177. Li, Z.; Li, L.; Zielke, H.R.; Cheng, L.; Xiao, R.; Crow, M.T.; Stetler-Stevenson, W.G.; Froehlich, J.; Lakatta, E.G. Increased expression of 72-kd type IV collagenase (MMP-2) in human aortic atherosclerotic lesions. *Am. J. Pathol.* **1996**, *148*, 121. [PubMed]
178. Molostvov, G.; Fletcher, S.; Bland, R.; Zehnder, D. Extracellular calcium-sensing receptor mediated signaling is involved in human vascular smooth muscle cell proliferation and apoptosis. *Cell. Physiol. Biochem.* **2008**, *22*, 413–422. [CrossRef] [PubMed]
179. Li, H.X.; Kong, F.J.; Bai, S.Z.; He, W.; Xing, W.J.; Xi, Y.H.; Li, G.W.; Guo, J.; Li, H.Z.; Wu, L.Y.; et al. Involvement of calcium-sensing receptor in oxLDL-induced MMP-2 production in vascular smooth muscle cells via PI3K/Akt pathway. *Mol. Cell. Biochem.* **2012**, *362*, 115–122. [CrossRef]
180. Tang, X.l.; Wang, Y.; Li, D.l.; Luo, J.; Liu, M.y. Orphan G protein-coupled receptors (GPCRs): biological functions and potential drug targets. *Acta Pharmacol. Sin.* **2012**, *33*, 363–371. [CrossRef]
181. Jetten, A.M.; Takeda, Y.; Slominski, A.; Kang, H.S. Retinoic acid-related orphan receptor γ (RORγ): connecting sterol metabolism to regulation of the immune system and autoimmune disease. *Curr. Opin. Toxicol.* **2018**, *8*, 66–80. [CrossRef]
182. Jeong, E.; Kim, Y.; Jeong, J.; Cho, Y. Structure of the class C orphan GPCR GPR158 in complex with RGS7-Gβ5. *Nat. Commun.* **2021**, *12*, 6805. [CrossRef] [PubMed]
183. Sutton, L.P.; Orlandi, C.; Song, C.; Oh, W.C.; Muntean, B.S.; Xie, K.; Filippini, A.; Xie, X.; Satterfield, R.; Yaeger, J.D.W.; et al. Orphan receptor GPR158 controls stress-induced depression. *eLife* **2018**, *7*, e33273. [CrossRef]
184. Khrimian, L.; Obri, A.; Ramos-Brossier, M.; Rousseaud, A.; Moriceau, S.; Nicot, A.S.; Mera, P.; Kosmidis, S.; Karnavas, T.; Saudou, F.; et al. Gpr158 mediates osteocalcin's regulation of cognition. *J. Exp. Med.* **2017**, *214*, 2859–2873. [CrossRef]
185. Insel, P.A.; Wilderman, A.; Zambon, A.C.; Snead, A.N.; Murray, F.; Aroonsakool, N.; McDonald, D.S.; Zhou, S.; McCann, T.; Zhang, L.; et al. G Protein–Coupled Receptor (GPCR) Expression in Native Cells: "Novel" endoGPCRs as Physiologic Regulators and Therapeutic Targets. *Mol. Pharmacol.* **2015**, *88*, 181–187. [CrossRef]
186. Civelli, O. Orphan GPCRs and Neuromodulation. *Neuron* **2012**, *76*, 12–21. [CrossRef] [PubMed]
187. Orlandi, C.; Posokhova, E.; Masuho, I.; Ray, T.A.; Hasan, N.; Gregg, R.G.; Martemyanov, K.A. GPR158/179 regulate G protein signaling by controlling localization and activity of the RGS7 complexes. *J. Cell Biol.* **2012**, *197*, 711–719. [CrossRef]
188. Hollinger, S.; Hepler, J.R. Cellular Regulation of RGS Proteins: Modulators and Integrators of G Protein Signaling. *Pharmacol. Rev.* **2002**, *54*, 527–559. [CrossRef] [PubMed]
189. Ross, E.M.; Wilkie, T.M. GTPase-Activating Proteins for Heterotrimeric G Proteins: Regulators of G Protein Signaling (RGS) and RGS-Like Proteins. *Annu. Rev. Biochem.* **2000**, *69*, 795–827. [CrossRef] [PubMed]

membranes

MDPI

Review

Molecular Dynamics Simulations of the Proteins Regulating Synaptic Vesicle Fusion

Maria Bykhovskaia

Neurology Department, Wayne State University, Detroit, MI 48202, USA; mbykhovs@med.wayne.edu

Abstract: Neuronal transmitters are packaged in synaptic vesicles (SVs) and released by the fusion of SVs with the presynaptic membrane (PM). An inflow of Ca^{2+} into the nerve terminal triggers fusion, and the SV-associated protein Synaptotagmin 1 (Syt1) serves as a Ca^{2+} sensor. In preparation for fusion, SVs become attached to the PM by the SNARE protein complex, a coiled-coil bundle that exerts the force overcoming SV-PM repulsion. A cytosolic protein Complexin (Cpx) attaches to the SNARE complex and differentially regulates the evoked and spontaneous release components. It is still debated how the dynamic interactions of Syt1, SNARE proteins and Cpx lead to fusion. This problem is confounded by heterogeneity in the conformational states of the prefusion protein–lipid complex and by the lack of tools to experimentally monitor the rapid conformational transitions of the complex, which occur at a sub-millisecond scale. However, these complications can be overcome employing molecular dynamics (MDs), a computational approach that enables simulating interactions and conformational transitions of proteins and lipids. This review discusses the use of molecular dynamics for the investigation of the pre-fusion protein–lipid complex. We discuss the dynamics of the SNARE complex between lipid bilayers, as well as the interactions of Syt1 with lipids and SNARE proteins, and Cpx regulating the assembly of the SNARE complex.

Keywords: synaptotagmin; SNARE complex; complexin; lipid bilayers; exocytosis; neuronal transmitters

Citation: Bykhovskaia, M. Molecular Dynamics Simulations of the Proteins Regulating Synaptic Vesicle Fusion. *Membranes* **2023**, *13*, 307. https://doi.org/10.3390/membranes13030307

Academic Editors: Yosuke Senju and Shiro Suetsugu

Received: 23 January 2023
Revised: 11 February 2023
Accepted: 1 March 2023
Published: 6 March 2023

1. Introduction

Neurons communicate by releasing neuronal transmitters into the synaptic gap. Transmitters are packed in synaptic vesicles (SVs) and released by the fusion of SVs with the presynaptic membrane (PM). The attachment of an SV to the PM is mediated by the SNARE complex [1–3], a coil-coiled four-helical bundle, which consists of the SV protein synaptobrevin (Sb) and the PM-associated proteins syntaxin 1A (Sx) and SNAP25, or t-SNARE. The assembly of the SNARE bundle enables overcoming the electrostatic and hydration repulsion between the SV and PM lipid bilayers [4].

Rapid synchronous fusion of SVs with PM is triggered by an influx of Ca^{2+} ions into the nerve terminal. An SV-associated protein Synaptotagmin 1 (Syt1) acts as a Ca^{2+} sensor, and evoked synaptic transmission is completely abolished in the absence of Syt1 [5,6]. Syt1 comprises two Ca^{2+} binding domains, C2A and C2B, which are attached to an SV by a transmembrane helix [7]. Each domain has two loops forming a Ca^{2+} binding pocket, and in each of the pockets, Ca^{2+} ions are chelated by five aspartic acids [8,9]. It is agreed that synergistic coordinated insertion of the tips of the C2 domains into the phospholipid membrane drives fusion [5,10,11], but other mechanistic details of Syt1 action are still debated.

Syt1 interacts with the SNARE complex, and multiple studies suggest an important role for Syt1–SNARE interactions during fusion [12–17]. However, other studies have argued against this possibility [18,19], and it is still debated how the SNARE–Syt1 complex is formed in vivo and what the role of Syt1–SNARE interactions is in the fusion process.

The fusion is tightly regulated by the cytosolic protein Complexin (Cpx), which attaches to the SNARE bundle [20] and serves as a positive regulator of synchronous release, promoting and accelerating evoked synaptic transmission [21–26]. The effect of Cpx on synchronous fusion is Ca^{2+}-dependent [21] and several studies suggested a functional [26–30] or molecular [31] interaction between Cpx and Syt1. Cpx deletion also produces a drastic increase in spontaneous Ca^{2+}-independent transmission [24], suggesting that the energetic barrier for SV fusion is reduced in the absence of Cpx [32,33]. It has been established that different domains of Cpx control evoked spontaneous transmission and that these two Cpx functions are decoupled [34,35]. The inhibitory role of Cpx in spontaneous transmission was extensively studied in vitro [36–39] and in vivo [24,34,40], and several competing models of the Cpx clamping function have been developed. However, it remains obscure how Cpx promotes and synchronizes the evoked transmission.

The proteins regulating synaptic fusion have been extensively studied with tools and perspectives of biochemistry and molecular biology, and tremendous progress has been achieved in understanding their interactions [2,18,41–45]. However, the atomistic details of the dynamic Syt1-SNARE-Cpx interactions are still debated, and a systematic approach to manipulating the fusion machinery and understanding disease-relevant mutations is still missing. One complication to this problem is that fusion occurs at a sub-millisecond timescale, and the underlying conformational transitions of the pre-fusion protein–lipid complex occur much faster, probably at a timescale of microseconds or tens of microseconds. Currently, such rapid conformational transitions cannot be monitored experimentally. However, they can be observed in silico employing molecular dynamics (MDs) simulations. In the present review, we discuss how MD simulations of proteins and protein–lipid complexes promoted our understanding of the protein dynamics regulating SV fusion.

2. The SNARE Complex Assembly

The fully assembled SNARE complex is a multicomponent molecular system, which encompasses a four-helical coil-coiled bundle, transmembrane (TM) domains of Sb and Sx, the palmitoylated loop of SNAP25, and the N-terminal domain of Sx, which attaches to PM [46]. Since zippering of the four-helical bundle is thought to provide the force to counterbalance the SV-PM repulsion, the mechanics and dynamics of the SNARE bundle assembly have been studied extensively. Crystallography studies [47] demonstrated that the bundle has distinct layers (Figure 1A), and the initial all-atom MD study (AAMD) [48] showed that a compact and stiff bundle has limited conformational dynamics. The latter study also showed that the bundle is largely stabilized by electrostatic forces, although the hydrophobic interactions add to the bundle rigidity.

Although the AAMD method was instrumental for the initial investigation of the dynamics of the bundle [48,49], as well as of the membrane insertion of the TM and linker domains of the SNARE proteins [50,51], the size of the molecular systems and timescales handled by the AAMD approach remained a limitation. Therefore, coarse-grain MD approaches (CGMD) were developed to simulate zippering of the SNARE complex between lipid bilayers. The initial CGMD simulations of the SNARE complex interacting with lipids [52–55] have been performed employing MARTINI force field [56]. This approach was employed to model fusion mediated by four SNARE complexes, starting from all the SNARE bundles being in a nearly assembled state (up Layer 5). These CGMD simulations enabled observation, in silico, of the final stages of SNARE zippering that trigger fusion, including lipid stalk formation and pore opening (Figure 1B) [57].

To investigate the assembly of the entire SNARE bundle and to understand how it depends on the number of the SNARE complexes attaching an SV to the PM, customized CGMD force fields were developed [58–61]. Indeed, it has been shown that the CGMD force fields, including MARTINI, are not suited to a wide range of applications, and they need to be refined and customized for specific molecular systems [62–66].

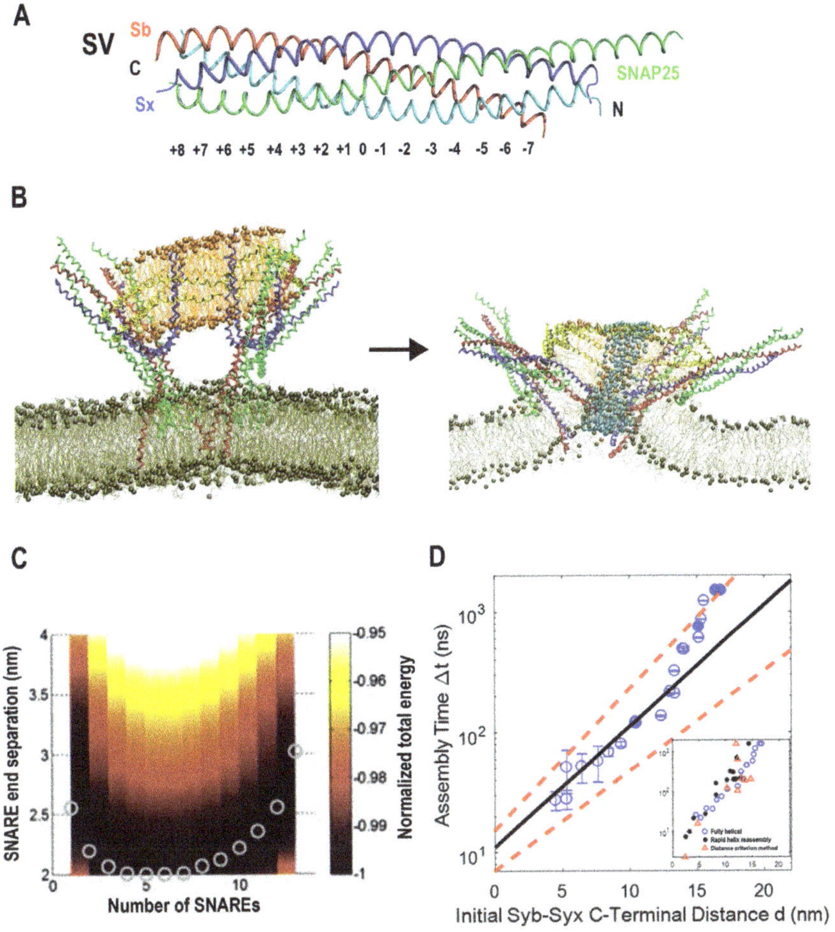

Figure 1. SNARE zippering. (**A**) The structure of the SNARE bundle with the denoted layers [47]. (**B**) The initial (left) and the final (right) states of the molecular system mimicking the SV and PM bilayers attached to each other by four SNARE bundles. Red: Sx, blue: Sb, green: SNAP25. Blue spheres denote water molecules diffusing through the open pore in the final state. Reproduced with permission from [57]. (**C**) The separation of an SV and the PM at equilibrium plotted against the number of the SNARE complexes mediating the SV-PM attachment [60]. Note a steep drop as the number of the SNARE complexes increases from one to two, and a further reduction in the SV-PM separation as the number of the SNARE complexes increases to three. Note also the plateau, as the number of the complexes increases further. (**D**) The assembly time of the SNARE complex depends exponentially on the initial separation of the Sb and Sx C-terminals [59]. The inset shows the results obtained using three different models of the helix assembly, which largely converge.

The customized CGMD approaches [58–61] modeled SNARE proteins as sequences of beads, each bead representing either a single amino acid [60] or a chain of four amino acids [61]. These studies did not model lipid bilayers explicitly but instead represented the PM and SV membranes as a continuum excreting electrostatic, hydration repulsion, and mechanical tension forces. Both models demonstrated, in silico, that SNARE zippering counterbalances membrane repulsion, and that increasing the number of SNARE complexes from one to three significantly accelerates fusion. Strikingly, both models also revealed that a further increase in the number of the SNARE complexes adds very little to the adhesive forces bringing together an SV and the PM [60,61] (Figure 1C). These findings were in agreement with experimental studies, which suggested that several SNARE complexes are likely to mediate synaptic fusion [67–69], even though under certain experimental conditions, a single SNARE complex may be sufficient [70].

Interestingly, the simulations of the kinetics of the SNARE zippering at various initial separations between the C-termini of Sb and Sx showed an exponential relationship between the number of the initially unraveled helical turns and the assembly times [59]. Notably, it was shown that the assembly of two or three membrane-proximal layers in the SNARE complex would take tens of nanoseconds, while the assembly of the entire bundle could take microseconds (Figure 1D).

Together, these findings suggest that three to four SNARE complexes in a nearly assembled state, with only several membrane-proximal layers being separated, would represent the most efficient prefusion complex, which could fully assemble at a sub-microsecond timescale.

3. Cpx as a Dynamic Fusion Clamp

The SV-PM fusion and release of transmitters can occur spontaneously, independently of Ca^{2+} influx. The spontaneous fusion can be clamped by Cpx [24,25,45], and it can be drastically promoted in Cpx-deleted synapses, which is especially prominent in invertebrates. Numerous studies suggested that the interaction of Cpx with the SNARE proteins inhibits the SNARE assembly [32–34,36,39,71,72]; however, the atomistic detail of this mechanism is still debated.

Cpx includes the central and accessory alpha helixes, as well as the C-terminal and N-terminal domains, which are largely unstructured, and the crystallography studies [20] demonstrated that Cpx binds the SNARE bundle via its central and accessory helixes (Figure 2A). Notably, it was also shown that the accessory helix predominantly contributes to the clamping mechanism [35]. Several competing models for the Cpx clamping function have been proposed, which implied that Cpx accessory helix either competes with Sb for the SNARE binding [33,36,39,71] or destabilizes the Cpx central helix [40].

To investigate, in silico, the role of Cpx in the SNARE assembly, the AAMD simulations of Cpx interacting with the partially unraveled SNARE bundle were performed [72]. Interestingly, this study revealed that the Cpx accessory helix could interact with the unstructured C-terminus of Sb, preventing it from zippering onto the core t-SNARE bundle, thus stabilizing the partially assembled structure of the SNARE complex with two or three of its C-terminal layers being unraveled (Figure 2B).

Subsequently, this model was extended to incorporate the SNARE-Cpx interactions with lipid bilayers mimicking the PM and an SV [73]. These AAMD simulations showed that the Cpx accessory helix could also act as a barrier between the SV and the SNARE bundle, thus hindering PM-SV fusion (Figure 2C). Importantly, this model enabled making several valid predictions for the poor-clamp and super-clamp mutations in Cpx and Sb [73,74].

Together, the AAMD simulations outlined above and coupled with in vivo studies [72–74] suggested that the Cpx accessory helix may simply act as a spacer between an SV and the SNARE bundle, in addition stabilizing the unstructured C-terminus of Sb, thus preventing spontaneous full SNARE assembly and PM-SV fusion.

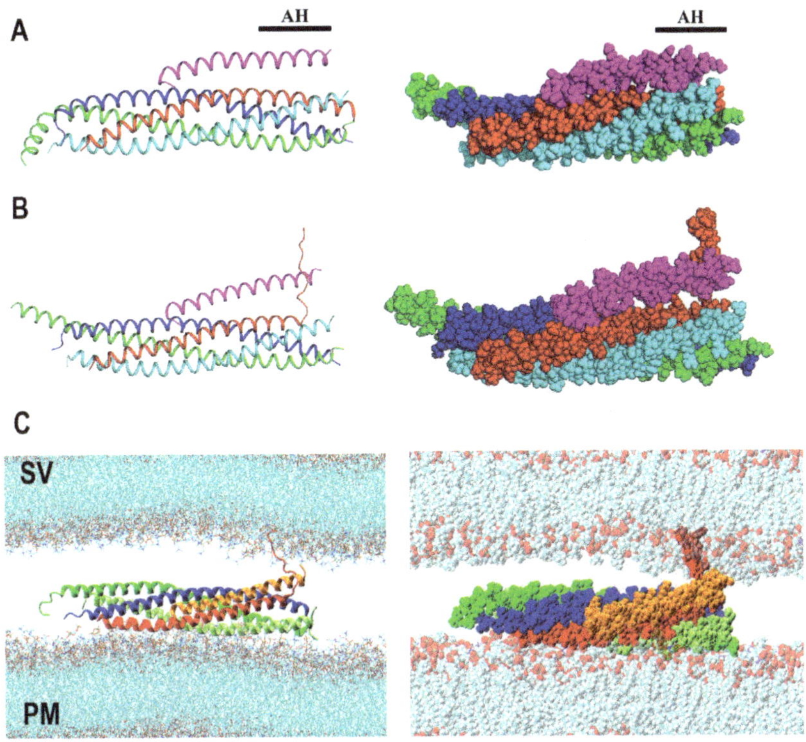

Figure 2. The model of Cpx clamping function driven by AAMD simulations. (**A**) Two representations of the SNARE-Cpx complex. The structure was obtained by crystallography and equilibrated by AAMD. Reproduced from [72]. Blue: Sx; red: Sb; green: SN1; cyan SN2; magenta: Cpx. AH: Accessory helix. (**B**) Cpx stabilizes a partially unraveled state of Sb (Layers 6–8). Reproduced from [75] with permission (license 5472810090395). (**C**) The partially unraveled SNARE complex between lipid bilayers mimicking an SV and the PM. Note that Cpx (orange) creates a barrier between the SNARE bundle and the SV via its accessory helix, in addition to stabilizing the partially unraveled state of Sb.

4. Syt1 and Its Interaction with Lipid Bilayers

Syt1 triggers fusion upon Ca^{2+} binding, presumably by inserting the Ca^{2+}-bound tips of its C2 domains into the lipid bilayer(s) [41]. However, the atomistic mechanics and dynamics of this process are still debated. The crystallography study showed that the C2A and C2B domains of Syt1 are tightly coupled and perpendicularly oriented [76]. However, optical studies suggested that in the solution, Syt1 may sample multiple conformations and the interactions with lipids would likely affect the Syt1 conformational ensemble [77–79].

The conformational space of Syt1 has been investigated employing AAMD coupled with Monte Carlo sampling of the C2 domain orientations [80]. This study identified several conformational states of the Syt1 C2AB tandem, all having tightly coupled C2 domains. Notably, this study also showed that Ca^{2+} binding decouples the C2 domains and allows them to rotate more freely, accelerating Syt1 conformational transitions.

Since the immersion into lipid bilayers is thought to be the major mechanism driving fusion [41], several studies employed AAMD to investigate the interactions of Syt1 C2 domains with lipids [81–83]. Initially, it was shown that binding the C2B module to the lipid bilayer drives lipid bending [81]. This finding supported the hypothesis that Syt1 drives formation of the stalk between lipid bilayers by promoting membrane curvature [11]. Subsequent studies [82,83] modeled the interactions of Syt1 domains with the PM bilayer by incorporating anionic lipids and phosphatidylinositol 4,5-biphosphate (PIP2), which is an essential component of the PM. These studies demonstrated that the C2B domain forms strong attachments to the PM via its Ca^{2+} binding loops and the polybasic motif (Figure 3A), in agreement with molecular biology and spectroscopy experiments [84–87]. Notably, both studies [82,83] demonstrated that the C2B domain does not robustly associate with the SV bilayer lacking PIP2, while the C2A domain does bind the SV bilayer via its Ca^{2+} binding loops (Figure 3B). These studies also demonstrated that the C2A domain robustly binds the PM bilayer (Figure 3C).

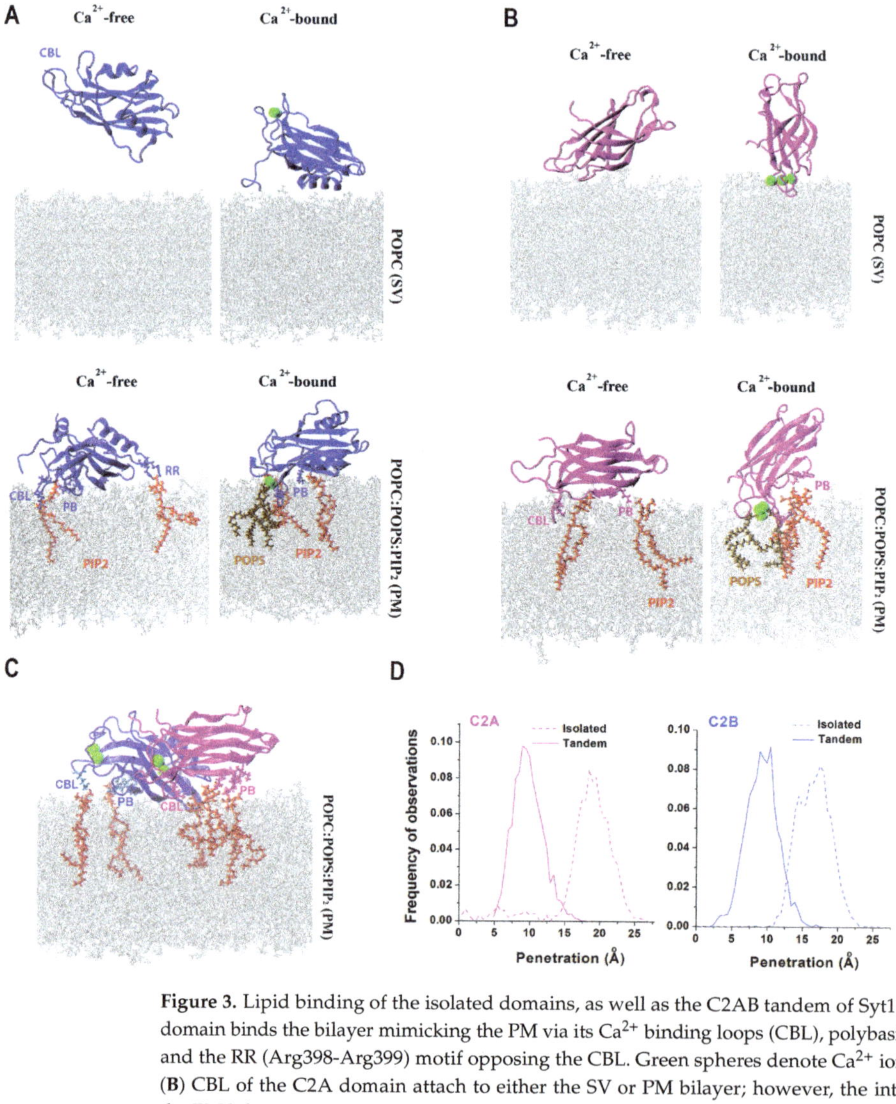

Figure 3. Lipid binding of the isolated domains, as well as the C2AB tandem of Syt1. (**A**) The C2B domain binds the bilayer mimicking the PM via its Ca^{2+} binding loops (CBL), polybasic stretch (PB), and the RR (Arg398-Arg399) motif opposing the CBL. Green spheres denote Ca^{2+} ions. Red: PIP2. (**B**) CBL of the C2A domain attach to either the SV or PM bilayer; however, the interaction with the PM bilayer is more extensive and the penetration into the PM is deeper. (**C**) Both C2 domains within Ca^{2+}C2AB tandem attach to the PM via their CBL and PB motifs. (**D**) The penetration into the PM bilayer is deeper for the isolated Ca^{2+}-bound C2 domains compared to the Ca^{2+}C2AB tandem. Reproduced from [83].

The AAMD studies cited above were consistent with two possibilities for Syt1 dynamics upon Ca^{2+} binding: (1) the C2AB tandem bridges the PM and an SV [78,88] and (2) both domains immerse into the PM, thus promoting PM curvature [11,89,90]. To discriminate between these possibilities, prolonged AAMD simulations of the C2AB tandem between lipid bilayers mimicking an SV and the PM were then performed at a microsecond scale [83]. This study demonstrated a conformational transition of the Syt1 C2AB tandem from the PM-SV bridging to the PM-attached conformation, suggesting that the second scenario is more likely.

Interestingly, a latter study [83] also showed that the C2 domains do not cooperate in penetrating into PM but rather preclude each other from deep immersion into lipids. Indeed, the isolated C2 domains immersed into the PM deeper than when being attached within the C2AB tandem (Figure 3D). These findings suggested that the C2 domains of Syt1 need to be decoupled within the prefusion protein complex, driving the hypothesis that the interactions with other components of the protein fusion machinery serve to uncouple the C2 domains of Syt1. One possibility is that the interactions of Syt1 with the SNARE bundle carry out this function.

5. The Prefusion Syt1-SNARE-Cpx Complex

Spin labeling studies demonstrated that the Syt1-SNARE complex samples multiple conformational states in the solution [91]. Consistently, multiple interfaces between the C2B domain and the SNARE complex were revealed by crystallography, including an extensive primary conserved interface [15]. Interestingly, a different C2B-SNARE interface was identified by the NMR approach [17]. These findings warranted systematic in silico studies of the Syt1-SNARE complex.

To sample the conformational space of the Syt1-SNARE complex, prolonged AAMD simulations were performed and coupled with in silico docking [83]. This study identified three different conformational states of the Syt1-SNARE-Cpx complex (Figure 4A), which were stable at a microsecond scale. The C2B-SNARE interface of the State 3 matched the primary conserved interface discovered by crystallography [15]. Interestingly, two of the three states had Syt1 directly interacting with Cpx (Figure 4A, States 1 and 2). The latter finding was in line with multiple experimental studies, which suggested a functional [26–30] or molecular [31] interaction between Cpx and Syt1 in vivo.

How does the association with the SNARE bundle affect the ability of the C2 domains of Syt1 to penetrate into lipid bilayers? The AAMD simulations [83] revealed that when the Syt1-SNARE-Cpx complex (State 3), with Syt1 being in its Ca^{2+}-bound state, interacted with the lipid bilayer mimicking PM (Figure 4B), the tips of both C2A and C2B domains penetrated into the PM significantly deeper than within the isolated C2AB tandem interacting with the PM (Figure 4C). In other words, the attachment of the C2B domain to the SNARE bundle uncoupled the C2A and C2B domains and, consequently, promoted the insertion of the tips of both C2 domains into the lipid bilayer mimicking the PM.

How does the Ca^{2+}Syt1-SNARE-Cpx pre-fusion complex trigger SV-PM fusion? To elucidate this question, AAMD simulations of the prefusion protein complex between lipid bilayers mimicking the PM and an SV were performed [83]. This study demonstrated, in silico, that this complex in its Ca^{2+}-bound form enables the insertion of the tips of C2 domain of Syt1 into the PM, thus promoting PM curvature and also firmly anchoring the t-SNARE bundle to the PM, acting synergistically with SNARE zippering and driving the SV-PM merging (Figure 4D,E). In contrast, the same molecular system in the absence of Ca^{2+} did not promote fusion (Figure 4E). In summary, this study [83] identified the conformation of the minimal protein machinery (Figure 4B,D) capable of driving SV-PM fusion.

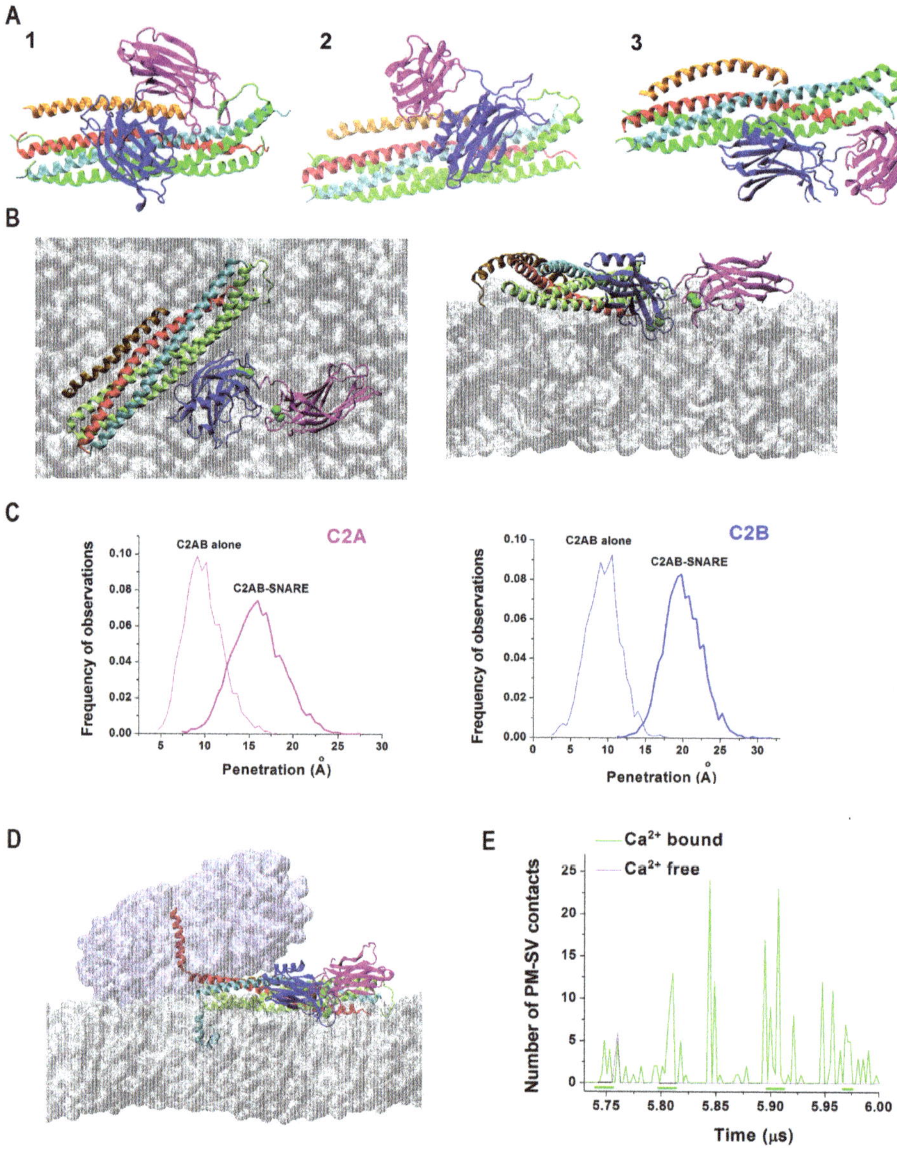

Figure 4. The prefusion Syt1-SNARE-Cpx complex. (**A**) Three conformational states of the Syt1-SNARE-Cpx complex obtained by AAMD simulations. Note that States 1 and 2 have Syt1 directly interacting with Cpx. (**B**) Two views of the prefusion Ca^{2+}Syt1-SNARE-Cpx complex attached to the PM. Note the Ca^{2+}-bound tips of C2 domains immersed into the PM. (**C**) The attachment of the C2B domain to the SNARE bundle decouples C2 domains and enables their deeper penetration into the PM. The graphs show the distributions of the penetration depths over respective 5 μs trajectories. (**D**) The prefusion Ca^{2+}Syt1-SNARE-Cpx complex drives the merging of the SV (top) and the PM (bottom) bilayers. (**E**) The number of SV-PM Van der Waals contacts for the Ca^{2+}-bound and Ca^{2+}-free prefusion Syt1-SNARE-Cpx complexes along respective trajectories. Note continuous stretches of the SV-PM attachment for the Ca^{2+}-bound complex (green lines). In contrast, for the Ca^{2+}-free complex, the PM and SV bilayers are not in contact for most of the trajectory. Reproduced from [83].

Subsequently, AAMD simulations were performed for the system containing several Syt1-SNARE-Cpx complexed between the PM and SV lipid bilayers (Figure 5A) [92]. Importantly, this study showed that even in the case of multiple SNARE complexes, the C2B domains of Syt1 robustly bind the SNARE bundles via their primary interfaces identified by crystallography [15]. In contrast, the interactions of the C2A domain showed some heterogeneity: in the end of the trajectory, two Syt1-SNARE-Cpx complexes had the C2A domains penetrating into the PM, while the other two complexes had the C2A domains bridging to the SV and interacting with Cpx (Figure 5B). This finding can be interpreted either as heterogeneity in Syt1-SNARE-Cpx conformational states within the prefusion protein–lipid complex or, alternatively, as dynamic intermediate states corresponding to the conformational transitions of the complexes to their final prefusion states. More prolonged AAMD simulations will be needed to discriminate between these possibilities.

Figure 5. SV-PM fusion mediated by four Syt1-SNARE-Cpx complexes. (**A**) The system at the initial (0 ns) and final (336 ns) points of the trajectory. (**B**) Each of the four complexes between the bilayers of the SV and PM in the end of the trajectory. Note that all the complexes have the C2B domains (navy) attached to the SNARE bundles. In contrast, the positions of the C2A domains (cyan) vary: two complexes (PC1 and PC4) have the C2A domain attached to the the PM, while the other two complexes (PC2 and PC3) have the C2A domains interacting with Cpx (yellow) and bridging to the SV. Reproduced from [92].

6. Conclusions and Further Directions

MD simulations of the SNARE proteins, Syt1, and Cpx elucidated the mechanistic detail of the final stages of SNARE zippering, enabled the development of the all-atom model of the fusion clamp, and revealed the atomistic detail of Syt1 immersing into lipid bilayers and triggering fusion (Table 1). As the developments in supercomputing enable more prolonged AAMD simulations of larger molecular systems [93–96], the dynamics of synaptic fusion will be further elucidated.

Table 1. A summary of the major MD studies of the synaptic fusion proteins.

Main Focus	Methodology	References
SNARE bundle dynamics	AAMD	[48,49,52]
SNARE TM domains embedded in lipids	AAMD	[50,51]
	CGMD, Martini force field	[53]
SNARE zippering under the forces exerted by the PM-SV repulsion	CGMD, Martini force field	[52,57]
	CGMD, customized force fields	[58–61]
Cpx fusion clamp	AAMD	[72–74]
Syt1 interdomain rotations	AAMD	[16,80]
Syt1 interactions with lipids	AAMD	[81–83]
Syt1-SNARE-Cpx prefusion complex between PM and SV	AAMD	[83,92]

Notably, in silico studies, in particular prolonged AAMD simulations at a microsecond scale, can elucidate how the fusion proteins transition to their pre-fusion states. Both in vitro [91] and in silico [92] studies suggest that the Syt1-SNARE complex is heterogeneous, which may reflect the dynamic conformational transitions of the pre-fusion Syt1-SNARE-Cpx complex to its final state triggering fusion, which occur in vivo. The timescale of such transitions would likely occur at a scale of microseconds or tens of microseconds, and, therefore, cannot be monitored experimentally. However, the dynamics of such conformational transitions can be captured in silico.

Importantly, the interactions of Syt1 and Cpx observed in silico [83] and in vitro [31] and also suggested by in vivo [26–30] studies may play a pivotal role in guiding the pre-fusion complex through the conformational transitions leading to fusion. Indeed, Cpx was shown to synchronize evoked release [22,24,26] by a mechanism which is distinct from clamping spontaneous fusion [34,35]. It is a plausible hypothesis that Cpx may synchronize fusion by accelerating the conformational transitions of the Syt1-SNARE complex, and in silico studies, such as AAMG or GCMD, can test this hypothesis directly.

Furthermore, the development of supercomputing capabilities makes it plausible to incorporate additional components of the protein fusion machinery and to develop the atomistic model of the pre-fusion protein dynamics beyond the minimal Syt-SNARE-Cpx complex. In particular, the Munc family of proteins was shown to orchestrate the assembly of the SNARE complex [1,2,43], with Minc18 possibly serving as a template, forming a tripartite complex with t-SNARE and Sb and stabilizing the half-zippered state of the SNARE bundle [97,98]. The in silico methods, such as AAMD or CGMD, could capture the dynamics of this process in the atomistic detail.

Finally, the CG [57,58,60] and AA [92] models of several SNARE bundles mediating fusion have set the stage for in silico studies of the SNARE self-organization, including the interactions and possible cooperation between multiple SNARE complexes. Indeed, competitive models for the interactions between SNARE bundles mediating fusion have been proposed [33,36,99], and the AAMD and GCMD methods could test the feasibility of these models in silico at the level of mechanics and dynamics of atomic interactions.

The mechanistic details outlined above could be the key for understanding numerous disease-relevant mutations in the fusion proteins, and they can be unraveled by further AAMD and GCMD studies.

Funding: This research received no external funding.

Institutional Review Board Statement: Not applicable.

Informed Consent Statement: Not applicable.

Conflicts of Interest: The authors declare no conflict of interest.

References

1. Sudhof, T.C.; Rothman, J.E. Membrane fusion: Grappling with SNARE and SM proteins. *Science* **2009**, *323*, 474–477. [CrossRef] [PubMed]
2. Sudhof, T.C. Neurotransmitter release: The last millisecond in the life of a synaptic vesicle. *Neuron* **2013**, *80*, 675–690. [CrossRef]
3. Kaeser, P.S.; Regehr, W.G. Molecular mechanisms for synchronous, asynchronous, and spontaneous neurotransmitter release. *Annu. Rev. Physiol.* **2014**, *76*, 333–363. [CrossRef] [PubMed]
4. Rizo, J.; Xu, J. The Synaptic Vesicle Release Machinery. *Annu. Rev. Biophys.* **2015**, *44*, 339–367. [CrossRef] [PubMed]
5. Chapman, E.R. How does synaptotagmin trigger neurotransmitter release? *Annu. Rev. Biochem.* **2008**, *77*, 615–641. [CrossRef]
6. Leitz, J.; Kavalali, E.T. Ca^{2+} Dependence of Synaptic Vesicle Endocytosis. *Neuroscientist* **2015**, *22*, 464–476. [CrossRef]
7. Perin, M.S.; Brose, N.; Jahn, R.; Sudhof, T.C. Domain structure of synaptotagmin (p65). *J. Biol. Chem.* **1991**, *266*, 623–629. [CrossRef]
8. Ubach, J.; Zhang, X.; Shao, X.; Sudhof, T.C.; Rizo, J. Ca^{2+} binding to synaptotagmin: How many Ca^{2+} ions bind to the tip of a C2-domain? *EMBO J.* **1998**, *17*, 3921–3930. [CrossRef]
9. Fernandez-Chacon, R.; Konigstorfer, A.; Gerber, S.H.; Garcia, J.; Matos, M.F.; Stevens, C.F.; Brose, N.; Rizo, J.; Rosenmund, C.; Sudhof, T.C. Synaptotagmin I functions as a calcium regulator of release probability. *Nature* **2001**, *410*, 41–49. [CrossRef]
10. Nyenhuis, S.B.; Karandikar, N.; Kiessling, V.; Kreutzberger, A.J.B.; Thapa, A.; Liang, B.; Tamm, L.K.; Cafiso, D.S. Conserved arginine residues in synaptotagmin 1 regulate fusion pore expansion through membrane contact. *Nat. Commun.* **2021**, *12*, 761. [CrossRef]
11. Martens, S.; Kozlov, M.M.; McMahon, H.T. How synaptotagmin promotes membrane fusion. *Science* **2007**, *316*, 1205–1208. [CrossRef] [PubMed]
12. Lynch, K.L.; Gerona, R.R.; Kielar, D.M.; Martens, S.; McMahon, H.T.; Martin, T.F. Synaptotagmin-1 utilizes membrane bending and SNARE binding to drive fusion pore expansion. *Mol. Biol. Cell* **2008**, *19*, 5093–5103. [CrossRef] [PubMed]
13. Schupp, M.; Malsam, J.; Ruiter, M.; Scheutzow, A.; Wierda, K.D.; Sollner, T.H.; Sorensen, J.B. Interactions Between SNAP-25 and Synaptotagmin-1 Are Involved in Vesicle Priming, Clamping Spontaneous and Stimulating Evoked Neurotransmission. *J. Neurosci.* **2016**, *36*, 11865–11880. [CrossRef] [PubMed]
14. Pang, Z.P.; Shin, O.H.; Meyer, A.C.; Rosenmund, C.; Sudhof, T.C. A gain-of-function mutation in synaptotagmin-1 reveals a critical role of Ca^{2+}-dependent soluble N-ethylmaleimide-sensitive factor attachment protein receptor complex binding in synaptic exocytosis. *J. Neurosci.* **2006**, *26*, 12556–12565. [CrossRef] [PubMed]
15. Zhou, Q.; Lai, Y.; Bacaj, T.; Zhao, M.; Lyubimov, A.Y.; Uervirojnangkoorn, M.; Zeldin, O.B.; Brewster, A.S.; Sauter, N.K.; Cohen, A.E.; et al. Architecture of the synaptotagmin-SNARE machinery for neuronal exocytosis. *Nature* **2015**, *525*, 62–67. [CrossRef] [PubMed]
16. Guan, Z.; Bykhovskaia, M.; Jorquera, R.A.; Sutton, R.B.; Akbergenova, Y.; Littleton, J.T. A synaptotagmin suppressor screen indicates SNARE binding controls the timing and Ca(2+) cooperativity of vesicle fusion. *Elife* **2017**, *6*, e28409. [CrossRef]
17. Brewer, K.D.; Bacaj, T.; Cavalli, A.; Camilloni, C.; Swarbrick, J.D.; Liu, J.; Zhou, A.; Zhou, P.; Barlow, N.; Xu, J.; et al. Dynamic binding mode of a Synaptotagmin-1-SNARE complex in solution. *Nat. Struct. Mol. Biol.* **2015**, *22*, 555–564. [CrossRef]
18. Jahn, R.; Fasshauer, D. Molecular machines governing exocytosis of synaptic vesicles. *Nature* **2012**, *490*, 201–207. [CrossRef]
19. Park, Y.; Seo, J.B.; Fraind, A.; Perez-Lara, A.; Yavuz, H.; Han, K.; Jung, S.R.; Kattan, I.; Walla, P.J.; Choi, M.; et al. Synaptotagmin-1 binds to PIP(2)-containing membrane but not to SNAREs at physiological ionic strength. *Nat. Struct. Mol. Biol.* **2015**, *22*, 815–823. [CrossRef]
20. Chen, X.; Tomchick, D.R.; Kovrigin, E.; Arac, D.; Machius, M.; Sudhof, T.C.; Rizo, J. Three-dimensional structure of the complexin/SNARE complex. *Neuron* **2002**, *33*, 397–409. [CrossRef]
21. Reim, K.; Mansour, M.; Varoqueaux, F.; McMahon, H.T.; Sudhof, T.C.; Brose, N.; Rosenmund, C. Complexins regulate a late step in Ca2+-dependent neurotransmitter release. *Cell* **2001**, *104*, 71–81. [CrossRef]
22. Xue, M.; Lin, Y.Q.; Pan, H.; Reim, K.; Deng, H.; Bellen, H.J.; Rosenmund, C. Tilting the balance between facilitatory and inhibitory functions of mammalian and Drosophila Complexins orchestrates synaptic vesicle exocytosis. *Neuron* **2009**, *64*, 367–380. [CrossRef]
23. Chang, S.; Reim, K.; Pedersen, M.; Neher, E.; Brose, N.; Taschenberger, H. Complexin stabilizes newly primed synaptic vesicles and prevents their premature fusion at the mouse calyx of held synapse. *J. Neurosci.* **2015**, *35*, 8272–8290. [CrossRef]
24. Huntwork, S.; Littleton, J.T. A complexin fusion clamp regulates spontaneous neurotransmitter release and synaptic growth. *Nat. Neurosci.* **2007**, *10*, 1235–1237. [CrossRef] [PubMed]
25. Trimbuch, T.; Rosenmund, C. Should I stop or should I go? The role of complexin in neurotransmitter release. *Nat. Rev. Neurosci.* **2016**, *17*, 118–125. [CrossRef] [PubMed]
26. Jorquera, R.A.; Huntwork-Rodriguez, S.; Akbergenova, Y.; Cho, R.W.; Littleton, J.T. Complexin controls spontaneous and evoked neurotransmitter release by regulating the timing and properties of synaptotagmin activity. *J. Neurosci.* **2012**, *32*, 18234–18245. [CrossRef]
27. Dhara, M.; Yarzagaray, A.; Schwarz, Y.; Dutta, S.; Grabner, C.; Moghadam, P.K.; Bost, A.; Schirra, C.; Rettig, J.; Reim, K.; et al. Complexin synchronizes primed vesicle exocytosis and regulates fusion pore dynamics. *J. Cell. Biol.* **2014**, *204*, 1123–1140. [CrossRef] [PubMed]
28. Xu, J.; Brewer, K.D.; Perez-Castillejos, R.; Rizo, J. Subtle Interplay between synaptotagmin and complexin binding to the SNARE complex. *J. Mol. Biol.* **2013**, *425*, 3461–3475. [CrossRef]

29. Tang, J.; Maximov, A.; Shin, O.H.; Dai, H.; Rizo, J.; Sudhof, T.C. A complexin/synaptotagmin 1 switch controls fast synaptic vesicle exocytosis. *Cell* **2006**, *126*, 1175–1187. [CrossRef]
30. Cao, P.; Yang, X.; Sudhof, T.C. Complexin activates exocytosis of distinct secretory vesicles controlled by different synaptotagmins. *J. Neurosci.* **2013**, *33*, 1714–1727. [CrossRef]
31. Zhou, Q.; Zhou, P.; Wang, A.L.; Wu, D.; Zhao, M.; Sudhof, T.C.; Brunger, A.T. The primed SNARE-complexin-synaptotagmin complex for neuronal exocytosis. *Nature* **2017**, *548*, 420–425. [CrossRef] [PubMed]
32. Giraudo, C.G.; Eng, W.S.; Melia, T.J.; Rothman, J.E. A clamping mechanism involved in SNARE-dependent exocytosis. *Science* **2006**, *313*, 676–680. [CrossRef] [PubMed]
33. Giraudo, C.G.; Garcia-Diaz, A.; Eng, W.S.; Chen, Y.; Hendrickson, W.A.; Melia, T.J.; Rothman, J.E. Alternative zippering as an on-off switch for SNARE-mediated fusion. *Science* **2009**, *323*, 512–516. [CrossRef] [PubMed]
34. Cho, R.W.; Kummel, D.; Li, F.; Baguley, S.W.; Coleman, J.; Rothman, J.E.; Littleton, J.T. Genetic analysis of the Complexin trans-clamping model for cross-linking SNARE complexes in vivo. *Proc. Natl. Acad. Sci. USA* **2014**, *111*, 10317–10322. [CrossRef]
35. Xue, M.; Reim, K.; Chen, X.; Chao, H.T.; Deng, H.; Rizo, J.; Brose, N.; Rosenmund, C. Distinct domains of complexin I differentially regulate neurotransmitter release. *Nat. Struct. Mol. Biol.* **2007**, *14*, 949–958. [CrossRef]
36. Krishnakumar, S.S.; Li, F.; Coleman, J.; Schauder, C.M.; Kummel, D.; Pincet, F.; Rothman, J.E.; Reinisch, K.M. Re-visiting the trans insertion model for complexin clamping. *Elife* **2015**, *4*, e04463. [CrossRef] [PubMed]
37. Krishnakumar, S.S.; Li, F.; Coleman, J.; Schauder, C.M.; Kummel, D.; Pincet, F.; Rothman, J.E.; Reinisch, K.M. Correction: Re-visiting the trans insertion model for complexin clamping. *Elife* **2017**, *6*, e31512. [CrossRef]
38. Krishnakumar, S.S.; Radoff, D.T.; Kummel, D.; Giraudo, C.G.; Li, F.; Khandan, L.; Baguley, S.W.; Coleman, J.; Reinisch, K.M.; Pincet, F.; et al. A conformational switch in complexin is required for synaptotagmin to trigger synaptic fusion. *Nat. Struct. Mol. Biol.* **2011**, *18*, 934–940. [CrossRef]
39. Kummel, D.; Krishnakumar, S.S.; Radoff, D.T.; Li, F.; Giraudo, C.G.; Pincet, F.; Rothman, J.E.; Reinisch, K.M. Complexin cross-links prefusion SNAREs into a zigzag array. *Nat. Struct. Mol. Biol.* **2011**, *18*, 927–933. [CrossRef]
40. Radoff, D.T.; Dong, Y.; Snead, D.; Bai, J.; Eliezer, D.; Dittman, J.S. The accessory helix of complexin functions by stabilizing central helix secondary structure. *Elife* **2014**, *3*, e04553. [CrossRef]
41. Chapman, E.R. A Ca(2+) Sensor for Exocytosis. *Trends. Neurosci.* **2018**, *41*, 327–330. [CrossRef]
42. Brunger, A.T.; Choi, U.B.; Lai, Y.; Leitz, J.; White, K.I.; Zhou, Q. The pre-synaptic fusion machinery. *Curr. Opin. Struct. Biol.* **2019**, *54*, 179–188. [CrossRef] [PubMed]
43. Rizo, J. Mechanism of neurotransmitter release coming into focus. *Protein. Sci.* **2018**, *27*, 1364–1391. [CrossRef] [PubMed]
44. Brunger, A.T.; Leitz, J.; Zhou, Q.; Choi, U.B.; Lai, Y. Ca(2+)-Triggered Synaptic Vesicle Fusion Initiated by Release of Inhibition. *Trends. Cell Biol.* **2018**, *28*, 631–645. [CrossRef] [PubMed]
45. Mohrmann, R.; Dhara, M.; Bruns, D. Complexins: Small but capable. *Cell. Mol. Life Sci.* **2015**, *72*, 4221–4235. [CrossRef] [PubMed]
46. Rizo, J. Molecular Mechanisms Underlying Neurotransmitter Release. *Annu. Rev. Biophys.* **2022**, *51*, 377–408. [CrossRef]
47. Sutton, R.B.; Fasshauer, D.; Jahn, R.; Brunger, A.T. Crystal structure of a SNARE complex involved in synaptic exocytosis at 2.4 A resolution. *Nature* **1998**, *395*, 347–353. [CrossRef] [PubMed]
48. Durrieu, M.P.; Lavery, R.; Baaden, M. Interactions between neuronal fusion proteins explored by molecular dynamics. *Biophys. J.* **2008**, *94*, 3436–3446. [CrossRef]
49. Bock, L.V.; Hutchings, B.; Grubmuller, H.; Woodbury, D.J. Chemomechanical regulation of SNARE proteins studied with molecular dynamics simulations. *Biophys. J.* **2010**, *99*, 1221–1230. [CrossRef]
50. Knecht, V.; Grubmuller, H. Mechanical coupling via the membrane fusion SNARE protein syntaxin 1A: A molecular dynamics study. *Biophys. J.* **2003**, *84*, 1527–1547. [CrossRef]
51. Blanchard, A.E.; Arcario, M.J.; Schulten, K.; Tajkhorshid, E. A highly tilted membrane configuration for the prefusion state of synaptobrevin. *Biophys. J.* **2014**, *107*, 2112–2121. [CrossRef] [PubMed]
52. Delalande, O.; Ferey, N.; Grasseau, G.; Baaden, M. Complex molecular assemblies at hand via interactive simulations. *J. Comput. Chem.* **2009**, *30*, 2375–2387. [CrossRef] [PubMed]
53. Lindau, M.; Hall, B.A.; Chetwynd, A.; Beckstein, O.; Sansom, M.S. Coarse-grain simulations reveal movement of the synaptobrevin C-terminus in response to piconewton forces. *Biophys. J.* **2012**, *103*, 959–969. [CrossRef] [PubMed]
54. Fang, Q.; Zhao, Y.; Herbst, A.D.; Kim, B.N.; Lindau, M. Positively Charged Amino Acids at the SNAP-25 C Terminus Determine Fusion Rates, Fusion Pore Properties, and Energetics of Tight SNARE Complex Zippering. *J. Neurosci.* **2015**, *35*, 3230–3239. [CrossRef] [PubMed]
55. Han, J.; Pluhackova, K.; Wassenaar, T.A.; Bockmann, R.A. Synaptobrevin Transmembrane Domain Dimerization Studied by Multiscale Molecular Dynamics Simulations. *Biophys. J.* **2015**, *109*, 760–771. [CrossRef]
56. Marrink, S.J.; Risselada, H.J.; Yefimov, S.; Tieleman, D.P.; de Vries, A.H. The MARTINI force field: Coarse grained model for biomolecular simulations. *J. Phys. Chem. B* **2007**, *111*, 7812–7824. [CrossRef]
57. Sharma, S.; Lindau, M. Molecular mechanism of fusion pore formation driven by the neuronal SNARE complex. *Proc. Natl. Acad. Sci. USA* **2018**, *115*, 12751–12756. [CrossRef]
58. Mostafavi, H.; Thiyagarajan, S.; Stratton, B.S.; Karatekin, E.; Warner, J.M.; Rothman, J.E.; O'Shaughnessy, B. Entropic forces drive self-organization and membrane fusion by SNARE proteins. *Proc. Natl. Acad. Sci. USA* **2017**, *114*, 5455–5460. [CrossRef]

59. Fortoul, N.; Bykhovskaia, M.; Jagota, A. Coarse-Grained Model for Zippering of SNARE from Partially Assembled States. *J. Phys. Chem. B.* **2018**, *122*, 10834–10840. [CrossRef]
60. Fortoul, N.; Singh, P.; Hui, C.-Y.; Bykhovskaia, M.; Jagota, A. Coarse-Grained Model of the Snare Complex Determines the Number of Snares Required for Docking. *Biophys. J.* **2015**, *108*, 154a. [CrossRef]
61. McDargh, Z.A.; Polley, A.; O'Shaughnessy, B. SNARE-mediated membrane fusion is a two-stage process driven by entropic forces. *FEBS. Lett.* **2018**, *592*, 3504–3515. [CrossRef] [PubMed]
62. Ayton, G.S.; Voth, G.A. Multiscale simulation of protein mediated membrane remodeling. *Semin. Cell Dev. Biol.* **2010**, *21*, 357–362. [CrossRef] [PubMed]
63. Ayton, G.S.; Tepper, H.L.; Mirijanian, D.T.; Voth, G.A. A new perspective on the coarse-grained dynamics of fluids. *J. Chem. Phys.* **2004**, *120*, 4074–4088. [CrossRef] [PubMed]
64. Ayton, G.S.; Voth, G.A. Systematic multiscale simulation of membrane protein systems. *Curr. Opin. Struct. Biol.* **2009**, *19*, 138–144. [CrossRef] [PubMed]
65. Mustafa, G.; Nandekar, P.P.; Yu, X.; Wade, R.C. On the application of the MARTINI coarse-grained model to immersion of a protein in a phospholipid bilayer. *J. Chem. Phys.* **2015**, *143*, 243139. [CrossRef]
66. Lamberg, A.; Taniguchi, T. Coarse-grained computational studies of supported bilayers: Current problems and their root causes. *J. Phys. Chem. B.* **2014**, *118*, 10643–10652. [CrossRef]
67. Bao, H.; Das, D.; Courtney, N.A.; Jiang, Y.; Briguglio, J.S.; Lou, X.; Roston, D.; Cui, Q.; Chanda, B.; Chapman, E.R. Dynamics and number of trans-SNARE complexes determine nascent fusion pore properties. *Nature* **2018**, *554*, 260–263. [CrossRef] [PubMed]
68. Shi, L.; Shen, Q.T.; Kiel, A.; Wang, J.; Wang, H.W.; Melia, T.J.; Rothman, J.E.; Pincet, F. SNARE proteins: One to fuse and three to keep the nascent fusion pore open. *Science* **2012**, *335*, 1355–1359. [CrossRef]
69. Sinha, R.; Ahmed, S.; Jahn, R.; Klingauf, J. Two synaptobrevin molecules are sufficient for vesicle fusion in central nervous system synapses. *Proc. Natl. Acad. Sci. USA* **2011**, *108*, 14318–14323. [CrossRef]
70. Van den Bogaart, G.; Holt, M.G.; Bunt, G.; Riedel, D.; Wouters, F.S.; Jahn, R. One SNARE complex is sufficient for membrane fusion. *Nat. Struct. Mol. Biol.* **2010**, *17*, 358–364. [CrossRef]
71. Zdanowicz, R.; Kreutzberger, A.; Liang, B.; Kiessling, V.; Tamm, L.K.; Cafiso, D.S. Complexin Binding to Membranes and Acceptor t-SNAREs Explains Its Clamping Effect on Fusion. *Biophys. J.* **2017**, *113*, 1235–1250. [CrossRef] [PubMed]
72. Bykhovskaia, M.; Jagota, A.; Gonzalez, A.; Vasin, A.; Littleton, J.T. Interaction of the complexin accessory helix with the C-terminus of the SNARE complex: Molecular-dynamics model of the fusion clamp. *Biophys. J.* **2013**, *105*, 679–690. [CrossRef] [PubMed]
73. Brady, J.; Vasin, A.; Bykhovskaia, M. The Accessory Helix of Complexin Stabilizes a Partially Unzippered State of the SNARE Complex and Mediates the Complexin Clamping Function In Vivo. *eNeuro* **2021**, *8*, 1–13. [CrossRef] [PubMed]
74. Vasin, A.; Volfson, D.; Littleton, J.T.; Bykhovskaia, M. Interaction of the Complexin Accessory Helix with Synaptobrevin Regulates Spontaneous Fusion. *Biophys. J.* **2016**, *111*, 1954–1964. [CrossRef]
75. Bykhovskaia, M. Molecular Dynamics Simulations of the SNARE Complex. *Methods. Mol. Biol.* **2019**, *1860*, 3–13. [CrossRef] [PubMed]
76. Fuson, K.L.; Montes, M.; Robert, J.J.; Sutton, R.B. Structure of human synaptotagmin 1 C2AB in the absence of Ca^{2+} reveals a novel domain association. *Biochemistry* **2007**, *46*, 13041–13048. [CrossRef]
77. Herrick, D.Z.; Kuo, W.; Huang, H.; Schwieters, C.D.; Ellena, J.F.; Cafiso, D.S. Solution and membrane-bound conformations of the tandem C2A and C2B domains of synaptotagmin 1: Evidence for bilayer bridging. *J. Mol. Biol.* **2009**, *390*, 913–923. [CrossRef]
78. Kuo, W.; Herrick, D.Z.; Cafiso, D.S. Phosphatidylinositol 4,5-bisphosphate alters synaptotagmin 1 membrane docking and drives opposing bilayers closer together. *Biochemistry* **2011**, *50*, 2633–2641. [CrossRef]
79. Nyenhuis, S.B.; Thapa, A.; Cafiso, D.S. Phosphatidylinositol 4,5 Bisphosphate Controls the cis and trans Interactions of Synaptotagmin 1. *Biophys. J.* **2019**, *117*, 247–257. [CrossRef]
80. Bykhovskaia, M. Calcium binding promotes conformational flexibility of the neuronal Ca(2+) sensor synaptotagmin. *Biophys. J.* **2015**, *108*, 2507–2520. [CrossRef]
81. Wu, Z.; Schulten, K. Synaptotagmin's role in neurotransmitter release likely involves Ca(2+)-induced conformational transition. *Biophys. J.* **2014**, *107*, 1156–1166. [CrossRef] [PubMed]
82. Prasad, R.; Zhou, H.X. Membrane Association and Functional Mechanism of Synaptotagmin-1 in Triggering Vesicle Fusion. *Biophys. J.* **2020**, *119*, 1255–1265. [CrossRef]
83. Bykhovskaia, M. SNARE complex alters the interactions of the Ca(2+) sensor synaptotagmin 1 with lipid bilayers. *Biophys. J.* **2021**, *120*, 642–661. [CrossRef] [PubMed]
84. Guillen, J.; Ferrer-Orta, C.; Buxaderas, M.; Perez-Sanchez, D.; Guerrero-Valero, M.; Luengo-Gil, G.; Pous, J.; Guerra, P.; Gomez-Fernandez, J.C.; Verdaguer, N.; et al. Structural insights into the Ca^{2+} and PI(4,5)P2 binding modes of the C2 domains of rabphilin 3A and synaptotagmin 1. *Proc. Natl. Acad. Sci. USA* **2013**, *110*, 20503–20508. [CrossRef] [PubMed]
85. Perez-Lara, A.; Thapa, A.; Nyenhuis, S.B.; Nyenhuis, D.A.; Halder, P.; Tietzel, M.; Tittmann, K.; Cafiso, D.S.; Jahn, R. PtdInsP2 and PtdSer cooperate to trap synaptotagmin-1 to the plasma membrane in the presence of calcium. *Elife* **2016**, *5*, e15886. [CrossRef]
86. Hui, E.; Bai, J.; Chapman, E.R. Ca^{2+}-triggered simultaneous membrane penetration of the tandem C2-domains of synaptotagmin I. *Biophys. J.* **2006**, *91*, 1767–1777. [CrossRef]
87. Wang, P.; Wang, C.T.; Bai, J.; Jackson, M.B.; Chapman, E.R. Mutations in the effector binding loops in the C2A and C2B domains of synaptotagmin I disrupt exocytosis in a nonadditive manner. *J. Biol. Chem.* **2003**, *278*, 47030–47037. [CrossRef]

88. Lin, C.C.; Seikowski, J.; Perez-Lara, A.; Jahn, R.; Hobartner, C.; Walla, P.J. Control of membrane gaps by synaptotagmin-Ca2+ measured with a novel membrane distance ruler. *Nat. Commun.* **2014**, *5*, 5859. [CrossRef]
89. Hui, E.; Johnson, C.P.; Yao, J.; Dunning, F.M.; Chapman, E.R. Synaptotagmin-mediated bending of the target membrane is a critical step in Ca(2+)-regulated fusion. *Cell.* **2009**, *138*, 709–721. [CrossRef]
90. McMahon, H.T.; Kozlov, M.M.; Martens, S. Membrane curvature in synaptic vesicle fusion and beyond. *Cell* **2010**, *140*, 601–605. [CrossRef]
91. Lai, A.L.; Huang, H.; Herrick, D.Z.; Epp, N.; Cafiso, D.S. Synaptotagmin 1 and SNAREs form a complex that is structurally heterogeneous. *J. Mol. Biol.* **2011**, *405*, 696–706. [CrossRef]
92. Rizo, J.; Sari, L.; Qi, Y.; Im, W.; Lin, M.M. All-atom molecular dynamics simulations of Synaptotagmin-SNARE-complexin complexes bridging a vesicle and a flat lipid bilayer. *Elife* **2022**, *11*, e76356. [CrossRef] [PubMed]
93. Shaw, D.E.; Dror, R.O.; Grossman, J.P.; Mackenzie, K.M.; Bank, J.A.; Young, C.; Deneroff, M.M.; Batson, B.; Bowers, K.J.; Chow, E.; et al. Millisecond-Scale Molecular Dynamics Simulations on Anton. In *Proceedings of the Conference on High Performance Computing Networking Storage and Analysis*; ACM: New York, NY, USA, 2009; Volume SC09.
94. Dror, R.O.; Jensen, M.O.; Shaw, D.E. Elucidating membrane protein function through long-timescale molecular dynamics simulation. *Annu. Int. Conf. IEEE Eng. Med. Biol. Soc.* **2009**, *2009*, 2340–2342. [PubMed]
95. Shaw, D.E.; Maragakis, P.; Lindorff-Larsen, K.; Piana, S.; Dror, R.O.; Eastwood, M.P.; Bank, J.A.; Jumper, J.M.; Salmon, J.K.; Shan, Y.; et al. Atomic-Level Characterization of the Structural Dynamics of Proteins. *Science* **2010**, *330*, 341–346. [CrossRef] [PubMed]
96. Dror, R.O.; Dirks, R.M.; Grossman, J.; Xu, H.; Shaw, D.E. Biomolecular Simulation: A Computational Microscope for Molecular Biology. *Annu. Rev. Biophys.* **2012**, *41*, 429–452. [CrossRef]
97. Baker, R.W.; Jeffrey, P.D.; Zick, M.; Phillips, B.P.; Wickner, W.T.; Hughson, F.M. A direct role for the Sec1/Munc18-family protein Vps33 as a template for SNARE assembly. *Science* **2015**, *349*, 1111–1114. [CrossRef]
98. Ma, L.; Rebane, A.A.; Yang, G.; Xi, Z.; Kang, Y.; Gao, Y.; Zhang, Y. Munc18-1-regulated stage-wise SNARE assembly underlying synaptic exocytosis. *Elife* **2015**, *4*, e09580. [CrossRef]
99. Montecucco, C.; Schiavo, G.; Pantano, S. SNARE complexes and neuroexocytosis: How many, how close? *Trends Biochem. Sci.* **2005**, *30*, 367–372. [CrossRef]

membranes

MDPI

Article

Precise Detection and Visualization of Nanoscale Temporal Confinement in Single-Molecule Tracking Analysis

Manon Westra and Harold D. MacGillavry *

Cell Biology, Neurobiology and Biophysics, Department of Biology, Faculty of Science, Utrecht University, 3584 CH Utrecht, The Netherlands; m.westra@uu.nl

* Correspondence: h.d.macgillavry@uu.nl

Abstract: The plasma membrane consists of a diverse mixture of molecules that dynamically assemble into a highly non-random organization. The formation of nanoscale domains in the membrane is of particular interest as these domains underlie critical cellular functions. Single-molecule tracking is a powerful method to detect and quantify molecular motion at high temporal and spatial resolution and has therefore been instrumental in understanding mechanisms that underlie membrane organization. In single-molecule trajectories, regions of temporal confinement can be determined that might reveal interesting biophysical interactions important for domain formation. However, analytical methods for the detection of temporal confinement in single-molecule trajectories depend on a variety of parameters that heavily depend on experimental factors and the influence of these factors on the performance of confinement detection are not well understood. Here, we present elaborate confinement analyses on simulated random walks and trajectories that display transient confined behavior to optimize the parameters for different experimental conditions. Furthermore, we demonstrate a heatmap visualization tool that allows spatial mapping of confinement hotspots relative to subcellular markers. Using these optimized tools, we reliably detected subdiffusive behavior of different membrane components and observed differences in the confinement behavior of two types of glutamate receptors in neurons. This study will help in further understanding the dynamic behavior of the complex membrane and its role in cellular functioning.

Keywords: single-molecule tracking; confinement; plasma membrane

Citation: Westra, M.; MacGillavry, H.D. Precise Detection and Visualization of Nanoscale Temporal Confinement in Single-Molecule Tracking Analysis. *Membranes* **2022**, *12*, 650. https://doi.org/10.3390/membranes12070650

Academic Editors: Shiro Suetsugu and Yosuke Senju

Received: 17 May 2022
Accepted: 22 June 2022
Published: 24 June 2022

1. Introduction

The plasma membrane is a highly complex and dynamic environment where a vast variety of transmembrane proteins are embedded in a mixture of over a hundred different types of lipids. Key to understanding membrane organization is determining how components are organized and move within the lateral plane of the membrane [1]. The development of single-molecule tracking techniques has been instrumental in quantifying the diffusion of membrane components in living cells and has provided important new insights into how the dynamic nanoscale organization of membrane components contributes to cellular functions. In contrast to other techniques, single-molecule tracking experiments provide trajectories that describe the motion of individual molecules, rather than the average, 'ensemble' behavior of a population of molecules. Careful analysis of single-molecule trajectories can therefore reveal a wealth of information on the dynamic behavior of molecules, the biophysical properties of the cellular environment, and the compartmentalization of molecules. Single-molecule trajectories can be described as directed, random (Brownian), or confined motion. Confinement zones, regions where a molecule remains longer than expected from a Brownian diffusant, are of particular biological interest. Confinement zones could indicate organizational hotspots where proteins undergo transient binding to intracellular scaffold molecules or are trapped by the underlying membrane cytoskeleton. Examples of such organizational hotspots that are relevant for biological processes are lipid domains [2,3], neurotransmitter receptor nanodomains [4–7], G-protein-coupled receptor hotspots [8] and ion channel nanoclusters [9,10]. There is thus a need for analytical tools that reliably detect, measure, and visualize confinement zones.

Several studies have reported on strategies to analyze and detect confined behavior in single-molecule trajectories [11–18]. In fact, Einstein formulated a theory about Brownian diffusants already in 1905, where he argued that the displacement of a Brownian particle is proportional to the square root of the elapsed time [19]. Confinement is defined as the portion of a trajectory that deviates from what a random walk would look like. Therefore, it is essential to know how long a Brownian diffusant would stay in a certain region. Saxton defined the probability that a molecule will stay in a region by the following equation: $\log \psi = 0.2048 - 2.5117(D \times t/R^2)$ where D is the diffusion coefficient, t the period of time, and R the radius of the region [20]. Simson et al. translated this probability into the confinement index, which is inversely related to the probability that a Brownian molecule will stay in a certain area [11]. When this probability during the trajectory becomes lower, the confinement index will increase, indicating a period of transient confinement within the trajectory. This confinement analysis has been used extensively by many labs to detect temporal confinement in single-molecule trajectories [21–33].

This confinement analysis depends on different manually defined parameters that need to be optimized to reliably detect confinement. These parameters include for instance the window of frames to analyze, threshold for the confinement index, and the minimal time a molecule should be in this state to be considered confined. Determining the optimal values for each of these parameters, however, is not trivial and varying individual parameters can have a large impact on the detection power of the analysis. Therefore, in this study, we set out to understand the influence of individual parameters on the performance of the confinement analysis and tested the robustness of this analysis on simulated random walks and trajectories that display transient confined behavior. This allowed us to optimize parameters and balance the detection of false-negatives and false-positives. Furthermore, we developed a tool to visualize confinement areas in heatmaps that allows spatial mapping of confinement hotspots relative to subcellular markers. To test the performance of this analysis, we applied our analysis on experimental data and reliably detected subdiffusive behavior for a variety of membrane components. Lastly, we found that two neuronal glutamate receptors, mGluR5 (metabotropic glutamate receptor 5) and AMPA-type glutamate receptors, reveal different confinement properties. This study will help in further understanding the dynamic behavior of membrane components and their role in membrane organization.

2. Materials and Methods

2.1. Simulations

To simulate 2D random walks, we tested two different simulation methods (Figure S1, see Supplementary Materials). In the first model (model 1), every consecutive coordinate is drawn from a Gaussian distribution based on the diffusion coefficient and the interval time between the steps: $step = \sqrt{2 \times D \times \Delta t} \times [x\ y]$ where x and y are normally distributed numbers, generated by the MATLAB function *randn*. In the second model (model 2), every step has the same step size derived from the set diffusion coefficient and interval time, however the angle and thus the direction of the step is random: $step = \sqrt{4 \times D \times \Delta t} \times [\cos(\theta)\ \sin(\theta)]$ where θ is a random angle between 0 and 2π, generated using MATLAB (MathWorks, Natick, MA, USA, R2021B) function *rand* (Figure S1A). The MSD curve of model 2 showed less variation in the curve, however the estimated diffusion coefficient is for both models highly similar to the set diffusion coefficient for the simulation. We applied subsampling to the simulated tracks so that every 100 steps the coordinates were saved in the trajectory. After subsampling, the distributions of the diffusion coefficients were almost identical for both models (Figure S1B) and for the rest of the simulations, we used model 2.

We also tested two different approaches for simulating transient confined trajectories. The principal idea behind the simulations is that the molecule cannot escape a simulated circle for as long as it is simulated to be confined. The first approach is the 'reflect' model, where the molecule bounces back from the simulated, circular confinement zone if the next step would be outside the circle (Figure S1C, left panel). The second approach is the 'stick' model, where the molecule sticks at the border at the place where it would otherwise escape the circle (Figure S1C, right panel). The 'reflective' model appeared to be more stable when varying the step sizes in comparison to the 'stick' model. When increasing the step size in the simulations, the plateau in the MSD curve reached higher values for the 'stick' model, while it remained the same for the 'reflect' model (Figure S1D). Therefore, we used the reflective model for the rest of the confinement simulations.

2.2. *Measure Power of Detection*

The power of detection was measured as the percentage of confined displacements, meaning all steps in all trajectories simulated that are in a period of detected confinement, i.e., above threshold *Lc* for period of time longer than *Tc*.

2.3. *Parameters Used in Simulations*

Simulations were performed to generate trajectories of 100 frames and confinement periods of 0, 15, or 50 frames, with diffusion coefficient of 0.05 $\mu m^2/s$ and interval time of 0.02 s unless reported otherwise. Basic simulation settings: D = 0.05 $\mu m^2/s$, *Nsteps* = 10,000, subsampling = 100, *dt* = 0.02 s, *dt_sim* = 0.0002 s. Basic confinement parameters: *Lcm* = 5.5, *Sm* = 15, *Smin* = 4, *Tc* = 0.2 s, α = 0.5, npointsMSDset = 10, *Rconf* = 0.04 µm (Table S1).

2.4. *Animals*

All animal experiments were performed in compliance with the guidelines for the welfare of experimental animals issued by the Government of the Netherlands (Wet op de Dierproeven, 1996) and European regulations (Guideline 86/609/EEC). All animal experiments were approved by the Dutch Animal Experiments Review Committee (Dier Experimenten Commissie; DEC), performed in line with the institutional guidelines of Utrecht University.

2.5. *Primary Neuron Cultures and Transfections*

Dissociated hippocampal cultures were prepared from embryonic day 18 (E18) Janvier Wistar rat brains of both genders [34]. Dissociated neurons were plated on Ø18-mm coverslips coated with poly-L-lysine (37.5 µg/mL, Sigma-Aldrich, St. Louis, MO, USA) and laminin (1.25 µg/mL, Roche Diagnostics, Mannheim, Germany) at a density of 100,000 neurons per well. Neurons were grown in Neurobasal medium (NB) supplemented with 1% penicillin/streptomycin (Gibco), 2% B27 (Gibco), and 0.5 mM L-glutamine (Gibco, Grand Island, NY, USA) at 37 °C in 5% CO_2. From days in vitro (DIV) 1 onward, medium was refreshed weekly by replacing half of the medium with BrainPhys Neuronal Medium (BP, STEMCELL Technologies, Vancouver, Canada) supplemented with 2% NeuroCult SM1 (STEMCELL Technologies, Vancouver, Canada) and 1% penicillin/streptomycin. At DIV14, neurons were transfected with indicated constructs using Lipofectamine 2000 (Invitrogen, Carlsbad, CA, USA). For each well, 1.8 µg DNA was mixed with 3.3 µL Lipofectamine 2000 in 200 µL BP, incubated for 30 min at room temperature (RT). Then, 400 µL conditioned medium was transferred to a new culture plate and replaced by 260 µL BP supplemented with 0.5 mM L-glutamine. The DNA mix was added to the neurons and incubated at 37 °C in 5% CO_2. After 1 to 2 h, neurons were briefly washed with BP and transferred to the new culture plate with conditioned medium supplemented with an additional 400 µL BP with SM1 and penicillin/streptomycin and kept at 37 °C in 5% CO_2. All experiments were performed using neurons at DIV18-22.

2.6. *DNA Plasmids*

All plasmids have been described before in previous studies: Homer1c-mCherry [35], SEP-GluA2 [4], SEP-mGluR5 [35]. The GFP-GPI and mHD-GT46 constructs are kind gifts from Dr. Ewers (Institute for Chemistry and Biochemistry, Free University Berlin, Germany; [36]).

2.7. Single-Molecule Tracking

Single-molecule tracking was performed on the Nanoimager S from ONI (Oxford Nanoimaging; ONI, Oxford, UK), equipped with a 100×/NA 1.4 oil immersion objective (Olympus Plan Apo, Hamburg, Germany), an XYZ closed-loop piezo stage, and with 471-, 561-, and 640-nm wavelength excitation lasers. Fluorescence emission was detected using a sCMOS camera (ORCA Flash 4, Hamamatsu, Hamamatsu City, Japan). Stacks of 5000 frames were acquired at 50 or 100 Hz with oblique illumination using the 640-nm laser. NimOS software (version 1.18.3) from ONI was used for localization and drift correction was performed. Neurons were imaged in extracellular imaging buffer containing 120 mM NaCl, 3 mM KCl, 10 mM HEPES, 2 mM CaCl2, 2 mM $MgCl_2$, and 10 mM glucose, pH adjusted to 7.35 with NaOH. The GFP/RFP-booster Atto647N (Chromotek, Martinsried, Germany) was added before image acquisition in a concentration of 1:200,000 in extracellular imaging buffer while blocking with 1% BSA. For the tracking of DOPE, we added Atto647N-DOPE (ATTO-TEC, Siegen, Germany) in complex with defatted BSA [37] before image acquisition at a final concentration of 6.5 nM. We used a minimum track length of 30 frames (10- or 20-ms interval) for visualization and quantification. Postsynaptic density (PSD) masks were created from a stack of 30 frames obtained for Homer1c-mCherry using the 561 nm excitation laser as described in [38].

2.8. Single-Molecule Tracking Analysis

Using MATLAB (version 2021b), molecules with a localization precision <50 nm were selected for analysis. Tracking was achieved using custom algorithms in MATLAB described previously with a tracking radius of <500 nm [39]. The first three points of the MSD with the addition of the value 0 at MSD(0) were used to fit the slope using a linear fit. Tracks with a negative slope were not used for further analysis. The diffusion coefficient was estimated based on the fit using the formula: $MSD = 4 \times D \times \Delta t$. Only tracks of at least 30 frames were selected for further analysis. Tracks were classified as immobile when the ratio between the radius of gyration and mean step size $(\frac{\sqrt{\pi/2} \cdot radius\ of\ gyration}{mean\ stepsize})$ was smaller than 2.11 [40]. The PSD mask was created based on the maximum intensity projection of Homer1c-mCherry. Peaks in intensity were detected, after which a FWHM-like boundary was defined for each PSD.

Transient confinement analysis on mobile trajectories was done in MATAB using slightly modified scripts from a previously published MATLAB implementation [21]. As modifications, we included the last segments of the trajectory shorter than Sm in the averaging to improve confinement detection at the end of a trajectory (Figure S2A). Furthermore, the point in the trajectory is the middle of the sliding window of the D over time instead of the beginning of the sliding window, for a better correlation with L over time (Figure S2B). Lastly, we corrected the ln into the log10 for defining confinement index L. As D, the maximum D_{inst} was used, estimated for each sub-trajectory of Δ10. Parameters used in the analysis are: Lc = 5.5, Sm = 15, α = 0.5, Tc = 0.2 s (Table S1). The confinement zones are further analyzed for size and duration of confinement and diffusion coefficient in and outside confinement zones. Confinement maps were created based on the detected confinement radius for each confinement zone. Each confinement zone was stored as a two-dimensional Gaussian distribution with the radius as FWHM. The final matrix was plotted with a color-code, where higher values indicate confinement hotspots because there are multiple Gaussians on top of each other.

2.9. Statistical Analysis

Statistical significance was tested using a Student's t-test when comparing two groups. Statistical tests with a p-value below 0.05 were considered significant. Significance is indicated as follows: $p < 0.05$ is indicated by *, $p < 0.01$ by **, and $p < 0.001$ by ***. Analysis was performed on neurons originating from three individual preparations of hippocampal neurons. The number of neurons used for analysis is indicated as n. Statistical analysis and graphs were prepared in GraphPad Prism (version 9, Graphpad Software, San Diego, CA, USA) and figures were generated in Adobe Illustrator 2022 (Adobe Systems, San Jose, CA, USA).

3. Results

3.1. Detection of Transient Confinement within Single-Molecule Trajectories

To detect transient confinement within single-molecule trajectories, Simson et al. [11] developed a statistical framework defining areas where a diffusing molecule remains significantly longer than expected if it would undergo Brownian motion. The probability ψ that a molecule stays within a bounded region for a set period of time has been defined by Saxton as:

$$\log \psi = 0.2048 - 2.5117\left(Dt/R^2\right) \tag{1}$$

with D being the diffusion coefficient, t the period of time, and R the radius of the region [20]. Simson et al. [11] used this probability to derive the confinement index L, which is inversely related to ψ:

$$L = \begin{cases} -\log(\psi) - 1) & \psi \leq 0.1 \\ 0 & \psi > 0.1 \end{cases} \tag{2}$$

High values of L are thus indicative of non-random confinement and $L = 0$ in cases of random diffusion. The calculation of L is inherently based on a number of variables that will influence the detection power of the confinement analysis. To correctly interpret the results of this analysis, here we aimed to understand the influence of these different variables on the detection power of the analysis. The analysis in this study is based on the MATLAB implementation from Ménchon et al. [21] with some modifications (see Section 2.8; Figure S2). To define the confinement index over time for individual trajectories, a trajectory is first divided into segments with minimum length *Smin* and maximum length *Sm* (Figure 1A) and for each individual segment, L is calculated as described above. As such, these parameters are thus expected to determine the spatial and temporal sensitivity of the analysis. The radius R of the segment is defined by the maximum distance between one of the points and the starting point of the segment. The diffusion coefficient D used in the formula is the same for all the segments in the trajectory and should represent the diffusion coefficient of the molecule if it would move freely. Then, for each point in the trajectory, L is averaged over all the segments that include that point. Note, that as a consequence, the averaged L for points at the beginning and the end of the track is averaged over fewer segments than points in the middle. As can be seen for a simulated trajectory with a short period of transient confinement (Figure 1B), L values are higher during the confined period, which is accompanied by a temporal decrease in the diffusion coefficient (Figure 1C). Confined periods within a trajectory can then be defined by selecting periods in which all points are above the critical minimum L (*Lcm*) for a duration longer than a critical period of time (*Tc*) (Figure 1C).

When increasing the diffusion coefficient D of simulated molecules, it can be seen that because for a molecule with a higher D it is less likely that it would stay in the same area for the same time, relatively higher L values are calculated (Figures 1D and S3A). On the other hand, when D remains constant but the region in which the molecule is observed is larger, the calculated confinement index becomes lower, because it is more likely that a freely moving molecule would stay in a larger area (Figures 1E and S3B). Lastly, if we only increase the observed time window (t)—which is represented by *Smin* and *Sm*—while keeping the other variables the same, a higher confinement index will be detected because it is less likely that a Brownian molecule with that diffusion coefficient would stay in the same area for a longer time (Figure 1F).

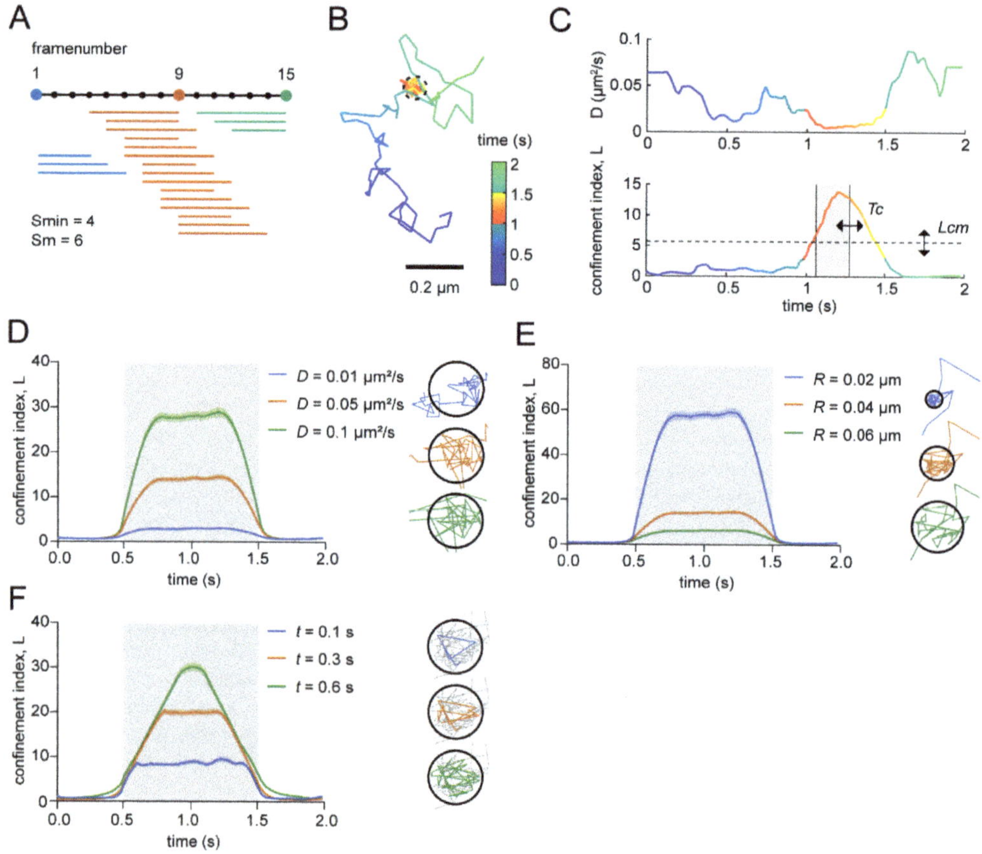

Figure 1. Influence of variables in confinement index formula. (**A**) Schematic diagram of how trajectories are divided into segments with different lengths. In this simplified example, the minimal segment length (*Smin*) is 4 frames and the maximum segment (*Sm*) is 6. All the segments are displayed that will be included in the confinement index of frame 1 (blue), 9 (orange), and 15 (green). (**B**) Simulated trajectory with a confinement zone with radius of 40 nm. Color-coded for time and confined period. Scale bar, 0.2 µm. (**C**) Diffusion coefficient (top) and confinement index (bottom) over time of the simulated trajectory in (**B**). Color-coded for time and confined period. *Tc* and *Lcm* indicate the critical time period and critical *L* threshold used to detect confinement periods. (**D**) Average confinement index over time for 100 simulated trajectories with three different diffusion coefficients. Simulated confined period is indicated with the gray box. Illustrations display for each condition one example segment of the track in the corresponding color. Simulated confinement radius (*Rconf*) is 0.04 µm and *Sm* is 15 frames. (**E**) Average confinement index over time for 100 simulated trajectories with three different simulated confinement radii. Simulated confined period is indicated with the gray box. Illustrations display for each condition one example segment of the track in the corresponding color. Diffusion coefficient is 0.05 $\mu m^2/s$ and *Sm* is 15 frames. (**F**) Average confinement index over time for 100 simulated trajectories with 3 different segment lengths (*Sm* = *Smin*). Simulated confined period is indicated with the gray box. Illustrations display for each condition one example segment of the track in the corresponding color. Diffusion coefficient is 0.05 $\mu m^2/s$ and *Rconf* is 0.04 µm. Data are represented as means ± SEM.

3.2. Optimizing Input Parameters for Accurate Transient Confinement Zone Detection in Single-Molecule Trajectories

After considering the influence of the different variables in the confinement index calculation, we aimed to optimize the user-defined parameters for the confinement detection to minimize the false positive rate while maximizing the true detection rate. False positives can be detected because *L* is not zero: even in random walks fluctuations in *L* could be inadvertently interpreted as confinement.

First, we wanted to define what effect the input parameters would have on incorrectly detecting confinement in random walks, therefore we simulated random walks with parameters comparable to experimental situations ($D = 0.05$ μm^2/s, $dt = 0.02$ s, 100 frames) and started off with a maximum window size (*Sm*) of 15 frames. Lowering the minimum *L* threshold (*Lcm*) increased the percentage of false positives detected in random walks, similar to decreasing the critical period of time (*Tc*) (Figure 2A). Changing the maximum window size (*Sm*) did not lead to a large change in the detection of false positives (Figure 2B). Thus, to achieve the lowest number of false positives, *Lcm* and *Tc* should be set relatively high, whereas the value of *Sm* is of less significance.

Next, we investigated the effect of the input parameters on the detection of true confinement. Therefore, we simulated trajectories with confined periods and parameters comparable to experimental situations ($D = 0.05$ μm^2/s, $dt = 0.02$ s, 100 frames, confined period of 15 or 50 frames and simulated confinement radius (*Rconf*) of 0.04–0.06 μm). Increasing *Lcm* lowered the detection power of true confinement (Figure 2C). Furthermore, in contrast to the effect on random walks, the time window plays an important role in the correct detection of true confinement. The power of detection increased with larger *Sm* when the confinement period was equally long, whereas the power of detection decreased if the confinement period was shorter than *Sm* (Figure 2D). Choosing this maximum time window will also affect the power of detecting multiple confinement zones after each other. Increasing the *Sm* to achieve increased power of detection could lead to combining multiple confinement zones because *L* will not go below the threshold in between the zones (Figure 2E, orange and green traces). Taken together, to achieve high power of detection, *Lcm* should not be too high and the value for *Sm* will affect the resolution at which confinement zones can be detected. While higher *Sm* values increase detection of confinement, it reduces the power to detect shorter periods of confinement that, at high *Sm* values, will be averaged out or combined with other confinement zones.

Because the diffusion coefficient can change over the course of a trajectory, Ménchon et al. introduced another *L* threshold where the threshold is defined by the average *L* of a trajectory multiplied by a factor α that ranges from 0 to 1 [21]. It is important to consider the effect of using this threshold definition as it can both increase and decrease the power of detection. For example, when the average *L* is high because of two highly confined regions, *Lc* will be higher than *Lcm*, resulting in not detecting a less confined third region (Figure 2F). On the other hand, when there are highly confined zones—possibly caused by a lower diffusion coefficient—shortly after each other, using the threshold *Lc* will separate the two zones, whereas *Lcm* would be too low (Figure 2G).

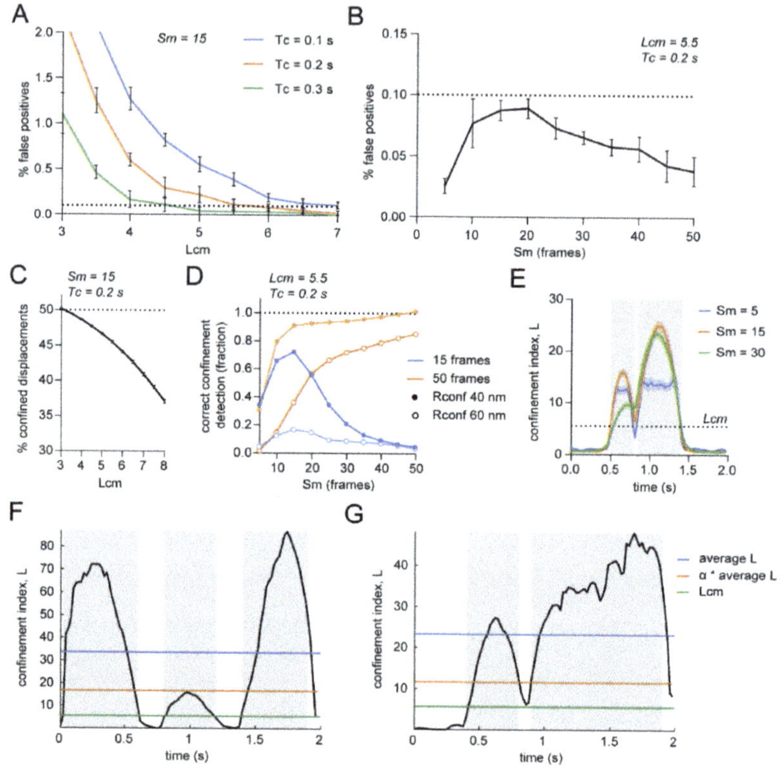

Figure 2. Influence of user-defined parameters on the confinement analysis. (**A**) Effect of minimal critical *L* (*Lcm*) and critical time (*Tc*) on percentage of false positives in simulated random walks. Dotted line indicates 0.1% false positives. *Sm* is 15. Five independent simulations of 100 trajectories. Data are represented as means ± SEM. (**B**) Effect of *Sm* on percentage of false positives in simulated random walks. Dotted line indicates 0.1% false positives. *Lcm* is 5.5 and *Tc* is 0.2 s. Five independent simulations of 1000 trajectories. Data are represented as means ± SEM. (**C**) Effect of *Lcm* on the percentage of detected confined displacements in trajectories simulated to be confined for 50 of the 100 frames. *Sm* = 15, *Tc* = 0.2 s, *Rconf* = 0.04 μm. Five independent simulations of 1000 trajectories. Data are represented as means ± SD. (**D**) Effect of *Sm* on correct confinement detection in trajectories simulated to be confined 15 or 50 of the 100 frames. *Lcm* = 5.5 and *Tc* = 0.2 s. One hundred simulated trajectories per condition. (**E**) Effect of *Sm* on the ability to discriminate between multiple confinement zones shortly after each other. *Lcm* = 5.5 and *Rconf* = 0.03 μm, 100 trajectories. Simulated confined periods are indicated with the gray boxes. Data are represented as means ± SEM. (**F**) Negative effect of 'α * average *L*' as *Lc*. Single simulated trajectory with three confinement zones. The middle confinement zone does not reach the threshold. *Rconf* = 0.015 μm and 0.03 μm and α = 0.5. Simulated confined periods are indicated with the gray boxes. (**G**) Positive effect of 'α * average *L*' as *Lc*. Single simulated trajectory with two confinement zones. With *Lcm* as threshold, both confinement zones would have been merged into one. *Rconf* = 0.03 μm and α = 0.5. Simulated confined periods are indicated with the gray boxes.

3.3. Estimated Diffusion Coefficient Can Be Influenced by Track Length

One important parameter in the confinement index formula is the diffusion coefficient. This is the diffusion coefficient of the molecule if it could move freely. However, setting this diffusion coefficient is not trivial. One approach is to use the maximum instantaneous diffusion coefficient (D_{inst}) per trajectory, assuming that the molecule undergoes Brownian diffusion during at least part of the trajectory [21]. To calculate the maximum D_{inst} of a trajectory, we need to determine how the diffusion coefficient changes along the trajectory. Therefore, the D_{inst} was estimated based on the linear part of the MSD versus time lag curve for each window of ten steps over the complete trajectory and the maximum D_{inst} was used as the D of that trajectory. For this short window, the uncertainty in estimating the diffusion coefficient could be relatively high, but it allows for the detection of changes in the diffusion coefficient within individual trajectories. Another approach would be to test against a set diffusion coefficient chosen a priori that would resemble the Brownian diffusion of the molecule of interest. This value could for instance be deduced from independent tracking experiments, taking the diffusion coefficient of a freely moving population of trajectories as a reference value [22,28,29].

We investigated the effect of the different approaches in defining the Brownian diffusion coefficient on the false positive rate and confinement detection. For simulations, it is possible to test against a set diffusion coefficient and we compared this approach (Dset) to using the maximum instantaneous diffusion coefficient (Dmax). An interesting observation was that the false positive rate depended on the length of the simulated tracks for the Dmax approach whereas this was stable with Dset (Figure 3A,B). This was caused by a higher Dmax for the longer tracks, probably because there is a higher chance that there is a peak in the diffusion coefficient over time when there are more steps in the trajectory (Figure 3C). Similarly, in confined tracks with the same confinement zone but a longer total length, the confinement index during the period of confinement was higher with the Dmax method, whereas with Dset, the confinement index was similar for all track lengths (Figure 3D,E). This difference could also be explained by the higher Dmax values for the longer tracks (Figure 3F). To improve the Dmax method, we tested the effect of including more steps in the sliding window for the estimation of D_{inst}. That resulted in a lower Dmax because the peaks are averaged out. However, Dmax still increases with longer track lengths (Figure 3G). A third option (Dinst) would be to use the estimated instantaneous diffusion coefficient per track as the D in the confinement formula as used by Simson et al. [11]. However, although based on only the first three points of the MSD versus time lag curve, Dinst is heavily influenced by confinement in the trajectory, whereas Dmax only declined when the largest part of the trajectory is confined (Figure 3H). Thus, using D_{inst} per trajectory negatively affects the ability to detect transient confinement. Particularly, when a track was confined for more than 50% of the time, the power of detection dropped dramatically (Figure 3I). To conclude, using a set diffusion coefficient as D in the confinement formula appears to be the most reliable method, however this is under the assumption that the Brownian diffusion coefficient of the molecule is known. The alternative method would be using the Dmax, where the diffusion of the whole trajectory is tested against the fastest segment in the trajectory, thus investigating changes in diffusion behavior over the course of the trajectory.

3.4. Detection Limits in the Confinement Analysis

Next, we wanted to understand the detection limits of the confinement analysis to interpret the results correctly. Can similarly sized confinement zones still be detected for molecules with different diffusion coefficients? Purely based on the theoretical formula behind the confinement analysis, it is only possible to detect a large confinement zone if the diffusion coefficient is high enough and the time is long enough (Figure 4A,B). Only then will the confinement index L still reach the threshold when the confinement radius is large. In other words, the ratio between the possible explored area of a random diffusing molecule ($D \times t$)—with given diffusion coefficient and time—and the observed explored area (R^2) should be high enough to be considered confined (Figure 4C). To study the limits of the confinement analysis in detecting confinement zones of specific sizes, we simulated confined tracks with varying diffusion coefficients (0.01–1 $\mu m^2/s$) and confinement radii (*Rconf*, 20–200 nm). Next, we ran the transient confinement analysis on these simulations and found the fraction of confinement zones that were detected and their detected confinement radius (Figure 4D,E). It is clear from the results that the larger confinement zones cannot be detected in tracks with low diffusion coefficients. Such a molecule with a low diffusion coefficient could reside in a larger area without deviating from a Brownian molecule in their behavior.

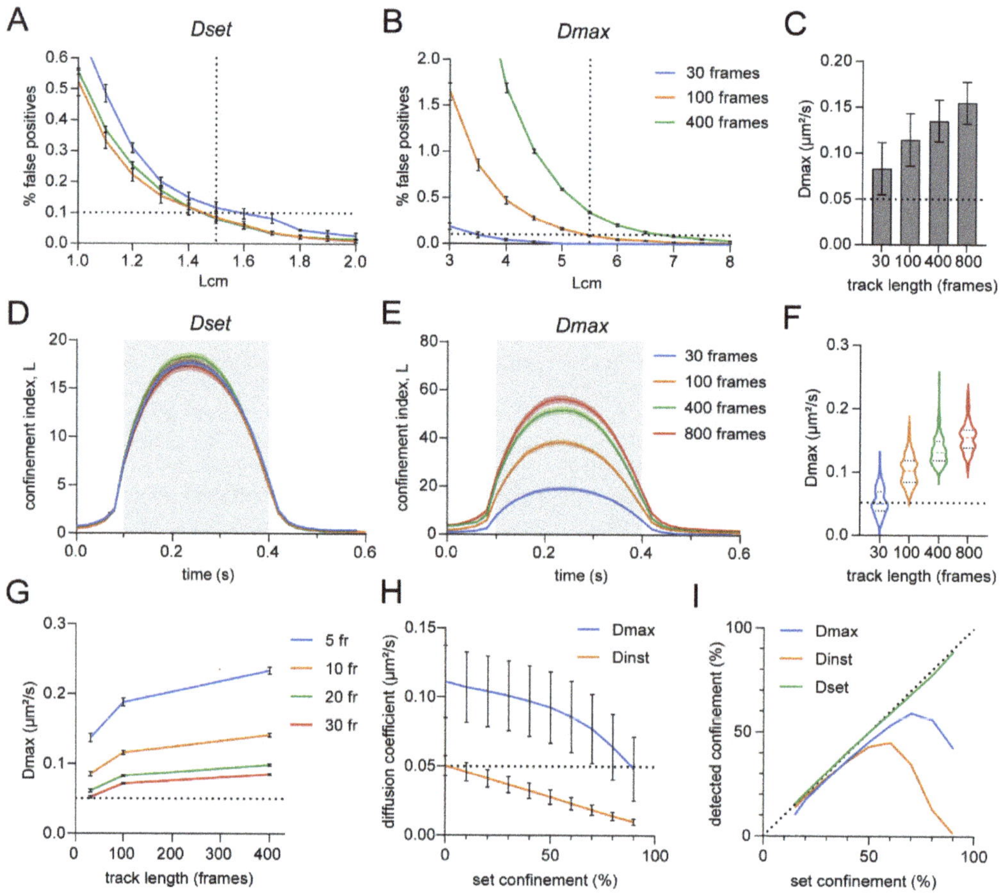

Figure 3. Effect of diffusion coefficient in the confinement index formula on confinement detection. (**A**,**B**) Effect of track length on the percentage of false positives in random walks using Dset (**A**) or Dmax (**B**) as diffusion coefficient in the confinement index formula. Five independent simulations of 1000 trajectories per condition. Data are represented as means ± SEM. (**C**) Effect of track length on Dmax. One hundred simulated random walks per condition. Data are represented as means ± SD. (**D**,**E**) Effect of track length on the confinement index over time using Dset (**D**) or Dmax (**E**) as diffusion coefficient in the confinement index formula. 100 confined trajectories per condition, $Rconf$ = 0.02 µm. Data are represented as means ± SEM. (**F**) Violin plots of Dmax values from (**E**). (**G**) Effect of the length of the segments used for estimating D_{inst} within trajectories on Dmax. (**H**) Effect of the percentage of confinement within a trajectory on the estimated diffusion coefficient of the whole trajectory (Dinst) or the Dmax. One thousand trajectories per condition, $Rconf$ = 0.04 µm. Data are represented as means ± SD. (**I**) Effect of the percentage of confinement within a trajectory on the detected confinement within the trajectory using Dmax, Dinst or Dset. Dotted line indicates set confinement = detected confinement. One thousand trajectories per condition, $Rconf$ = 0.04 µm.

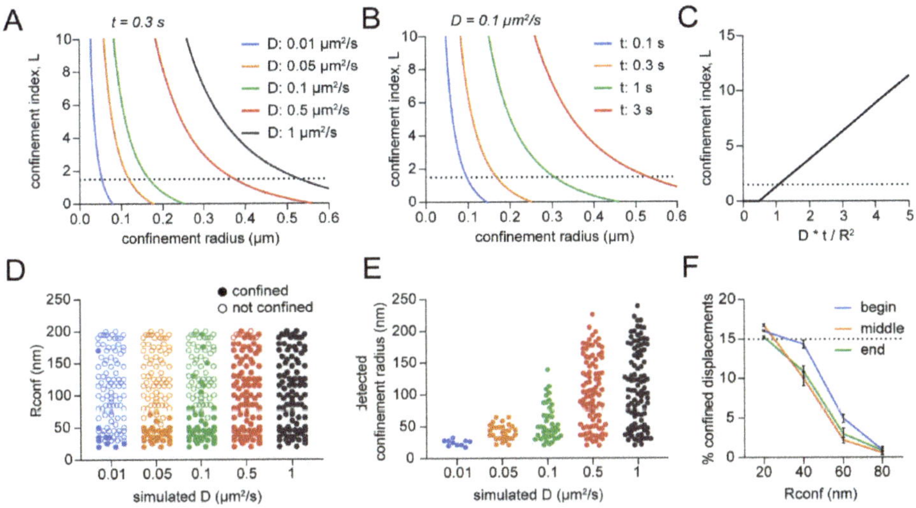

Figure 4. Detection limits of the confinement analysis. (**A**) Theoretical effect of the confinement radius and diffusion coefficient on the confinement index that can be detected. Observing time = 16 frames or 0.3 s. Dotted line indicates *Lcm* of 1.5 as used with Dset. (**B**) Theoretical effect of the confinement radius and observing time on the confinement index that can be detected. $D = 0.1\ \mu m^2/s$. (**C**) Theoretical relation between $D \times t/R^2$ and the confinement index. (**D**) Effect of the diffusion coefficient in trajectories on the ability to detect a range of different simulated confinement radii (*Rconf*). (**E**) Effect of the diffusion coefficient in trajectories on the detected confinement radius given the same range of confinement radii in (**D**) as input for the simulations. (**F**) Effect of the timing of simulated confinement periods and *Rconf* on the correct detection of confinement. Fifteen frames confined of total 100 frames per simulated track. Five independent simulations of 100 trajectories per condition. Dotted line indicates correct confinement detection. Data are represented as means ± SD.

Additionally, the first and last points in the confinement profile of a track are averaged over a lower number of segments. Therefore, we tested whether the timing of a confinement zone would affect the power of detection. We simulated confinement zones with varying radii and duration of 15 or 50 frames at the beginning, middle, and end of the track, and observed no considerable differences in the detected confined displacements (Figures 4F and S4). Thus, the detection of confinement in this analysis is not limited by the timing of the confinement periods.

3.5. Influence of Localization Error and Frame Rate on Confinement Detection Accuracy

In single-molecule tracking analysis, the first step is to localize the sub-pixel position of fluorescence emission events of single molecules. Generally, the point spread function of single-molecule emission spots can be fitted with a two-dimensional Gaussian function. The uncertainty inherent to the fitting routine, or localization error, influences the accuracy of the detected single-molecule trajectories. To investigate the effect of the localization error on the performance of the confinement detection, we simulated random walks and confined tracks with no error, 20 nm, and 50 nm localization error. We found that in random walks, a larger localization error resulted in fewer false positives (Figures 5A and S5A). Thus, the *Lcm* threshold could be lowered to achieve the same detection precision. This can be explained by R^2 being, on average, larger with a higher localization error. However, a higher localization error also resulted in a dramatic decrease in the detection of both short and longer confinement periods, even with the lowered Lcm threshold (Figures 5B,C and S5B,C).

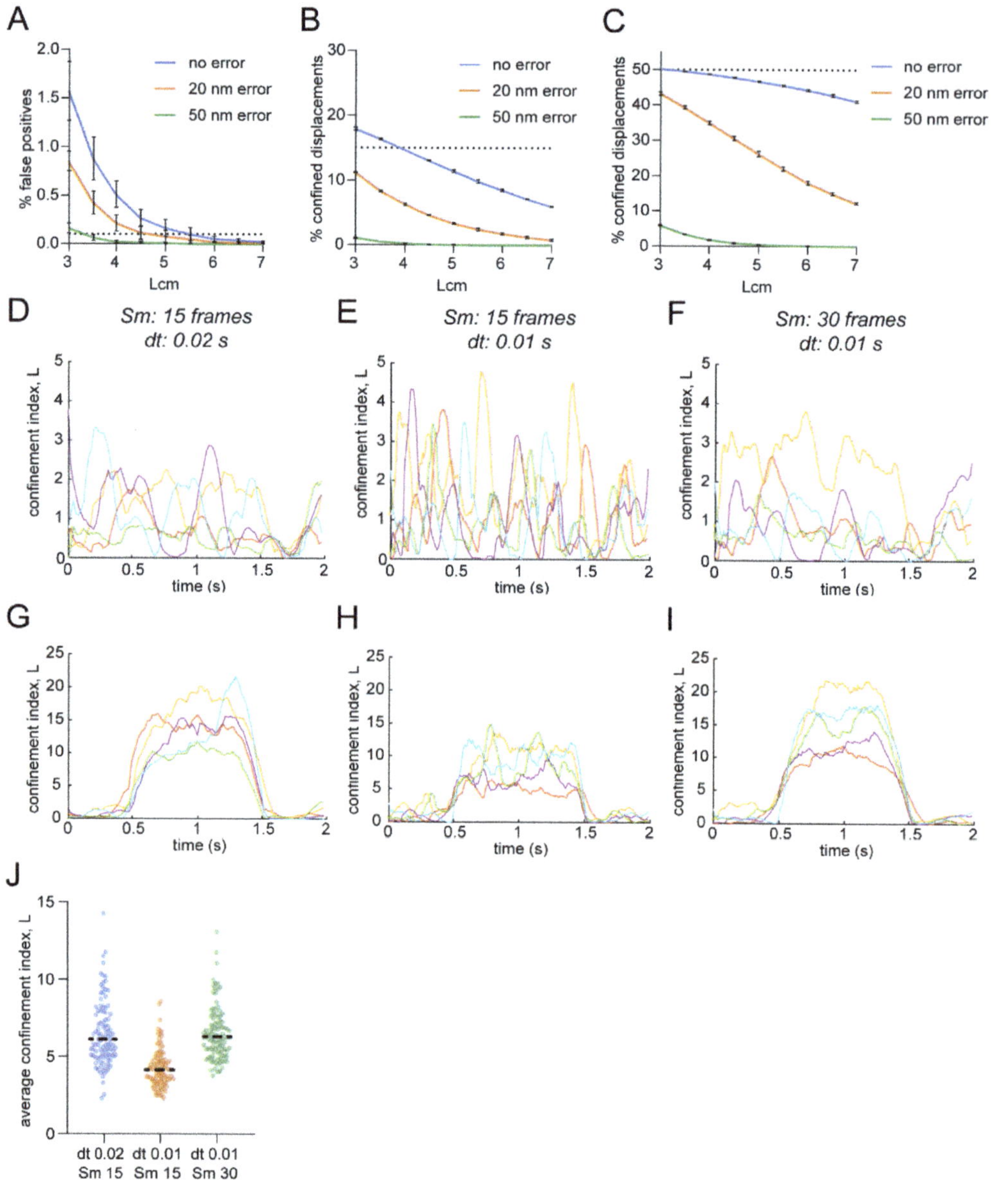

Figure 5. Influence of experimental parameters on confinement detection. (**A**) Effect of localization error on the percentage of false positives detected in random walks. Three independent simulations of 1000 trajectories per condition. Data are represented as means ± SD. (**B**,**C**) Effect of localization error on the percentage of detected confined displacements in tracks simulated to be confined for 15 (**B**) or 50 (**C**) of the 100 frames. *Rconf* = 0.04 µm. Three independent simulations of 1000 trajectories. Data are represented as means ± SD. (**D**–**F**) Confinement index of 5 random trajectories over time with (**D**) *Sm* = 15, *dt* = 0.02 s and 100 frames, (**E**) *Sm* = 15, *dt* = 0.01 s and 200 frames, (**F**) *Sm* = 30, *dt* = 0.01 s and 200 frames. (**G**–**I**) Confinement index of 5 confined trajectories over time with (**G**) *Sm* = 15, *dt* = 0.02 s and 100 frames, (**H**) *Sm* = 15, *dt* = 0.01 s and 200 frames, (**I**) *Sm* = 30, *dt* = 0.01 s and 200 frames. *Rconf* = 0.04 µm. (**J**) Corresponding average confinement index values of (**G**–**I**).

Another experimental parameter we considered is the frame rate of the acquisition that determines the time resolution in the single-molecule trajectories. It is important to adjust the parameters of the analysis accordingly, especially when comparing different experiments with different experimental parameters. Using the same maximum segment window with a smaller interval time will lead to narrower peaks in random walks (Figure 5D,E). Moreover, for a similar long period of confinement, the average confinement index will be lower with a smaller interval time and the same segment window (Figure 5G,H). Changing the interval time can be corrected by adjusting the maximum segment window (*Sm*) accordingly to obtain a more similar confinement index over time profile (Figure 5F,I,J).

3.6. Spatial Mapping of Transient Confinement of Membrane Probes

Next, we wanted to test the robustness of the confinement analysis on experimental single-molecule tracking data. We therefore focused on the neuronal membrane, where the dynamic behavior of individual membrane components collectively contributes to the efficient transfer of synaptic signals [41]. Subsynaptic domains enriched in glutamate receptors and scaffolding molecules were found to be aligned with the presynaptic glutamate release to optimize synaptic transmission, which emphasizes the importance of the heterogeneous organization of the neuronal membrane [42,43]. First, we developed a visualization tool to map the detected confinement zones resulting in a spatial heatmap of confinement hotspots. Such a heatmap can be displayed together with other cellular markers to locate confinement hotspots relative to specific subcellular domains. To create such a heatmap, we plotted every confinement zone as a Gaussian with amplitude 1 and the radius of the confinement zone as the FWHM (Figure 6A). Overlapping confinement zones would result in higher amplitude values, resulting in a heatmap that is color-coded for the amplitude values. Neurons were transfected with GFP-GPI, and we performed single-molecule tracking experiments using an anti-GFP nanobody coupled to Atto647N (Figure 6B). After tracking the molecules, we performed the transient confinement analysis on the mobile trajectories to obtain information on the location and timing of confinement zones in experimental trajectories (Figure 6C,D). Finally, we plotted the trajectories and confinement heatmap in combination with a marker of excitatory synapses (Homer1c-mCherry) (Figure 6E–J). We observed confinement of GFP-GPI in the axon, dendrite, and spines, where confinement does not seem to be specifically enriched at synaptic locations. Overall, this visualization tool allows for the detection of areas with high levels of transient confinement in a cellular context, by mapping the confinement zones relative to cellular markers.

To address the effect of changing parameters in the confinement analysis on the resulting confinement measures in experimental datasets, we applied the confinement analysis on a single-molecule tracking acquisition of GFP-GPI in the neuronal membrane. Lowering the *Lcm* increased the detected confinement radius, the diffusion coefficient inside confinement zones, and resulted in more trajectories with detected confinement (number of tracks with confinement for *Lcm* 3: 381, *Lcm* 4: 309, *Lcm* 5.5: 219, Figures 7A and S6A). On the other hand, varying *Tc* did not affect the average detected confinement radius or the diffusion coefficient inside confinement zones, although with higher *Tc*, fewer trajectories with detected confinement were found because of the stricter thresholds (*Tc* 0.1: 311, *Tc* 0.2: 219, *Tc* 0.3: 139, Figures 7B and S6B). Lastly, a longer time window resulted in the detection of larger confinement zones and a higher diffusion coefficient inside confinement zones, but did not noticeably affect the number of trajectories with detected confinement (*Sm* 5: 181, *Sm* 15: 219, *Sm* 30: 177, Figures 7C and S6C).

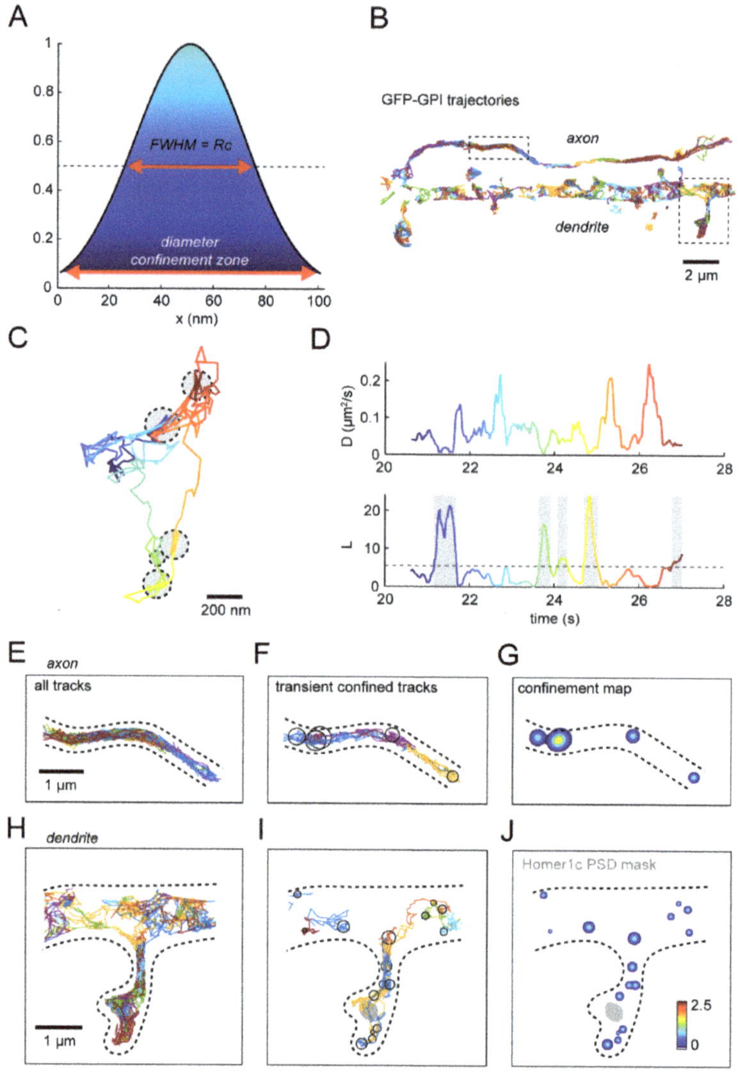

Figure 6. Spatial visualization of confinement. (**A**) Gaussian function used in confinement map with amplitude 1. FWHM is the confinement radius. Color-coded for amplitude. (**B**) GFP-GPI anti-GFP nanobody Atto647N trajectories in axon and dendrite. Boxes mark location of zooms in (**E–J**). Scale bar, 2 µm. (**C**) Example experimental GPI trajectory with 5 confinement zones indicated with the grey circles. Color-coded for time. Scale bar, 200 nm. (**D**) Diffusion coefficient (top) and confinement index (bottom) over time of the experimental trajectory in (**C**). Color-coded for time and the detected confined periods are indicated with the gray boxes. (**E–J**) Zoom of trajectories in the axon (**E–G**) and dendrite (**H–J**). (**E,H**) All trajectories displayed in different colors. (**F,I**) Trajectories with confined periods. Confinement zones are indicated with the black circles. (**G,J**) Confinement map indicating areas with multiple confinement zones close to or on top of each other. Homer1c PSD mask is indicated in gray in the dendrite. Scalebar, 1 µm.

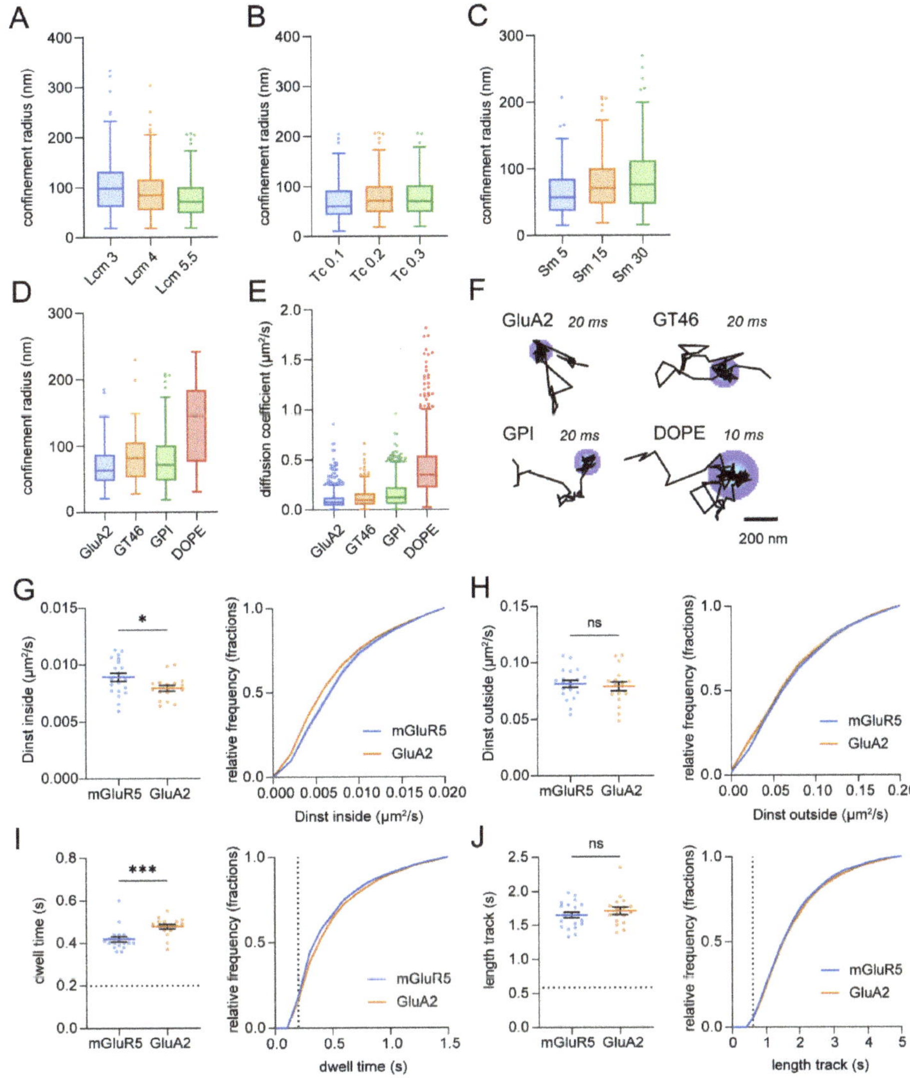

Figure 7. Confinement analysis on experimental trajectories. (**A–C**) Effect of user-defined parameters on the detected confinement radius. (**A**) Vary critical *L*, *Lcm*. Number of trajectories with confinement: *Lcm* 3: 381, *Lcm* 4: 309, *Lcm* 5.5: 219. (**B**) Vary critical time, *Tc*. Number of trajectories with confinement: *Tc* 0.1: 311, *Tc* 0.2: 219, *Tc* 0.3: 139. (**C**) Vary maximum segment length, *Sm*. Number of trajectories with confinement: *Sm* 5: 181, *Sm* 15: 219, *Sm* 30: 177. (**D–F**) Confinement zones detected for membrane probes: GluA1, GT46, GPI, and DOPE. (**D**) Detected confinement radius for each probe. (**E**) Estimated diffusion coefficients for each probe. (**F**) Example trajectories with their detected confinement zone. Interval times during the acquisition are indicated. Scale bar, 200 nm. (**G–J**) (**G**) Average estimated diffusion coefficient inside confinement zones, (**H**) average estimated diffusion coefficient outside confinement zones, (**I**) average dwell time, (**J**) average total track length for mGluR5 ($n = 20$) and GluA2 ($n = 17$; unpaired t-test) (left), and corresponding cumulative frequency distribution (right). (**I**) Dotted line indicates dwell time threshold. (**J**) Dotted line indicates minimum track length threshold. Data are represented as means ± SEM. * $p < 0.05$, *** $p < 0.001$, ns, $p > 0.05$.

Furthermore, we wanted to investigate the possibility to compare confinement of different membrane probes with different diffusional properties. Therefore, we compared the confinement behavior of (1) the AMPAR subunit GluA2 with (2) an artificial protein with single-spanning transmembrane domain (GT46), (3) GFP-GPI, and (4) the phospholipid DOPE in neurons. We detected transient confinement in the trajectories of all four membrane probes, but considerably larger confinement zones were detected for DOPE in comparison to GluA2 (Figure 7D). This difference could be explained by their difference in diffusion as DOPE has a higher diffusion coefficient (Figure 4, Figures 7E,F, and S6D). The difference in size of the confinement zones related to their difference in diffusion coefficient for all probes, except GPI which diffused slightly faster than GT46 but revealed slightly smaller confinement zones. In conclusion, comparison of confinement zone properties should be approached with caution when the diffusion coefficients of the probes differ considerably.

Lastly, we aimed to compare the transient confinement properties of two glutamate receptors that have comparable diffusion coefficients: the metabotropic glutamate receptor 5 (mGluR5) and the AMPA receptor subunit GluA2 (Figure S6E). These receptors both reside in the postsynaptic membrane, but GluA2 is found to be more enriched in the synapse whereas mGluR5 is concentrated in an area around the synapse [44,45]. We found that the diffusion coefficient of GluA2 inside confinement zones was lower than for mGluR5 (mGluR5: 0.00891 ± 0.00035 $\mu m^2/s$ and GluA2: 0.00793 ± 0.00025 $\mu m^2/s$; Figure 7G), while the diffusion coefficient outside confinement zones was not significantly different (mGluR5: 0.0813 ± 0.0030 $\mu m^2/s$ and GluA2: 0.0789 ± 0.0040 $\mu m^2/s$; Figure 7H). In addition, the time GluA2 resided in confinement zones was longer than mGluR5 (mGluR5: 0.419 ± 0.012 s and GluA2: 0.477 ± 0.010 s; Figure 7I), while the total track length was not significantly different (mGluR5: 1.65 ± 0.043 s and GluA2: 1.71 ± 0.054 s; Figure 7J). These results suggest that there are different mechanisms responsible for the confinement of mGluR5 and GluA2.

4. Discussion

Single-molecule tracking is a powerful approach to investigate the heterogeneous organization of cellular membranes. However, precise and reliable analysis of single-molecule tracking data remains challenging. This study set out to gain a better understanding of the detection of transient confinement in single-molecule trajectories.

We found that a considerable factor in the confinement analysis is the diffusion coefficient that is used in the formula to define the diffusion coefficient of the molecule when it would move freely. However, defining this free diffusion coefficient a priori is not straightforward. We therefore compared three different methods: using the maximum instantaneous diffusion coefficient per trajectory based on shorter segments [21], the instantaneous diffusion coefficient per trajectory [11], and defining a constant diffusion coefficient for the molecule of interest considered to be the free diffusion coefficient [22,25,26]. It is important to note the limitations of all these methods and their consequences on the results. The Dmax could be an overestimation of the free diffusion coefficient and therefore lead to an overestimation of the detected confinement in that trajectory and in addition, the value of Dmax seems to depend on the length of the track. However, it does allow for detecting changes in diffusion behavior over the trajectory. Estimating the instantaneous diffusion coefficient per trajectory could be heavily influenced by the percentage of confinement in the trajectory and therefore lead to reduced power of detection. Using a set diffusion coefficient is another approach of using the confinement analysis as with this method not only transient confinement would be detected but also completely confined tracks, as all the tracks are tested against the same diffusion coefficient and thus no longer for a temporal deviation within the trajectory. This heavily depends on choosing the free diffusion coefficient in a correct way. Previous studies used this approach, but this was under the assumption that the diffusion of these molecules was mainly Brownian [22,25,26]. This should be done with caution as we showed that the percentage of confinement in a track could influence the Dinst. It is important to note that when using this approach in the case of comparing confinement after, for instance, an experimental treatment, that the difference may solely come from a change in the diffusion coefficient and does not perse reflect a true difference in the confinement behavior.

One of the interesting quantitative outcomes of the confinement analysis is the size of a confinement zone. We found, however, that the maximal size that can be detected, depends on the diffusion coefficient of the molecule. Thus, comparing confinement of molecules with varying diffusion coefficients could lead to erroneous conclusions about differences in their confinement zone sizes. It is also not directly evident from the data which biological mechanism underlies these confined periods, and different mechanisms might also result in different confinement measures. Phase separation into liquid-ordered and liquid-disordered phases in model membranes arises mainly from the combination of lipid composition and temperature. Biological membranes are even more complex, both in the heterogeneity of lipids and involvement of membrane-associated proteins that affect the thermodynamics. Neurons and more specifically synapses are enriched in cholesterol and sphingolipids [46,47]. These lipids comprise the more ordered phase in the membrane and especially add up to the heterogeneous organization of biological membranes [48–50]. Furthermore, characteristics of confinement zones could be explained by compartmentalization by the actin-based membrane skeleton, molecular crowding, or binding to scaffold proteins [37–39].

Interestingly, we detected differences in the confinement behavior of glutamate receptors mGluR5 and GluA2, which suggests that different mechanisms are responsible for their confinement. It is known that these glutamate receptors show different distribution patterns within the postsynaptic membrane, GluA2 is concentrated in the postsynaptic density (PSD) whereas mGluR5 seems to be surrounding the PSD [51]. As our results reveal that GluA2 is trapped for longer time and moves slower, this could be an indication that GluA2 is trapped at a specific location mainly by scaffold molecules, supported by the finding that the actin cytoskeleton is absent from the PSD [52]. On the other hand, mGluR5 might be slowed down in its diffusion by other mechanisms such as steric hindrance or interactions with the cytoskeleton. Other methods for confinement analysis might help in detecting specific underlying mechanisms. For example, Meilhac et al. modified the method from Simson et al. to be able to detect jumps between compartments in the membrane [11,12]. Another way to characterize confinement are potential wells, describing a field of force resulting from molecular interactions, which has been suggested for synaptic receptors [18,40,41]. Lastly, this method could also be applied to less complex membranes to get a better estimate of the mechanism behind detected confinement behavior. For example, to study the effect of temperature on confined behavior and thus membrane organization.

5. Conclusions

This study provides a systematic analysis of the influence of the different parameters that are used for detecting temporal confinement in single-molecule trajectories and proposes a visualization tool to map confinement zones in the cellular context. To demonstrate the validity of the presented approach, we studied the diffusion of two glutamate receptor types in neurons and found that mGluR5 and GluA2 differ in their confinement behavior. We believe the presented results can guide future studies in the correct detection and interpretation of confinement analysis. As such, this study will help in better understanding the complex organization of the membrane by reliably detecting and spatially mapping confinement.

Supplementary Materials: The following supporting information can be downloaded at: https://www.mdpi.com/article/10.3390/membranes12070650/s1, Table S1: Confinement parameters and their effect on the power of detection; Figure S1: Random walk and confinement simulations; Figure S2: Improvements in the confinement analysis; Figure S3: Raw simulated transient confined trajectories from Figure 1; Figure S4: Effect of timing confinement on power of detection; Figure S5:

Effect of localization error on confinement detection using Dset; Figure S6: Confinement analysis on experimental trajectories. Reference [21] is cited in Supplementary Materials.

Author Contributions: Conceptualization, M.W. and H.D.M.; methodology, M.W. and H.D.M.; software, M.W. and H.D.M.; validation, M.W. and H.D.M.; formal analysis, M.W. and H.D.M.; investigation, M.W. and H.D.M.; resources, H.D.M.; data curation, M.W. and H.D.M.; writing—original draft preparation, M.W. and H.D.M.; writing—review and editing, M.W. and H.D.M.; visualization, M.W. and H.D.M.; supervision, H.D.M.; project administration, H.D.M.; funding acquisition, H.D.M. All authors have read and agreed to the published version of the manuscript.

Funding: This work was funded by the Netherlands Organization for Scientific Research (ALW-VIDI 171.029 to H.D.M.).

Institutional Review Board Statement: All experiments were approved by the Dutch Animal Experiments Committee (Dier Experimenten Commissie (DEC), work protocol project number: AVD1080020173404), performed in line with institutional guidelines of Utrecht University, and conducted in agreement with Dutch law (Wet op de Dierproeven, 1996) and European regulations (Directive 2010/63/EU).

Informed Consent Statement: Not applicable.

Data Availability Statement: The data that support the findings of this study are available on request from the corresponding author, H.D.M.

Acknowledgments: We would like to thank Marijn Siemons for his help with the simulations and all members of the MacGillavry lab for helpful discussions.

Conflicts of Interest: The authors declare no conflict of interest.

References

1. Jacobson, K.; Liu, P.; Lagerholm, B.C. The Lateral Organization and Mobility of Plasma Membrane Components. *Cell* **2019**, *177*, 806–819. [CrossRef] [PubMed]
2. Pike, L.J. Rafts defined: A report on the Keystone symposium on lipid rafts and cell function. *J. Lipid Res.* **2006**, *47*, 1597–1598. [CrossRef] [PubMed]
3. Kusumi, A.; Fujiwara, T.K.; Tsunoyama, T.A.; Kasai, R.S.; Liu, A.; Hirosawa, K.M.; Kinoshita, M.; Matsumori, N.; Komura, N.; Ando, H.; et al. Defining raft domains in the plasma membrane. *Traffic* **2020**, *21*, 106–137. [CrossRef] [PubMed]
4. MacGillavry, H.D.; Song, Y.; Raghavachari, S.; Blanpied, T.A. Nanoscale Scaffolding Domains within the Postsynaptic Density Concentrate Synaptic AMPA Receptors. *Neuron* **2013**, *78*, 615–622. [CrossRef] [PubMed]
5. Nair, D.; Hosy, E.; Petersen, J.D.; Constals, A.; Giannone, G.; Choquet, D.; Sibarita, J.-B. Super-Resolution Imaging Reveals That AMPA Receptors Inside Synapses Are Dynamically Organized in Nanodomains Regulated by PSD95. *J. Neurosci.* **2013**, *33*, 13204–13224. [CrossRef]
6. Crosby, K.C.; Gookin, S.E.; Garcia, J.D.; Hahm, K.M.; Dell'Acqua, M.L.; Smith, K.R. Nanoscale Subsynaptic Domains Underlie the Organization of the Inhibitory Synapse. *Cell Rep.* **2019**, *26*, 3284–3297.e3. [CrossRef]
7. Specht, C.G.; Izeddin, I.; Rodriguez, P.C.; ElBeheiry, M.; Rostaing, P.; Darzacq, X.; Dahan, M.; Triller, A. Quantitative nanoscopy of inhibitory synapses: Counting gephyrin molecules and receptor binding sites. *Neuron* **2013**, *79*, 308–321. [CrossRef]
8. Sungkaworn, T.; Jobin, M.-L.; Burnecki, K.; Weron, A.; Lohse, M.J.; Calebiro, D. Single-molecule imaging reveals receptor-G protein interactions at cell surface hot spots. *Nature* **2017**, *550*, 543–547. [CrossRef]
9. Akin, E.J.; Solé, L.; Johnson, B.; El Beheiry, M.; Masson, J.-B.; Krapf, D.; Tamkun, M.M. Single-Molecule Imaging of Nav1.6 on the Surface of Hippocampal Neurons Reveals Somatic Nanoclusters. *Biophys. J.* **2016**, *111*, 1235–1247. [CrossRef]
10. Heck, J.; Parutto, P.; Ciuraszkiewicz, A.; Bikbaev, A.; Freund, R.; Mitlöhner, J.; Andres-Alonso, M.; Fejtova, A.; Holcman, D.; Heine, M. Transient Confinement of CaV2.1 Ca2+-Channel Splice Variants Shapes Synaptic Short-Term Plasticity. *Neuron* **2019**, *103*, 66–79.e12. [CrossRef]
11. Simson, R.; Sheets, E.D.; Jacobson, K. Detection of temporary lateral confinement of membrane proteins using single-particle tracking analysis. *Biophys. J.* **1995**, *69*, 989–993. [CrossRef]
12. Meilhac, N.; Le Guyader, L.; Salomé, L.; Destainville, N. Detection of confinement and jumps in single-molecule membrane trajectories. *Phys. Rev. E* **2006**, *73*, 011915. [CrossRef] [PubMed]
13. Renner, M.; Wang, L.; Levi, S.; Hennekinne, L.; Triller, A. A Simple and Powerful Analysis of Lateral Subdiffusion Using Single Particle Tracking. *Biophys. J.* **2017**, *113*, 2452–2463. [CrossRef] [PubMed]
14. Weihs, D.; Gilad, D.; Seon, M.; Cohen, I. Image-based algorithm for analysis of transient trapping in single-particle trajectories. *Microfluid. Nanofluidics* **2012**, *12*, 337–344. [CrossRef]
15. Rajani, V.; Carrero, G.; Golan, D.E.; de Vries, G.; Cairo, C.W. Analysis of Molecular Diffusion by First-Passage Time Variance Identifies the Size of Confinement Zones. *Biophys. J.* **2011**, *100*, 1463–1472. [CrossRef]

16. Sikora, G.; Wyłomańska, A.; Gajda, J.; Solé, L.; Akin, E.J.; Tamkun, M.M.; Krapf, D. Elucidating distinct ion channel populations on the surface of hippocampal neurons via single-particle tracking recurrence analysis. *Phys. Rev. E* **2017**, *96*, 062404. [CrossRef]
17. Lanoiselée, Y.; Grimes, J.; Koszegi, Z.; Calebiro, D. Detecting Transient Trapping from a Single Trajectory: A Structural Approach. *Entropy* **2021**, *23*, 1044. [CrossRef]
18. Hoze, N.; Nair, D.; Hosy, E.; Sieben, C.; Manley, S.; Herrmann, A.; Sibarita, J.-B.; Choquet, D.; Holcman, D. Heterogeneity of AMPA receptor trafficking and molecular interactions revealed by superresolution analysis of live cell imaging. *Proc. Natl. Acad. Sci. USA* **2012**, *109*, 17052–17057. [CrossRef]
19. Einstein, A. Über die von der molekularkinetischen Theorie der Wärme geforderte Bewegung von in ruhenden Flüssigkeiten suspendierten Teilchen. *Ann. Phys.* **1905**, *322*, 549–560. [CrossRef]
20. Saxton, M.J. Lateral diffusion in an archipelago. Single-particle diffusion. *Biophys. J.* **1993**, *64*, 1766–1780. [CrossRef]
21. Menchón, S.A.; Martín, M.G.; Dotti, C.G. APM_GUI: Analyzing particle movement on the cell membrane and determining confinement. *BMC Biophys.* **2012**, *5*, 4. [CrossRef] [PubMed]
22. Sergé, A.; Fourgeaud, L.; Hémar, A.; Choquet, D. Receptor Activation and Homer Differentially Control the Lateral Mobility of Metabotropic Glutamate Receptor 5 in the Neuronal Membrane. *J. Neurosci.* **2002**, *22*, 3910–3920. [CrossRef] [PubMed]
23. Orr, G.; Hu, D.; Özçelik, S.; Opresko, L.K.; Wiley, H.S.; Colson, S.D. Cholesterol Dictates the Freedom of EGF Receptors and HER2 in the Plane of the Membrane. *Biophys. J.* **2005**, *89*, 1362–1373. [CrossRef] [PubMed]
24. Wu, H.M.; Lin, Y.H.; Yen, T.C.; Hsieh, C.L. Nanoscopic substructures of raft-mimetic liquid-ordered membrane domains revealed by high-speed single-particle tracking. *Sci. Rep.* **2016**, *6*, 20542. [CrossRef]
25. Kovtun, O.; Torres, R.; Ferguson, R.S.; Josephs, T.; Rosenthal, S.J. Single Quantum Dot Tracking Unravels Agonist Effects on the Dopamine Receptor Dynamics. *Biochemistry* **2021**, *60*, 1031–1043. [CrossRef]
26. Neupert, C.; Schneider, R.; Klatt, O.; Reissner, C.; Repetto, D.; Biermann, B.; Niesmann, K.; Missler, M.; Heine, M. Regulated dynamic trafficking of neurexins inside and outside of synaptic terminals. *J. Neurosci.* **2015**, *35*, 13629–13647. [CrossRef]
27. Dietrich, C.; Yang, B.; Fujiwara, T.; Kusumi, A.; Jacobson, K. Relationship of Lipid Rafts to Transient Confinement Zones Detected by Single Particle Tracking. *Biophys. J.* **2002**, *82*, 274–284. [CrossRef]
28. Notelaers, K.; Rocha, S.; Paesen, R.; Smisdom, N.; De Clercq, B.; Meier, J.C.; Rigo, J.M.; Hofkens, J.; Ameloot, M. Analysis of $\alpha 3$ GlyR single particle tracking in the cell membrane. *Biochim. Biophys. Acta Mol. Cell Res.* **2014**, *1843*, 544–553. [CrossRef]
29. Meier, J.; Vannier, C.; Sergé, A.; Triller, A.; Choquet, D. Fast and reversible trapping of surface glycine receptors by gephyrin. *Nat. Neurosci.* **2001**, *4*, 253–260. [CrossRef]
30. Bürli, T.; Baer, K.; Ewers, H.; Sidler, C.; Fuhrer, C.; Fritschy, J.M. Single Particle Tracking of $\alpha 7$ Nicotinic AChR in Hippocampal Neurons Reveals Regulated Confinement at Glutamatergic and GABAergic Perisynaptic Sites. *PLoS ONE* **2010**, *5*, e11507. [CrossRef]
31. Syed, A.; Zhu, Q.; Smith, E.A. Ligand binding affinity and changes in the lateral diffusion of receptor for advanced glycation endproducts (RAGE). *Biochim. Biophys. Acta-Biomembr.* **2016**, *1858*, 3141–3149. [CrossRef] [PubMed]
32. Borgdorff, A.J.; Choquet, D. Regulation of AMPA receptor lateral movements. *Nature* **2002**, *417*, 649–653. [CrossRef] [PubMed]
33. Wong, W.C.; Juo, J.-Y.; Lin, C.-H.; Liao, Y.-H.; Cheng, C.-Y.; Hsieh, C.-L. Characterization of Single-Protein Dynamics in Polymer-Cushioned Lipid Bilayers Derived from Cell Plasma Membranes. *J. Phys. Chem. B* **2019**, *123*, 6492–6504. [CrossRef] [PubMed]
34. Kapitein, L.C.; Yau, K.W.; Hoogenraad, C.C. *Microtubule Dynamics in Dendritic Spines. Methods in Cell Biology*, 1st ed.; Elsevier Inc.: Amsterdam, The Netherlands, 2010; Volume 97, pp. 111–132. ISBN 9780123813497.
35. Scheefhals, N.; Catsburg, L.A.E.; Westerveld, M.L.; Blanpied, T.A.; Hoogenraad, C.C.; MacGillavry, H.D. Shank Proteins Couple the Endocytic Zone to the Postsynaptic Density to Control Trafficking and Signaling of Metabotropic Glutamate Receptor 5. *Cell Rep.* **2019**, *29*, 258–269.e8. [CrossRef]
36. Albrecht, D.; Winterflood, C.M.; Ewers, H. Dual color single particle tracking via nanobodies. *Methods Appl. Fluoresc.* **2015**, *3*, 024001. [CrossRef]
37. Eggeling, C.; Ringemann, C.; Medda, R.; Schwarzmann, G.; Sandhoff, K.; Polyakova, S.; Belov, V.N.; Hein, B.; Von Middendorff, C.; Schönle, A.; et al. Direct observation of the nanoscale dynamics of membrane lipids in a living cell. *Nature* **2009**, *457*, 1159–1162. [CrossRef]
38. Willems, J.; de Jong, A.P.H.; Scheefhals, N.; Mertens, E.; Catsburg, L.A.E.; Poorthuis, R.B.; de Winter, F.; Verhaagen, J.; Meye, F.J.; MacGillavry, H.D. Orange: A Crispr/Cas9-based genome editing toolbox for epitope tagging of endogenous proteins in neurons. *PLOS Biol.* **2020**, *18*, e3000665. [CrossRef]
39. Lu, H.E.; MacGillavry, H.D.; Frost, N.A.; Blanpied, T.A. Multiple Spatial and Kinetic Subpopulations of CaMKII in Spines and Dendrites as Resolved by Single-Molecule Tracking PALM. *J. Neurosci.* **2014**, *34*, 7600–7610. [CrossRef]
40. Golan, Y.; Sherman, E. Resolving mixed mechanisms of protein subdiffusion at the T cell plasma membrane. *Nat. Commun.* **2017**, *8*, 15851. [CrossRef]
41. Westra, M.; Gutierrez, Y.; MacGillavry, H.D. Contribution of Membrane Lipids to Postsynaptic Protein Organization. *Front. Synaptic Neurosci.* **2021**, *13*, 790773. [CrossRef]
42. Tang, A.H.; Chen, H.; Li, T.P.; Metzbower, S.R.; MacGillavry, H.D.; Blanpied, T.A. A trans-synaptic nanocolumn aligns neurotransmitter release to receptors. *Nature* **2016**, *536*, 210–214. [CrossRef] [PubMed]

43. Li, S.; Raychaudhuri, S.; Lee, S.A.; Brockmann, M.M.; Wang, J.; Kusick, G.; Prater, C.; Syed, S.; Falahati, H.; Ramos, R.; et al. Asynchronous release sites align with NMDA receptors in mouse hippocampal synapses. *Nat. Commun.* **2021**, *12*, 677. [CrossRef] [PubMed]
44. Luján, R.; Nusser, Z.; Roberts, J.D.B.; Shigemoto, R.; Somogyi, P. Perisynaptic location of metabotropic glutamate receptors mGluR1 and mGluR5 on dendrites and dendritic spines in the rat hippocampus. *Eur. J. Neurosci.* **1996**, *8*, 1488–1500. [CrossRef] [PubMed]
45. Nusser, Z.; Mulvihill, E.; Streit, P.; Somogyi, P. Subsynaptic segregation of metabotropic and ionotropic glutamate receptors as revealed by immunogold localization. *Neuroscience* **1994**, *61*, 421–427. [CrossRef]
46. Fitzner, D.; Bader, J.M.; Penkert, H.; Bergner, C.G.; Su, M.; Weil, M.-T.; Surma, M.A.; Mann, M.; Klose, C.; Simons, M. Cell-Type- and Brain-Region-Resolved Mouse Brain Lipidome. *Cell Rep.* **2020**, *32*, 108132. [CrossRef]
47. Breckenridge, W.C.; Gombos, G.; Morgan, I.G. The lipid composition of adult rat brain synaptosomal plasma membranes. *Biochim. Biophys. Acta Biomembr.* **1972**, *266*, 695–707. [CrossRef]
48. Valsecchi, M.; Mauri, L.; Casellato, R.; Prioni, S.; Loberto, N.; Prinetti, A.; Chigorno, V.; Sonnino, S. Ceramide and sphingomyelin species of fibroblasts and neurons in culture. *J. Lipid Res.* **2007**, *48*, 417–424. [CrossRef]
49. Mainali, L.; Raguz, M.; Subczynski, W.K. Phase-Separation and Domain-Formation in Cholesterol-Sphingomyelin Mixture: Pulse-EPR Oxygen Probing. *Biophys. J.* **2011**, *101*, 837–846. [CrossRef]
50. Balleza, D.; Mescola, A.; Marín–Medina, N.; Ragazzini, G.; Pieruccini, M.; Facci, P.; Alessandrini, A. Complex Phase Behavior of GUVs Containing Different Sphingomyelins. *Biophys. J.* **2019**, *116*, 503–517. [CrossRef]
51. Scheefhals, N.; MacGillavry, H.D. Functional organization of postsynaptic glutamate receptors. *Mol. Cell. Neurosci.* **2018**, *91*, 82–94. [CrossRef]
52. Sidenstein, S.C.; D'Este, E.; Böhm, M.J.; Danzl, J.G.; Belov, V.N.; Hell, S.W. Multicolour multilevel STED nanoscopy of actin/spectrin organization at synapses. *Sci. Rep.* **2016**, *6*, 26725. [CrossRef] [PubMed]

membranes

MDPI

Article

Stabilization of *Cereibacter sphaeroides* Photosynthetic Reaction Center by the Introduction of Disulfide Bonds

Georgii Selikhanov [1,2], Anastasia Atamas [1], Diana Yukhimchuk [1], Tatiana Fufina [2], Lyudmila Vasilieva [2] and Azat Gabdulkhakov [1,*]

1 Institute of Protein Research, Russian Academy of Sciences, Institutskaya 4, 142290 Pushchino, Moscow Region, Russia
2 Federal Research Center Pushchino Scientific Center for Biological Research PSCBR, Institute of Basic Biological Problems, Russian Academy of Sciences, Institutskaya 2, 142290 Pushchino, Moscow Region, Russia
* Correspondence: azat@vega.protres.ru

Abstract: The photosynthetic reaction center of the purple nonsulfur bacterium *Cereibacter sphaeroides* is a useful model for the study of mechanisms of photoinduced electron transfer and a promising component for photo-bio-electrocatalytic systems. The basic research and technological applications of this membrane pigment-protein complex require effective approaches to increase its structural stability. In this work, a rational design approach to genetically modify the reaction centers by introducing disulfide bonds is used. This resulted in significantly increasing the thermal stability of some of the mutant pigment-protein complexes. The formation of the S-S bonds was confirmed by X-ray crystallography as well as SDS-PAGE, and the optical properties of the reaction centers were studied. The genetically modified reaction centers presented here preserved their ability for photochemical charge separation and could be of interest for basic science and biotechnology.

Keywords: protein stabilization; disulfide bonds; photosynthetic reaction center; integral membrane protein; site-directed mutagenesis; crystal structure

Citation: Selikhanov, G.; Atamas, A.; Yukhimchuk, D.; Fufina, T.; Vasilieva, L.; Gabdulkhakov, A. Stabilization of *Cereibacter sphaeroides* Photosynthetic Reaction Center by the Introduction of Disulfide Bonds. *Membranes* **2023**, *13*, 154. https://doi.org/10.3390/membranes13020154

Academic Editors: Yosuke Senju and Shiro Suetsugu

Received: 29 December 2022
Revised: 19 January 2023
Accepted: 23 January 2023
Published: 25 January 2023

1. Introduction

The photosynthetic reaction center (RC) of the purple bacterium *Cereibacter sphaeroides* (recently renamed from *Rhodobacter sphaeroides*) is a photosensitive membrane pigment-protein complex that serves as a convenient and informative test model for studying the mechanisms of electron transfer, photosynthesis, and pigment-protein interactions. It is a well-studied, relatively stable integral membrane protein (IMP) with established protocols for its expression, isolation from membranes, and purification. The RC from *C. sphaeroides* consists of three protein subunits and ten cofactors arranged in two membrane-spanning branches, A and B. The RC contains two bacteriochlorophylls (BChl), P_A and P_B, combined into a special pair P, two monomeric BChls, B_A and B_B, two bacteriopheophytins (BPhe), H_A and H_B, two quinones, a non-heme Fe atom, and a spheroidene molecule [1]. This bacterial reaction center shares considerable similarities with the photosystem II of plants, algae, and cyanobacteria [2] and has therefore served for many years as a structural and functional model for the study of the more complex photosystem II.

One of the most common methods to study the mechanisms of photochemical processes in the reaction centers is via amino acid substitutions. This method allows new data to be obtained, but such substitutions can often lead to decreased structural stability of the protein [3]. Studies on mutant forms of the reaction center with reduced stability prove difficult due to the higher denaturation rate of the purified complexes, especially when X-ray protein crystallography is to be used, where the stability of the macromolecule under study is one of the main factors for successful crystallization [4]. Increasing the stability of complexes by introducing compensatory mutations that do not affect their function is one of the possible methods to study such objects.

Another aspect where the overall stability of the reaction center is of great importance is the potential technological applications in the field of photo-bioelectrocatalysis. Both isolated reaction centers and intact cells of purple bacteria are promising catalysts that convert light energy into chemical energy in charge-separated states. Primary charge separation and subsequent electron transfer across photosynthetic membranes occur with a quantum yield approaching 100% [5]. However, to make photo-bioelectrochemical systems appealing to industry, it is critical to improve the stability of their components to enable long-term application and to develop cost-effective systems (both in terms of their assembly and operation) [6].

In this regard, it is important to have effective approaches to improving the structural stability of reaction centers to facilitate their basic studies and industrial applications. Recently, the effect of various detergents and osmolytes on the thermal stability of bacterial RC complexes has been studied [7]. Sodium cholate has been shown to have a significant stabilizing effect on the structure of native and genetically modified RCs. However, it is not always possible to change the storage conditions of the reaction centers due to the specifics of the experiment or biotechnological process in which the complex is used.

H-bond networks have been shown to play an important role in the stability of reaction centers [8]. For *C. sphaeroides* RC, the H-bond of the acetyl group of each of the bacteriochlorophylls in the dimer ensures the stability of the complex at elevated temperature and pressure [9,10]. The introduction of hydrogen bonds to stabilize the RC is possible but has two major drawbacks: (1) The complexity of hydrogen bond interactions in membrane complexes makes it difficult to design appropriate substitutions and predict their influence on the RC structure; (2) There are hydrogen bonding networks located near the electron transport cofactors of the RC, and their manipulations may affect the functional activity of the complex.

Here we present an approach for stabilization of the photosynthetic reaction center of *Cereibacter sphaeroides*, which is the formation of disulfide bonds between its α-helices. Investigations were undertaken into the possibility of S-S bond formation at different sites and also the effects of their introduction on the thermal stability and functional activity of the complex.

2. Materials and Methods

2.1. Mutagenesis

A genetic system for site-directed mutagenesis consisting of the *C. sphaeroides* DD13 strain [11] deficient in RC and antennae systems synthesis and the pRK plasmids described elsewhere [12] was used. Mutations were introduced into *puf*-operone using PCR oligonucleotides via the QuikChange plasmid mutagenesis protocol as described in [13]. The nucleotide changes were confirmed by DNA sequencing. The altered pufL and pufM genes were transferred into a broad-host-range vector: a derivative of pRK415 that contained a 4.2 kb EcoRI-HindIII restriction fragment and included the pufLMX genes [14]. The resulting plasmids were introduced into *C. sphaeroides* strain DD13 by conjugative crossing to produce transconjugant strains with RC-only phenotypes [11].

2.2. Bacterial Growth and Protein Samples Preparation

Growth of wild-type and mutant bacterial strains under dark, semiaerobic conditions was performed as previously described [14,15]. Cells were harvested and disrupted by ultrasonication; membranes for RC purification were then pelleted by ultracentrifugation. Reaction centers were solubilized from membranes with lauryldimethylamine oxide (LDAO) (Sigma-Aldrich, St. Louis, MO, USA) and purified using a polyhistidine tag (6X-His) attached to the carboxy terminus of the RC M subunit [16]. The purity of the reaction centers was estimated by absorbance spectroscopy, measuring the ratio of protein absorbance at 280 nm to bacteriochlorophyll absorbance at 802 nm (A280/A802; [17]). If the value of A280/A802 was less than 1.4, the RC sample was considered sufficiently pure for crystallization. Absorption spectra were recorded using a Shimadzu UV-1800 spectrophotometer (Shimadzu Corporation, Kyoto, Japan).

2.3. Studies on the Properties of Reaction Centers

Pigment extraction and pigment composition analysis of RCs were performed as previously described [18]. Thermal stability was investigated at 48 °C according to previous methodology [19], with the difference that 0.1% LDAO (Sigma-Aldrich, St. Louis, MO, USA) was used as detergent instead of 0.1% Triton X-100. The number of intact RCs in the sample was estimated by the absorption of monomeric BChl at 804 as reported [9]. The construction of the curves of absorption changes was carried out using the Origin software package (OriginLab Corporation, Northampton, MA, USA). Differential (light minus dark) absorption spectra were recorded at constant illumination with SZS-22 and KS-19 crossed light filters using a Shimadzu UV-1800 spectrophotometer (Shimadzu Corporation, Kyoto, Japan).

2.4. Polyacrylamide Gel Electrophoresis

Tris/MES SDS-polyacrylamide gel electrophoresis was chosen to confirm the presence of a disulfide bond between subunits in the mutant forms of RC [20], modified from [15]. It differs from conventional SDS polyacrylamide gel electrophoresis [21] by the addition of a higher percentage of polymer (acrylamide), the addition of MES (2-(N-morpholino)ethanesulfonic acid) (Sigma-Aldrich, St. Louis, MO, USA) to the buffer solution, and urea to the gels. These method alterations afford an increase in resolution. Otherwise, the protocol for setting up the experiment (polymerization of the gels, introduction of the samples, electrophoresis, staining, and washing of the gels) does not differ from the classical method. Comparatively mild denaturation conditions (30 °C, 60 min) for protein samples were used.

2.5. Crystallization and X-ray Diffraction Analysis

Protein crystallization was performed using vapor diffusion in a hanging drop with the addition of detergent or by *in meso* approach using a lipid sponge phase following the conditions used previously [22,23]. For the mutant forms L37Cys+L99Cys and L53Cys+L64Cys, we obtained trigonal crystals, space group $P3_121$; for the crystals of mutants L172Cys+L246Cys and M19Cys+L214Cys, space group $P4_12_12$; and for the crystals of the reaction center M84Cys+L278Cys, space group C2.

Samples of photosynthetic reaction centers suitable for crystallization were prepared as described [14,15]. Sample purity A280/A800 was <1.4. RC solutions with a protein concentration of 25–30 mg/mL were used.

Diffraction data for L37Cys+L99Cys, M19Cys+L214Cys, and M84Cys+L278Cys crystals were collected at the ID30A-3 beamline at the European Synchrotron Radiation Facility (ESRF), Grenoble, France [24], equipped with a Pilatus 6M detector (Dectris AG, Baden—Daettwill, Switzerland). Data collection was controlled by the MxCuBE system [25], and the strategy was calculated by BEST [26]. Data were processed and scaled using the XDS package [27].

Diffraction data for L53Cys+L64Cys and L172Cys+L246Cys crystals were collected using Proteum X8 (Bruker, Billerica, MA, USA) and XtaLAB Synergy-S (Rigaku Corporation, Tokyo, Japan) diffractometers, respectively.

Structures were solved by molecular replacement with Phaser [28], using the structure of the photosynthetic reaction center of *C. sphaeroides* strain RV [18] (PDB ID 3V3Y) as a search model. Water molecules were removed from the model. The initial model was refined using REFMAC5 [29]. Manual rebuilding of the model was performed in Coot [30]. Data statistics are summarized in Tables 1 and 2. The figures were prepared using PyMOL [31].

Table 1. Data collection and processing.

	L37Cys+L99Cys	L53Cys+L64Cys	L172Cys+L246Cys	M19Cys+L214Cys	M84Cys+L278Cys
Diffraction source	ESRF, beamline ID30A-3	Proteum X8 (Bruker)	XtaLAB Synergy-S (Rigaku)	ESRF, beamline ID30A-3	ESRF, beamline ID30A-3
Wavelength (Å)	0.9677	1.54178	1.54178	0.9677	0.9677
Temperature (K)	100	110	120	100	100
Detector	DECTRIS Eiger X 4M	PLATINUM135 CCD	HyPix-6000C	DECTRIS Eiger X 4M	DECTRIS Eiger X 4M
Rotation range per image (°)	0.15	0.5	0.5	0.05	0.1
Total rotation range (°)	150	180	120	50	200
Space group	$P3_121$	$P3_121$	$P4_12_12$	$P4_12_12$	C2
a, b, c (Å), α, β, γ, °	139.8 139.8 186.5 90 90 120	139.6 139.6 185.0 90 90 120	99.7 99.7 239.1 90 90 90	100.91 100.91 237.0 90 90 90	253.1 75.9 65.8 90.0 95.5 90.0
Resolution range (Å)	30.00–2.60 (2.67–2.60)	30.00–2.85 (2.95–2.85)	30.00–2.30 (2.40–2.30)	30.00–2.75 (2.82–2.75)	30.00–2.60 (2.67–2.60)
Total No. of reflections	564 510 (43 474)	283 148 (16 729)	366 770 (30 085)	115 943 (8 973)	150 052 (11 638)
No. of unique reflections	65 303 (4 784)	49 267 (4 765)	53 136 (5 315)	32 119 (2 369)	38 037 (2 832)
Completeness (%)	99.9 (100.0)	99.7 (97.6)	97.3 (82.7)	98.1 (99.9)	98.7 (99.3)
Redundancy	8.6 (9.1)	5.7 (3.5)	6.9 (5.7)	3.6 (3.8)	3.9 (4.1)
$\langle I/\sigma(I)\rangle$	10.13 (1.42)	5.42 (1.79)	8.26 (1.10)	7.92 (1.06)	8.15 (1.02)
$R_{r.i.m.}$‡	17.3 (161.4)	27.1 (58.2)	11.7 (65.8)	11.9 (124.4)	9.8 (134.9)
$CC_{1/2}$	99.7 (57.9)	99.6 (59.4)	96.6 (60.2)	99.6 (40.1)	99.7 (58.6)

Table 2. Structure solution and refinement.

	L37Cys+L99Cys	L53Cys+L64Cys	L172Cys+L246Cys	M19Cys+L214Cys	M84Cys+L278Cys
Resolution range (Å)	30.00–2.60 (2.64–2.60)	30.00–2.85 (2.95–2.85)	30.00–2.30 (2.40–2.30)	46.0–2.75 (2.82–2.75)	41.00–2.60 (2.67–2.60)
Completeness (%)	99.9 (100.0)	99.5 (96.1)	99.3 (99.0)	98.0 (96.0)	98.6 (99.0)
No. of reflections, working set	65 301 (2 567)	48 500 (2 565)	44 919 (2 614)	32 091 (3 403)	38 001 (3 296)
No. of reflections, test set	3 351 (153)	2 435 (131)	2 273 (140)	1 282 (141)	1 519 (138)
R_{cryst}	18.56 (25.95)	22.85 (33.77)	25.86 (26.08)	19.91 (34.63)	19.27 (40.40)
R_{free}	20.74 (30.34)	27.90 (38.49)	30.82 (32.56)	27.06 (42.24)	24.93 (40.01)
R.m.s. deviations					
Bonds (Å)	0.009	0.010	0.010	0.010	0.009
Angles (°)	1.122	1.314	1.354	1.234	1.159
PDB ID	8C5X	8C6K	8C87	8C88	8C7C

3. Results

3.1. The Design and Introduction of Disulfide Bonds

Of the three protein subunits of *C. sphaeroides* RC, M and L, are pseudosymmetric with respect to each other and have 5 transmembrane α-helices, whereas subunit H has only one transmembrane α-helix, which is mostly located outside the membrane on the cytoplasmic side. This structural organization allows the introduction of disulfide bonds within one L or M subunit or between them. In the wild-type reaction center, there are several natural cysteine residues that could potentially be used for the introduction of disulfide bonds. However, because the cysteines themselves are highly reactive amino acid residues that are often important for the proper function of proteins [32], they were not used so as not to interfere with their potential role in the reaction center. In addition, it has been shown that native cysteines can be used to unidirectionally bind electrochemically active proteins to metal electrodes [33], which could be important for the development of biooptoelectronic materials and devices.

In our study, the rational design approach was employed. Using a high-resolution crystal structure of *C. sphaeroides* RC (PDB ID 6Z1J, [23]), several positions that had good geometric parameters were selected in anticipation of forming the intra- or inter-subunit S-S bonds if cysteines were placed there. When designing the mutations, the following rules were adhered to: (1) the side-chain volume of the introduced amino acids should not be significantly different from the side-chain volume of the substituted residues; (2) the side groups of the introduced cysteines should face each other to increase the probability of disulfide bond formation; and (3) in the case of proximity to electron transfer cofactors, the side groups of the substituted amino acid residues should not be turned in their direction to avoid direct influence of the cysteines on the RC function. Another important feature of the introduced substitutions was the depth of their immersion in the membrane.

Finally, the following mutation pairs were designed and obtained (Figure 1):

(1) V(M84)C+G(L278)C, periplasmic surface, intra-subunit S-S bond;
(2) A(L53)C+I(L64)C, periplasmic surface, inter-subunit S-S bond;
(3) A(L172)C+L(L246)C, membrane zone closer to periplasm, inter-subunit S-S bond;
(4) A(L37)C+S(L99)C, hydrophobic zone near BPheo H_A, inter-subunit S-S bond;
(5) G(M19)C+T(L214)C, cytoplasmic surface, intra-subunit S-S bond.

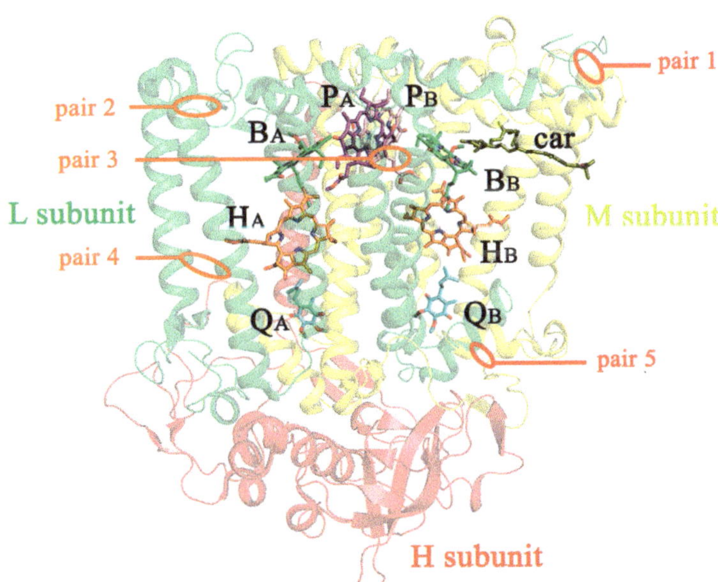

Figure 1. Locations of cysteine substitutions in the *C. sphaeroides* RC complex. Pairs of mutations within the subunit and between subunits are shown in orange and red, respectively. P_A and P_B are BChls of the special pair; B_A and B_B are monomeric BChl; H_A and H_B are monomeric BPheo; Q_A and Q_B are ubiquinones; car is a carotenoid. In this model, the tails of the cofactors are truncated.

3.2. Confirmation of Disulfide Bond Formation by X-ray Crystallography

In our work, X-ray crystallography was used as the main tool to observe disulfide bonds. All mutant forms were successfully crystallized, diffraction data collected, and the structures solved. Data collection and processing statistics are shown in Table 1. Apart from the sites of amino acid substitutions, no other significant changes were found in the structure of the mutant RCs compared with the wild-type. The locations of amino acid substitutions in the spatial structures of the RC mutant forms are shown in Figure 2.

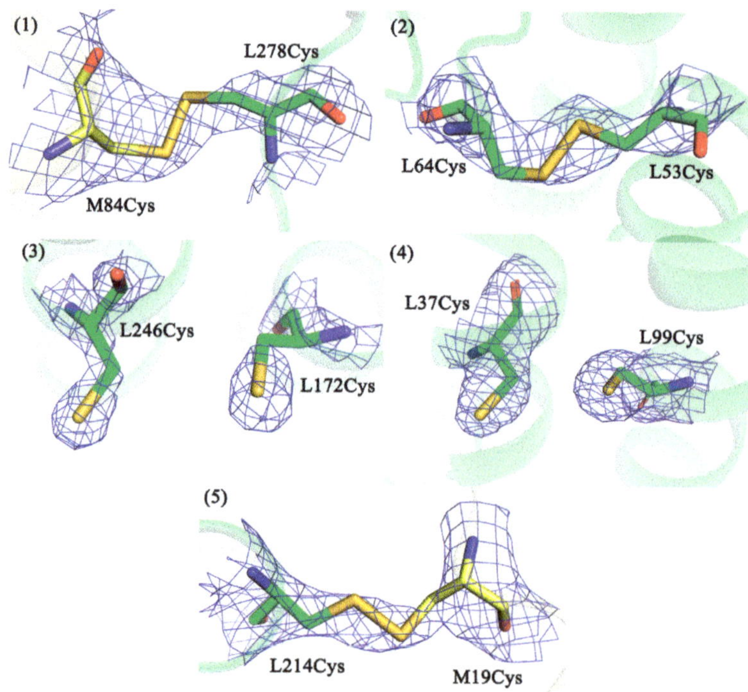

Figure 2. Fragments of the *2Fo-Fc* electron density maps for the crystal structures of the RC mutant forms at the sites of amino acid substitutions: (pair 1) V(M84)C+G(L278)C mutant form, 2.6 Å resolution, 0.7 σ; (pair 2) A(L53)C+I(L64)C mutant form, 2.85 Å resolution, 2.0 σ; (pair 3) A(L172)C+L(L246)C mutant form, 2.3 Å resolution, 2.1 σ; (pair 4) A(L37)C+S(L99)C mutant form, 2.6 Å resolution, 2.0 σ; (pair 5) G(M19)C+T(L214)C mutant form, 2.75 Å resolution, 1.2 σ.

Figure 2 shows that in all mutant RCs, the electron density for the introduced cysteine residues is clearly visible. It can be seen that in the cases where the mutant pairs are located closer to the protein surface, disulfide bonds are formed between the cysteines, regardless of the cytoplasmic or periplasmic side (pairs 1, 2, and 5). In the case where the mutant pairs are located deeper in the membrane part of the RC (pairs 3 and 4), no bonds are observed, and the side groups of the amino acid residues are turned away from each other.

3.3. Confirmation of the Formation of Disulfide Bonds by Polyacrylamide Gel Electrophoresis (PAGE)

In the inter-subunit mutants V(M84)C+G(L278)C and G(M19)C+T(L214)C (pairs 1 and 5, respectively), the electron density for the S-S bonds was visible but of poor quality and detectable only with a low cutoff. At the same time, no alternative conformations for these cysteine residues were visible either. The reasons for this could be increased mobility of the side groups of the amino acid residues at the interaction interface between the L and

M subunits and/or radiation damage during the imaging process. To provide additional confirmation of the formation of these disulfide bonds, SDS-PAGE was used.

The electrophoretic method for separating the subunits of the RC has some peculiarities. The first is related to the membrane property of the protein. The bands of the RC subunits do not migrate in a gel according to their molecular weight. The characteristic letter designations of the subunits are not based on their actual molecular weight but on their electrophoretic mobility in the gel. L–light (31.4 kDa), M–medium (34.5 kDa), and H–heavy (28 kDa) subunits separate as if their molecular weights were 21, 24, and 28 kDa, respectively [17]. This is explained by the fact that the binding of charged SDS molecules to hydrophobic regions of membrane proteins is higher than the binding to amphiphilic regions of globular proteins, which affects electrophoretic mobility [34].

The second feature is that the bands of the L and M subunits disappear and the intensity of the H band increases when the RC samples are heated at 100 °C for two minutes or longer in the presence of SDS and beta-mercaptoethanol [35]. In addition, the disappearance of the bands corresponding to the L and M subunits upon heating has been described [36]. This effect was shown to be due to the aggregation of the subunits induced by beta-mercaptoethanol–they form high molecular weight aggregates that cannot penetrate the gel. It was also shown that the interactions holding the LM complex together are not disulfidic in nature.

Keeping in mind the abovementioned details, comparatively mild denaturation conditions were used, with a temperature of 30 °C and a denaturation time of 60 min. The results of Tris/MES SDS PAGE are shown in Figure 3.

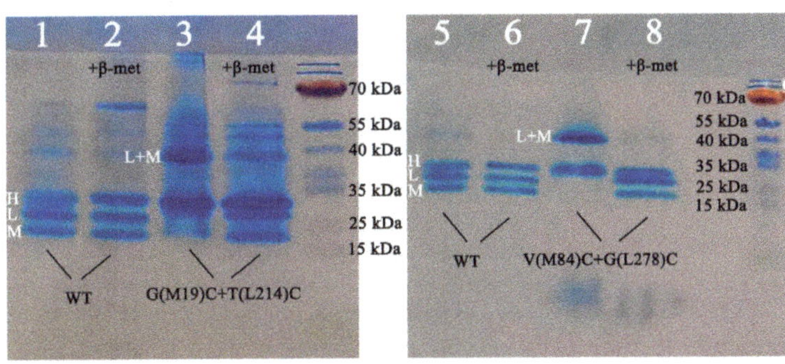

Figure 3. Tris/MES SDS PAGE of G(M19)C+T(L214)C and V(M84)C+G(L278)C intra-subunit disulfide mutant reaction centers. β-Mercaptoethanol (β-met) was added to the samples in lines 2, 4, 6, and 8.

In the wild-type RC, the formation of three bands corresponding to the RC subunits is observed (lines 1, 2, 5, and 6). In the samples containing beta-mercaptoethanol (lines 2 and 6), the band boundaries are more pronounced than without the reducing agent (lines 1 and 5).

In the G(M19)C+T(L214)C mutant RC, the formation of the LM complex is observed in the gel without beta-mercaptoethanol (line 3). Upon addition of beta-mercaptoethanol, this complex breaks down and three bands become visible, each corresponding to one of the RC subunits (line 4).

For the mutant form V(M84)C+G(L278)C, the same picture is obtained: in the gel without beta-mercaptoethanol, the formation of the LM complex is observed (line 7), and in the presence of beta-mercaptoethanol, the complex breaks down and three bands, each corresponding to one of the RC subunits, become visible (line 8).

Summarizing the results of the electrophoresis, it can be concluded that the breakdown of the LM complex in the mutant forms (lines 4 and 8) is due to the addition of beta-mercaptoethanol being associated with the breaking of the disulfide bond that appeared as a result of the cysteine pair introduction.

3.4. Pigment Content and Photochemical Properties of Mutant RCs

The wild-type reaction center contains two BPheo molecules and four BChl molecules. The pigment composition remained unchanged in all investigated RC mutant forms according to the performed pigment analysis.

The absorption spectra of the isolated reaction centers are shown in Figure 4.

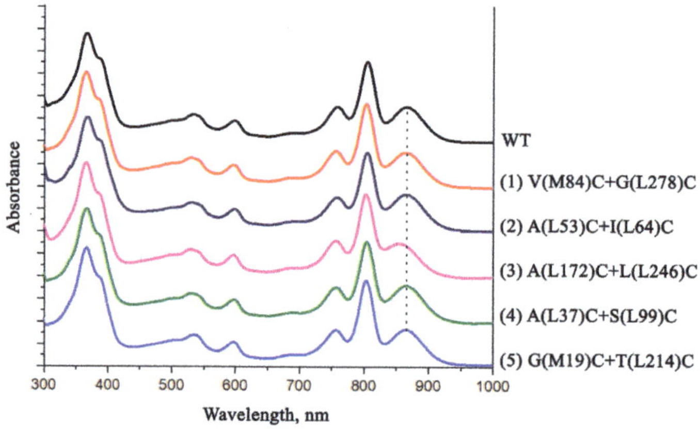

Figure 4. Electronic ground-state absorption spectra of the wild-type *C. sphaeroides* reaction center and its mutant forms. Spectra were measured at room temperature and normalized at the Q_Y H absorption band.

In the absorption spectrum of the isolated wild-type RC, shown in Figure 4, a long-wavelength band of Q_Y P with a maximum peak at 865 nm is accounted by the absorption of bacteriochlorophyll dimer P. The Q_Y B band with the maximum peak at 804 nm represents an absorption of BChl monomers and also a high-energy transition in the primary electron donor molecule. A Q_Y H band with a maximum at 760 nm corresponds to the absorption of bacteriopheophytin molecules. In the short wavelength region of the spectrum, the band at 599 nm reflects Q_X transitions in BChl molecules. At 532 nm is the Q_X H maximum, which corresponds to the absorption band of BPheo molecules in active and inactive electron transport chains. The shoulder at 500 nm is assigned to the carotenoid molecule. In the short-wavelength region of the absorption spectrum of isolated RCs, there is a Soret band with a maximum at 363 nm and a shoulder at the long-wavelength slope of the band at 390 nm, which reflects the absorption of all bacteriochlorins of RCs.

As can be seen from the absorption spectra shown, the introduced mutations have no significant effect on the position and amplitude of the absorption bands of the pigments. The only exception is the A(L172)C+L(L246)C mutant form, in the absorption spectrum of which a short-wavelength shift of the long-wavelength maximum of the special pair absorption band is observed. It is likely that the reason for these changes may be associated with the location of one of the introduced cysteines (L172) in close proximity to the histidine amino acid residue L173, which acts as a ligand for the magnesium atom of BChl P_A.

In the differential spectra of the wild-type RC, light-induced formation of the state $P^+Q_A^-$ causes bleaching of the Q_Y P band at 865 nm, a short-wavelength shift of the Q_YB band, and a long-wavelength shift of the Q_Y H band (Figure 5). Similar spectral changes were observed upon illumination of all mutant reaction centers, demonstrating effective electron transfer from the primary donor P to the acceptor Q_A.

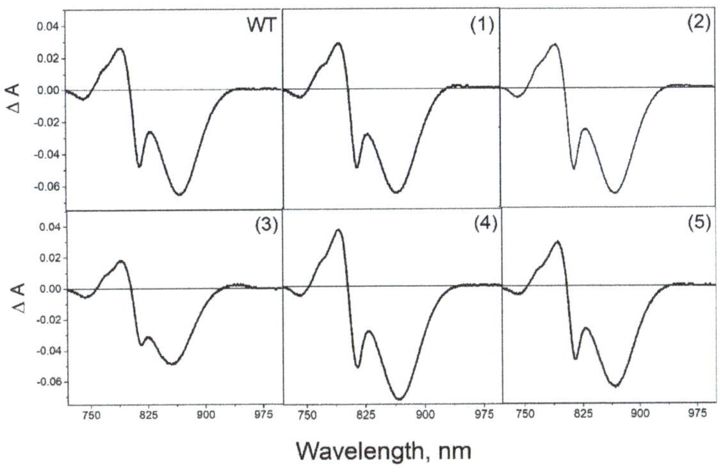

Figure 5. Differential (light minus dark) absorption spectra of wild-type RC *C. sphaeroides* and mutant reaction centers: (1) V(M84)C+G(L278)C, (2) A(L53)C+I(L64)C, (3) A(L172)C+L(L246)C, (4) A(L37)C+S(L99)C, and (5) G(M19)C+T(L214)C. Spectra were measured at room temperature and normalized at the Q_Y H absorption band.

As can be seen from the differential (light minus dark) absorption spectra, the introduced mutations have no significant effect on the photochemical properties of mutant RCs. Similar to what was observed in the absorption spectra, the amplitude of the Q_Y P band was noticeably reduced in the RC A(L172)C+L(L246)C. As mentioned above the possible reason for these changes may be associated with the location of one of the introduced cysteines (L172) in close proximity to a ligand for the magnesium atom of BChl P_A.

3.5. Thermostability of the Mutant RCs

Since the main goal of this research was to stabilize the RCs, the thermal stability of the isolated mutant pigment-protein complexes was investigated (Figure 6).

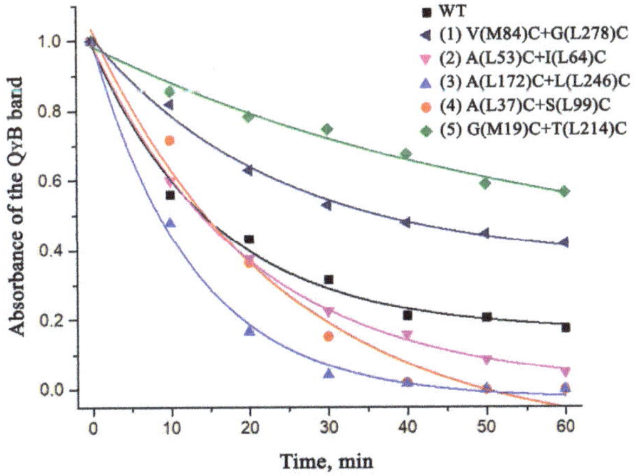

Figure 6. Thermodependent changes in the long-wavelength absorption band of monomeric BChls (at an incubation temperature of 48 °C).

According to the data obtained, the mutations made for the formation of intra-subunit disulfide bridges (A(L37)C+S(L99)C, A(L172)C+L(L246)C, and A(L53)C+I(L64)C) did not contribute positively to the stabilization of the RC but, on the contrary, had a destabilizing effect on the structure of the complex. However, it was observed that the mutations between the L and M subunits (G(M19)C+T(L214)C and V(M84)C+G(L278)C) significantly contributed to the stabilization of the RC. After 60 min of incubation, more than 50% of the native complexes remained in the sample in the case of G(M19)C+T(L214)C and more than 40% in the case of V(M84)C+G(L278)C, which was significantly more compared to about 20% of the undenatured wild-type RCs under the same conditions.

4. Discussion

The introduction of disulfide bonds into protein molecules was previously reported within the literature as a method to increase their overall structural stability [37–39]. A disulfide bond formed between the thiol groups of two spatially close cysteine residues is often important for protein folding, stability, and function [40,41]. Due to conformational entropy, native disulfide bonds stabilize the conformation of protein molecules [42], whereas removal of native disulfides can result in decreased stability of the target protein [43]. Previous studies have shown that the proper introduction of disulfide bonds can stabilize the flexible region of target proteins and reduce conformational entropy by fixing the protein in a single desired conformation [44].

This approach is more commonly used with globular proteins but is also possible with integral membrane proteins. The major developments of this method are in the area of stabilization of G protein-coupled receptors (GCPRs), also known as seven-(pass)-transmembrane domain receptors. For example, a double cysteine mutant of the opsin form of rhodopsin was obtained in which the formation of a disulfide bond between the introduced amino acid residues was observed [45], which was later confirmed by the crystal structure [46]. The formation of the S-S bond resulted in an increase in thermal stability and had a minor effect on the functional properties of the protein. Another example of a GCPR whose thermal stability was improved by the introduction of a disulfide bond is the serotonin 5-HT_{2C} receptor [47].

To our knowledge, the possibility of stabilization of the photosynthetic reaction center by disulfide bridges has not been previously studied. The reaction centers, unlike GCPRs, are complexes with multiple subunits. Therefore, it was chosen to introduce mutations both within one subunit and between two subunits.

In two intra-subunit mutant forms A(L172)C+L(L246)C and A(L37)C+S(L99)C (pairs 3 and 4, respectively), the formation of disulfide bonds did not occur. This can be attributed to the fact that these cysteine pairs were too deeply immersed in the membrane and were therefore inaccessible to the bacterial disulfide bond formation systems [48] that promote the oxidation process. Considering these data and the fact that in rhodopsin, successful S-S bond formation also occurred near the surface [46], it was confirmed that proximity to the extramembranous part of the integral membrane protein is essential for S-S bond formation.

In three mutant forms V(M84)C+G(L278)C, A(L53)C+I(L64)C and G(M19)C+T(L214)C (pairs 1, 2, and 5, respectively), in which the introduced cysteines were close to the protein surface, disulfide bonds were formed regardless of whether these amino acid residues were on the cytoplasmic side or on the periplasmic surface. It should be noted that no specific oxidizing agents were used during bacterial growth or protein purification to stimulate S-S bond formation.

All inter-subunit mutant forms were less thermally stable than wild-type RCs. However, the cysteine pair L53Cys-L64Cys (pair 2), which formed a disulfide bond, had the least destabilizing effect on the structure of the complex compared with cysteine pairs that did not form S-S bridges.

In the case of mutant RCs with inter-subunit S-S bonds V(M84)C+G(L278)C and G(M19)C+T(L214)C (pairs 1 and 5, respectively), both complexes exhibited increased thermal stability compared with wild-type RCs. It is assumed that such strengthening of the RC structure occurs due to the strengthening of the complex at the level of the quaternary structure of the protein.

The genetically modified reaction centers described in this work retained the ability for photoinduced electron transfer from the primary electron donor molecule (P) to the electron acceptor molecules (Q). This implies that the mutant RCs V(M84)C+G(L278)C and G(M19)C+T(L214)C with enhanced thermal stability could potentially be used in photo-bioelectrochemical systems where wild-type reaction centers are used. Examples of which include the sunlight-driven online sensing of various toxic compounds [49] and bio-hybrid systems that have been shown to be effective transducers of solar radiation [50,51]. Bio-hybrids can also be used as materials for biooptoelectronics [52,53], functionally integrated into devices [54], and used as active elements in bio-photonic energy cells [6]. Due to their increased structural stability, the V(M84)C+G(L278)C and G(M19)C+T(L214)C mutants could be more effective for the processes mentioned above, which is only a speculation for now and requires further research and confirmation.

In summary, the data obtained in this work demonstrate the possibility of introducing disulfide bonds into bacterial photosynthetic reaction centers to increase their thermal stability without loss of functional activity. The reaction center is a representative of integral membrane proteins with an α-helical structure. This is the most common type of transmembrane protein. In humans, for example, an estimated 27% of all proteins are α-helical membrane proteins [55]. This class includes many other IMPs of interest to the scientific community, where the introduction of disulfide bonds may be an effective stabilization mechanism. We believe that our results may be useful for the future development of multisubunit α-helical integral membrane protein complexes with enhanced stability.

Author Contributions: Conceptualization, G.S., L.V., and A.G.; methodology, G.S., A.A., and D.Y.; investigation, G.S., A.A., D.Y. and T.F.; data curation, L.V. and A.G.; writing—original draft preparation, G.S.; writing—review and editing, A.A., D.Y., T.F., L.V., and A.G.; visualization, G.S. and T.F.; supervision, L.V. and A.G. All authors have read and agreed to the published version of the manuscript.

Funding: Russian Foundation for Basic Research (grant No. 18-02-40008).

Institutional Review Board Statement: Not applicable.

Data Availability Statement: MDPI Research Data Policies.

Acknowledgments: We acknowledge the European Synchrotron Radiation Facility for provision of synchrotron radiation facilities, and we would like to thank Alexander Popov for assistance in using beamline ID30A-3.

Conflicts of Interest: The authors declare no conflict of interest.

References

1. Allen, J.P.; Feher, G.; Yeates, T.O.; Komiya, H.; Rees, D.C. Structure of the Reaction Center from Rhodobacter Sphaeroides R-26: The Cofactors. *Proc. Natl. Acad. Sci. USA* **1987**, *84*, 5730–5734. [CrossRef]
2. Kálmán, L.; Williams, J.C.; Allen, J.P. Comparison of Bacterial Reaction Centers and Photosystem II. *Photosynth. Res.* **2008**, *98*, 643–655. [CrossRef] [PubMed]
3. Pakula, A.A.; Sauer, R.T. Genetic analysis of protein stability and function. *Annu. Rev. Genet.* **1989**, *23*, 289–310. [CrossRef] [PubMed]
4. McPherson, A.; Gavira, J.A. Introduction to Protein Crystallization. *Acta Cryst. F* **2014**, *70*, 2–20. [CrossRef]
5. Wraight, C.A.; Clayton, R.K. The Absolute Quantum Efficiency of Bacteriochlorophyll Photooxidation in Reaction Centres of Rhodopseudomonas Spheroides. *Biochim. Biophys. Acta BBA–Bioenerg.* **1974**, *333*, 246–260. [CrossRef] [PubMed]
6. Grattieri, M.; Beaver, K.; Gaffney, E.M.; Dong, F.; Minteer, S.D. Advancing the Fundamental Understanding and Practical Applications of Photo-Bioelectrocatalysis. *Chem. Commun.* **2020**, *56*, 8553–8568. [CrossRef] [PubMed]
7. Fufina, T.Y.; Vasilieva, L.G. Effect of Detergents and Osmolytes on Thermal Stability of Native and Mutant Rhodobacter Sphaeroides Reaction Centers. *Biochem. Mosc.* **2021**, *86*, 517–524. [CrossRef]
8. Selikhanov, G.; Fufina, T.; Guenther, S.; Meents, A.; Gabdulkhakov, A.; Vasilieva, L. X-ray Structure of the *Rhodobacter Sphaeroides* Reaction Center with an M197 Phe→His Substitution Clarifies the Properties of the Mutant Complex. *IUCrJ* **2022**, *9*, 261–271. [CrossRef]

9. Holden-Dye, K.; Crouch, L.I.; Williams, C.M.; Bone, R.A.; Cheng, J.; Böhles, F.; Heathcote, P.; Jones, M.R. Opposing Structural Changes in Two Symmetrical Polypeptides Bring about Opposing Changes to the Thermal Stability of a Complex Integral Membrane Protein. *Arch. Biochem. Biophys.* **2011**, *505*, 160–170. [CrossRef]
10. Kangur, L.; Jones, M.R.; Freiberg, A. Hydrogen Bonds in the Vicinity of the Special Pair of the Bacterial Reaction Center Probed by Hydrostatic High-Pressure Absorption Spectroscopy. *Biophys. Chem.* **2017**, *231*, 27–33. [CrossRef]
11. Jones, M.R.; Visschers, R.W.; Van Grondelle, R.; Hunter, C.N. Construction and Characterization of a Mutant of Rhodobacter Sphaeroides with the Reaction Center as the Sole Pigment-Protein Complex. *Biochemistry* **1992**, *31*, 4458–4465. [CrossRef] [PubMed]
12. Vasilieva, L.G.; Bolgarina, T.I.; Khatynov, R.A.; Shkuropatov, A.Y.; Miyake, J.; Shuvalov, V.A. Substitution of Valine-157 Residue by Tyrosine in the L-Subunit of the Rhodobacter Sphaeroides Reaction Center. *Dokl. Biochem. Biophys.* **2001**, *376*, 46–49. [CrossRef] [PubMed]
13. Liu, H.; Naismith, J.H. An Efficient One-Step Site-Directed Deletion, Insertion, Single and Multiple-Site Plasmid Mutagenesis Protocol. *BMC Biotechnol.* **2008**, *8*, 91. [CrossRef] [PubMed]
14. Khatypov, R.A.; Vasilieva, L.G.; Fufina, T.Y.; Bolgarina, T.I.; Shuvalov, V.A. Substitution of Isoleucine L177 by Histidine Affects the Pigment Composition and Properties of the Reaction Center of the Purple Bacterium Rhodobacter Sphaeroides. *Biochemistry* **2005**, *70*, 1256–1261. [CrossRef]
15. Fufina, T.Y.; Vasilieva, L.G.; Khatypov, R.A.; Shkuropatov, A.Y.; Shuvalov, V.A. Substitution of Isoleucine L177 by Histidine in *Rhodobacter Sphaeroides* Reaction Center Results in the Covalent Binding of P_A Bacteriochlorophyll to the L Subunit. *FEBS Lett.* **2007**, *581*, 5769–5773. [CrossRef]
16. Goldsmith, J.O.; Boxer, S.G. Rapid Isolation of Bacterial Photosynthetic Reaction Centers with an Engineered Poly-Histidine Tag. *Biochim. Biophys. Acta BBA–Bioenerg.* **1996**, *1276*, 171–175. [CrossRef]
17. Okamura, M.Y.; Steiner, L.A.; Feher, G. Characterization of Reaction Centers from Photosynthetic Bacteria. I. Subunit Structure of the Protein Mediating the Primary Photochemistry in Rhodopseudomonas Spheroides R-26. *Biochemistry* **1974**, *13*, 1394–1403. [CrossRef]
18. Vasilieva, L.G.; Fufina, T.Y.; Gabdulkhakov, A.G.; Leonova, M.M.; Khatypov, R.A.; Shuvalov, V.A. The Site-Directed Mutation I(L177)H in Rhodobacter Sphaeroides Reaction Center Affects Coordination of PA and BB Bacteriochlorophylls. *Biochim. Biophys. Acta BBA–Bioenerg.* **2012**, *1817*, 1407–1417. [CrossRef]
19. Fufina, T.Y.; Vasilieva, L.G.; Shuvalov, V.A. Examination of Stability of Mutant Photosynthetic Reaction Center of Rhodobacter Sphaeroides I(L177)H and Determination of Location of Bacteriochlorophyll Covalently Bound to the Protein. *Biochem. Mosc.* **2010**, *75*, 208–213. [CrossRef]
20. Kashino, Y.; Koike, H.; Satoh, K. An Improved Sodium Dodecyl Sulfate-Polyacrylamide Gel Electrophoresis System for the Analysis of Membrane Protein Complexes. *Electrophoresis* **2001**, *22*, 1004–1007. [CrossRef]
21. Laemmli, U.K. Cleavage of Structural Proteins during the Assembly of the Head of Bacteriophage T4. *Nature* **1970**, *227*, 680–685. [CrossRef] [PubMed]
22. Gabdulkhakov, A.G.; Fufina, T.Y.; Vasilieva, L.G.; Mueller, U.; Shuvalov, V.A. Expression, Purification, Crystallization and Preliminary X-Ray Structure Analysis of Wild-Type and L(M196)H-Mutant *Rhodobacter Sphaeroides* Reaction Centres. *Acta Crystallogr. Sect. F Struct. Biol. Cryst. Commun.* **2013**, *69*, 506–509. [CrossRef] [PubMed]
23. Selikhanov, G.; Fufina, T.; Vasilieva, L.; Betzel, C.; Gabdulkhakov, A. Novel Approaches for the Lipid Sponge Phase Crystallization of the *Rhodobacter Sphaeroides* Photosynthetic Reaction Center. *IUCrJ* **2020**, *7*, 1084–1091. [CrossRef]
24. Nurizzo, D.; Mairs, T.; Guijarro, M.; Rey, V.; Meyer, J.; Fajardo, P.; Chavanne, J.; Biasci, J.-C.; McSweeney, S.; Mitchell, E. The ID23-1 Structural Biology Beamline at the ESRF. *J. Synchrotron Radiat.* **2006**, *13*, 227–238. [CrossRef]
25. Gabadinho, J.; Beteva, A.; Guijarro, M.; Rey-Bakaikoa, V.; Spruce, D.; Bowler, M.W.; Brockhauser, S.; Flot, D.; Gordon, E.J.; Hall, D.R.; et al. *MxCuBE*: A Synchrotron Beamline Control Environment Customized for Macromolecular Crystallography Experiments. *J. Synchrotron Radiat.* **2010**, *17*, 700–707. [CrossRef]
26. Bourenkov, G.P.; Popov, A.N. Optimization of Data Collection Taking Radiation Damage into Account. *Acta Crystallogr. Sect. D Biol. Crystallogr.* **2010**, *66*, 409–419. [CrossRef] [PubMed]
27. Kabsch, W. *XDS*. *Acta Crystallogr. Sect. D Biol. Crystallogr.* **2010**, *66*, 125–132. [CrossRef]
28. McCoy, A.J.; Grosse-Kunstleve, R.W.; Adams, P.D.; Winn, M.D.; Storoni, L.C.; Read, R.J. *Phaser* Crystallographic Software. *J. Appl. Crystallogr.* **2007**, *40*, 658–674. [CrossRef]
29. Murshudov, G.N.; Skubák, P.; Lebedev, A.A.; Pannu, N.S.; Steiner, R.A.; Nicholls, R.A.; Winn, M.D.; Long, F.; Vagin, A.A. *REFMAC* 5 for the Refinement of Macromolecular Crystal Structures. *Acta Crystallogr. Sect. D Biol. Crystallogr.* **2011**, *67*, 355–367. [CrossRef]
30. Emsley, P.; Lohkamp, B.; Scott, W.G.; Cowtan, K. Features and Development of *Coot*. *Acta Crystallogr. Sect. D Biol. Crystallogr.* **2010**, *66*, 486–501. [CrossRef]
31. Delano, W.L. The PyMOL Molecular Graphics System. 2022. Available online: http://www.pymol.org (accessed on 8 November 2022).
32. Netto, L.E.S.; de Oliveira, M.A.; Monteiro, G.; Demasi, A.P.D.; Cussiol, J.R.R.; Discola, K.F.; Demasi, M.; Silva, G.M.; Alves, S.V.; Faria, V.G.; et al. Reactive Cysteine in Proteins: Protein Folding, Antioxidant Defense, Redox Signaling and More. *Comp. Biochem. Physiol. Part C Toxicol. Pharmacol.* **2007**, *146*, 180–193. [CrossRef] [PubMed]

33. Reiss, B.D.; Hanson, D.K.; Firestone, M.A. Evaluation of the Photosynthetic Reaction Center Protein for Potential Use as a Bioelectronic Circuit Element. *Biotechnol. Prog.* **2007**, *23*, 985–989. [CrossRef] [PubMed]
34. Rath, A.; Glibowicka, M.; Nadeau, V.G.; Chen, G.; Deber, C.M. Detergent Binding Explains Anomalous SDS-PAGE Migration of Membrane Proteins. *Proc. Natl. Acad. Sci. USA* **2009**, *106*, 1760–1765. [CrossRef]
35. Clayton, R.K.; Haselkorn, R. Protein Components of Bacterial Photosynthetic Membranes. *J. Mol. Biol.* **1972**, *68*, 97–105. [CrossRef] [PubMed]
36. Shepherd, W.D.; Kaplan, S. Effect of Heat and 2-Mercaptoethanol on Intracytoplasmic Membrane Polypeptides of Rhodopseudomonas Sphaeroides. *J. Bacteriol.* **1978**, *135*, 656–667. [CrossRef]
37. Liu, H.-X.; Li, L.; Yang, X.-Z.; Wei, C.-W.; Cheng, H.-M.; Gao, S.-Q.; Wen, G.-B.; Lin, Y.-W. Enhancement of Protein Stability by an Additional Disulfide Bond Designed in Human Neuroglobin. *RSC Adv.* **2019**, *9*, 4172–4179. [CrossRef]
38. van Beek, H.L.; Wijma, H.J.; Fromont, L.; Janssen, D.B.; Fraaije, M.W. Stabilization of Cyclohexanone Monooxygenase by a Computationally Designed Disulfide Bond Spanning Only One Residue. *FEBS Open Biol.* **2014**, *4*, 168–174. [CrossRef]
39. Sanchez-Romero, I.; Ariza, A.; Wilson, K.S.; Skjøt, M.; Vind, J.; De Maria, L.; Skov, L.K.; Sanchez-Ruiz, J.M. Mechanism of Protein Kinetic Stabilization by Engineered Disulfide Crosslinks. *PLoS ONE* **2013**, *8*, e70013. [CrossRef]
40. Creighton, T.E.; Zapun, A.; Darby, N.J. Mechanisms and Catalysts of Disulphide Bond Formation in Proteins. *Trends Biotechnol.* **1995**, *13*, 18–23. [CrossRef]
41. Fass, D. Disulfide Bonding in Protein Biophysics. *Annu. Rev. Biophys.* **2012**, *41*, 63–79. [CrossRef]
42. Dill, K.A. Dominant Forces in Protein Folding. *Biochemistry* **1990**, *29*, 7133–7155. [CrossRef] [PubMed]
43. Liu, D.; Cowburn, D. Combining Biophysical Methods to Analyze the Disulfide Bond in SH2 Domain of C-Terminal Src Kinase. *Biophys. Rep.* **2016**, *2*, 33–43. [CrossRef] [PubMed]
44. Matsumura, M.; Signor, G.; Matthews, B.W. Substantial Increase of Protein Stability by Multiple Disulphide Bonds. *Nature* **1989**, *342*, 291–293. [CrossRef] [PubMed]
45. Xie, G.; Gross, A.K.; Oprian, D.D. An Opsin Mutant with Increased Thermal Stability. *Biochemistry* **2003**, *42*, 1995–2001. [CrossRef]
46. Standfuss, J.; Xie, G.; Edwards, P.C.; Burghammer, M.; Oprian, D.D.; Schertler, G.F.X. Crystal Structure of a Thermally Stable Rhodopsin Mutant. *J. Mol. Biol.* **2007**, *372*, 1179–1188. [CrossRef]
47. Popov, P.; Peng, Y.; Shen, L.; Stevens, R.C.; Cherezov, V.; Liu, Z.-J.; Katritch, V. Computational Design of Thermostabilizing Point Mutations for G Protein-Coupled Receptors. *eLife* **2018**, *7*, e34729. [CrossRef]
48. Landeta, C.; Boyd, D.; Beckwith, J. Disulfide Bond Formation in Prokaryotes. *Nat. Microbiol.* **2018**, *3*, 270–280. [CrossRef]
49. Grattieri, M.; Minteer, S.D. Self-Powered Biosensors. *ACS Sens.* **2018**, *3*, 44–53. [CrossRef]
50. Kornienko, N.; Zhang, J.Z.; Sakimoto, K.K.; Yang, P.; Reisner, E. Interfacing Nature's Catalytic Machinery with Synthetic Materials for Semi-Artificial Photosynthesis. *Nat. Nanotech.* **2018**, *13*, 890–899. [CrossRef]
51. Grattieri, M. Purple Bacteria Photo-Bioelectrochemistry: Enthralling Challenges and Opportunities. *Photochem. Photobiol. Sci.* **2020**, *19*, 424–435. [CrossRef]
52. Hasan, K.; Grippo, V.; Sperling, E.; Packer, M.A.; Leech, D.; Gorton, L. Evaluation of Photocurrent Generation from Different Photosynthetic Organisms. *ChemElectroChem* **2017**, *4*, 412–417. [CrossRef]
53. Milano, F.; Punzi, A.; Ragni, R.; Trotta, M.; Farinola, G.M. Photonics and Optoelectronics with Bacteria: Making Materials from Photosynthetic Microorganisms. *Adv. Funct. Mater.* **2019**, *29*, 1805521. [CrossRef]
54. Kruse, O.; Rupprecht, J.; Mussgnug, J.H.; Dismukes, G.C.; Hankamer, B. Photosynthesis: A Blueprint for Solar Energy Capture and Biohydrogen Production Technologies. *Photochem. Photobiol. Sci.* **2005**, *4*, 957–970. [CrossRef] [PubMed]
55. Almén, M.S.; Nordström, K.J.; Fredriksson, R.; Schiöth, H.B. Mapping the Human Membrane Proteome: A Majority of the Human Membrane Proteins Can Be Classified According to Function and Evolutionary Origin. *BMC Biol.* **2009**, *7*, 50. [CrossRef] [PubMed]

membranes MDPI

Review

Interfacial Enzymes Enable Gram-Positive Microbes to Eat Fatty Acids

Christopher D. Radka

Department of Infectious Diseases, St. Jude Children's Research Hospital, Memphis, TN 38105, USA; christopher.radka@stjude.org; Tel.: +1-901-595-5616

Abstract: Exogenous fatty acid (eFA) activation and utilization play key roles in bacterial physiology and confer growth advantages by bypassing the need to make fatty acids for lipid synthesis. In Gram-positive bacteria, eFA activation and utilization is generally carried out by the fatty acid kinase (FakAB) two-component system that converts eFA to acyl phosphate, and the acyl-ACP:phosphate transacylase (PlsX) that catalyzes the reversible conversion of acyl phosphate to acyl–acyl carrier protein. Acyl–acyl carrier protein is a soluble format of the fatty acid that is compatible with cellular metabolic enzymes and can feed multiple processes including the fatty acid biosynthesis pathway. The combination of FakAB and PlsX enables the bacteria to channel eFA nutrients. These key enzymes are peripheral membrane interfacial proteins that associate with the membrane through amphipathic helices and hydrophobic loops. In this review, we discuss the biochemical and biophysical advances that have established the structural features that drive FakB or PlsX association with the membrane, and how these protein–lipid interactions contribute to enzyme catalysis.

Keywords: interfacial enzymes; fatty acid metabolism; peripheral membrane proteins; membrane binding; mechanism

Citation: Radka, C.D. Interfacial Enzymes Enable Gram-Positive Microbes to Eat Fatty Acids. *Membranes* **2023**, *13*, 423. https://doi.org/10.3390/membranes13040423

Academic Editors: Shiro Suetsugu and Yosuke Senju

Received: 7 March 2023
Revised: 8 April 2023
Accepted: 9 April 2023
Published: 10 April 2023

1. Introduction

1.1. Protein–Membrane Association

The bacterial cytoplasmic membrane envelopes the cell and separates the cytosol from the extracellular environment. The cytoplasmic membrane is loaded with proteins that functionalize the membrane for a variety of processes including energy generation, import and export, and signaling. Bacterial membranes can be made up of as much as 50% proteins, thus demonstrating their importance to the membrane composition [1].

Integral membrane proteins are completely embedded in the hydrophobic core of the membrane and span the lipid bilayer. The membrane-spanning segments of these proteins are rich in hydrophobic amino acids whose side chains interact with the hydrophobic core and transmembrane structures and generally consist of one or multiple α-helices or a rolled-up β-sheet (i.e., β-barrel). Hydrophilic segments on either side of the bilayer enable integral membrane proteins to recognize and respond to soluble intracellular or extracellular factors. Monotopic proteins are a subset of integral membrane proteins that are permanently fixed to one leaflet of the membrane through amphipathic α-helices or hydrophobic loops.

Lipid-anchored proteins contain covalently attached lipids or fatty acids. The hydrophobic segment of the lipid or fatty acid is embedded in one of the membrane leaflets and anchors the water-soluble protein to the bilayer, whereas the polypeptide chain typically plays a negligible role in entering the bilayer. Both Gram-positive and Gram-negative bacteria have lipoproteins with covalently attached fatty acids on their amino termini [2,3]. Bacterial lipoproteins are potent agonists that stimulate the immune system, but when the fatty acids are hydrolyzed from the lipoprotein, the immunostimulatory property of the deacylated protein is inactivated [4].

1.2. Interfacial Enzymes

Peripheral membrane proteins indirectly associate with the membrane through noncovalent interactions with integral or lipid-anchored proteins, or directly through electrostatic or ionic interactions with the bilayer lipid head groups. Two of the most common structural elements that drive peripheral membrane protein association with the membrane are amphipathic helices and hydrophobic loops [5–8]. Interfacial enzymes whose substrates are lipids must bind to the membrane surface to carry out their function. Interfacial enzymes are soluble proteins that gather on a membrane surface by interfacial adsorption. This process can have the effect of concentrating and stabilizing protein on the membrane, and hyper-activating the enzyme to enhance its specific activity (also referred to as interfacial activation). Interfacial adsorption is crucial for interfacial enzymes because their hydrophobic substrates are water-insoluble and partition into the membrane. Therefore, the enzyme must interact with the membrane to entice the substrate into its active site. Membrane binding induces a conformational change that makes a productive enzyme-substrate complex and therefore enables extraction of the substrate from the membrane. Lipases are a well-studied example of water-soluble enzymes that reversibly bind the membrane and catalyze a hydrolysis reaction at the lipid–water interface [9].

Since they are soluble, interfacial enzymes are readily used to catalyze unnatural reactions to modify hydrophobic substrates and analogs that are delivered by artificial bilayers (i.e., liposomes and vesicles), detergents, or carrier proteins (i.e., albumin) as surrogates for natural membranes. However, natural membranes are complex, having diverse lipid compositions, and are integral for membrane-associated proteins that may contribute to the overall interfacial enzyme–membrane binding event. It is unclear if detergents or protein carrier membrane surrogates elicit the full physiological change in interfacial enzymes that natural membranes do. Thus, caution should be used when inferring natural membrane binding properties from studies using membrane surrogates, particularly those with detergent or carrier proteins. Artificial bilayers are an optimal reagent for biophysical studies of the mode of membrane binding (i.e., insertion depth, angular orientation, electrostatic interaction).

1.3. Membrane Binding Experimental Methods

A common biophysical technique to measure lipid affinity for interfacial enzymes is surface plasmon resonance (SPR) [10]. Lipids can be immobilized to an SPR L1 carboxydextran-coated chip surface, and the enzyme of interest can be injected in increasing concentrations to measure binding affinity. Interfacial enzymes adsorb to the lipid and then desorb under denaturing conditions (i.e., sodium hydroxide wash) to reuse the immobilized lipids and repeat the binding process at another enzyme concentration or buffer composition. Lipids are removed from the chip using detergents to regenerate the L1 chip with a new lipid composition and take advantage of the reversibility of the immobilization technique.

Other biophysical tools used to study lipid binding include nuclear magnetic resonance (NMR) and electron paramagnetic resonance (EPR). EPR focuses on the interaction of an external magnetic field with an unpaired electron spin in a molecule, while NMR focuses on the interaction of an external magnetic field with isotopic nuclei of the individual atom [11]. Site-directed spin labeling of amino acid side chains in EPR experiments enables the determination of properties or protein–membrane interactions, such as the topology of the protein with respect to the membrane, and information about local secondary structure in the membrane and degree of membrane insertion [11,12]. In NMR, isotropic chemical shifts and spin exchange signals are converted to torsion angles and interatomic distances, and dipolar couplings and anisotropic chemical shifts are converted to bond orientation restraints to provide dynamic atomic information about the location of amino acids [13]. NMR signals are sensitive to local environments and can be used to study the structure–activity correlations of proteins in detergent micelles versus detergent-free lipids [13], the binding events of ligand-induced conformational rearrangements, and the effect of allostery on the equilibrium of conformational exchange [14].

Molecular dynamics (MDs) is a useful computational technique for hypothesis generation through visualization of hypothetical membrane binding. MD simulations enable identification of possible key residues that may be responsible for tight binding and candidates for mutagenesis and validation of the simulations. During MD simulations, a substantial portion of the computational resources are spent on simulating the dynamics of lipid acyl chains, which are not expected to play a major role in interfacial adsorption. To reduce this computational cost and shorten simulation times for higher throughput, the acyl chains can be truncated and the empty space filled with an oil layer of fewer atoms ("accelerated membrane models" or "highly mobile membrane mimetic") [15]. This method has been validated to be reliable and yields results comparable to simulations with full acyl chains ("full membrane/full lipid") [16] and will be a useful tool to study the mechanisms of interfacial enzyme–membrane binding.

2. Bacterial Fatty Acid Metabolism

Bacterial fatty acid synthesis (FASII) is an essential energy-intensive process that produces the fatty acids required for lipid synthesis and membrane biogenesis [17]. Discrete monofunctional FASII enzymes catalyze the activation of short chain acyl-coenzyme A (acyl-CoA) to acyl–acyl carrier protein (acyl-ACP), which undergoes successive rounds of condensation, reduction, dehydration, and reduction to extend the acyl chain of acyl-ACP by two carbons with each elongation cycle [18]. Fatty acids are insoluble molecules that partition in the membrane compartment of cells and are linked to CoA and ACP for solubility to make them compatible with the soluble FASII enzymes. Nature has evolved a repertoire of antibiotics inhibiting different aspects of FASII, hence validating these enzymes as potential targets for new antibiotic discovery and development [19]. FASII inhibitors also have clinical potential, shown by novel therapeutic candidates advancing through clinical trials, with encouraging antimicrobial efficacy [20,21]. Some bacteria can acquire exogenous fatty acids (eFAs) from their environment and bypass the need to expend energy for de novo synthesis of substrates for membrane biogenesis. eFA can be acquired as free unesterified fatty acids [22] (although this form is generally in low abundance), or extracellular lipases can hydrolyze esterified fatty acids from abundant host lipids and liberate monomeric fatty acids [23]. eFA acquisition has gained attention as a possible FASII inhibition resistance mechanism in the treatment of Gram-positive pathogens, with a major focus on *Staphylococcus aureus* [24–26], although heterogeneity in how bacteria eat eFA suggests the viability of FASII inhibition as an antimicrobial strategy may be pathogen-specific [19,27]. *S. aureus* is a leading cause of skin infection and leading cause of death from antibiotic-resistant infection [28], highlighting the need for new therapies in the clinic. Studying the eFA acquisition pathway may uncover new features that can be inhibited and yield novel compounds to enhance the efficacy of FASII inhibitors against many Gram-positive organisms. To this end, this review will discuss advancements made in understanding how Gram-positive bacteria eat eFA.

2.1. Bacterial Phospholipid Synthesis

Phosphatidic acid is the central intermediate of phospholipid synthesis and the most widely distributed pathway for phospholipid synthesis in bacteria is the PlsXYC pathway [17,29–32]. The first step is catalyzed by acyl-ACP:phosphate transacylase (PlsX), which catalyzes the reversible conversion of acyl-ACP from FASII to acyl phosphate. The second step is catalyzed by glycerol-phosphate acyltransferase (PlsY), which utilizes acyl phosphate to convert glycerol-3-phosphate to lysophosphatidic acid. The third step is catalyzed by 1-acyl-sn-glycerol-3-phosphate acyltransferase (PlsC), which utilizes acyl-ACP to convert lysophosphatidic acid to phosphatidic acid. Fatty acid kinase is a two-component system consisting of a fatty acid binding protein (FakB) that obtains eFA from the bilayer and presents the eFA to the kinase (FakA) for phosphorylation to acyl phosphate, and then FakB exchanges the acyl phosphate with eFA in the bilayer and the cycle repeats [33]. The acyl phosphates made by FakAB can be used by PlsY or converted to acyl-ACP by PlsX for

FASII modification and/or PlsC utilization (Figure 1). Thus, through PlsX, FakAB products can be used for both steps of phospholipid synthesis. PlsY is an integral membrane protein with a seven-transmembrane helix fold [34], and PlsC is a monotopic integral membrane protein anchored by a hydrophobic/aromatic amino-terminal two-helix motif [35] (Figure 2). PlsX and FakB are soluble proteins that must solve the topological problem of exchanging their insoluble substrates with the membrane, and recent studies have shed light onto these processes. The FakAB system is a Gram-positive strategy to activate eFA for cell metabolism, whereas Gram-negative bacteria generally use acyl-CoA and acyl-ACP synthetases to activate eFA.

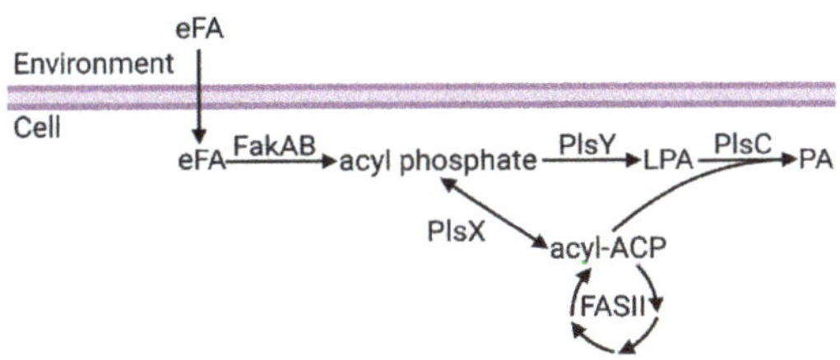

Figure 1. Channeling of exogenous fatty acids (eFA) into phospholipid synthesis. eFA enters the cell and is activated to acyl phosphate by the FakAB two-component system. Acyl phosphate can either be utilized by PlsY to make lysophosphatidic acid (LPA) or converted to acyl-ACP by PlsX. Acyl-ACP can be modified through bacterial fatty acid biosynthesis (FASII) or utilized with LPA by PlsC to make phosphatidic acid (PA). Figure created using BioRender.com.

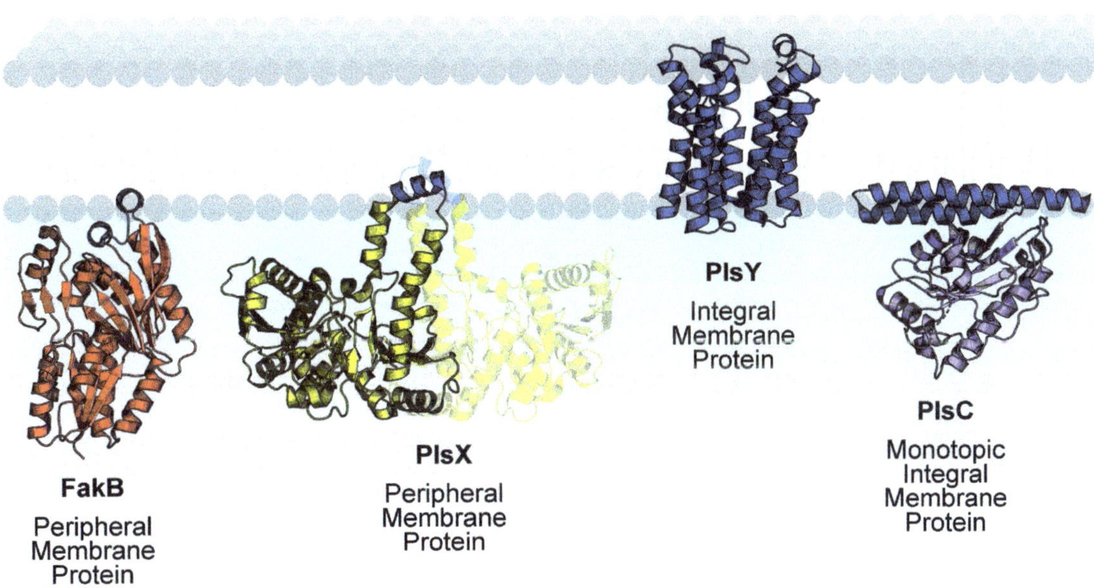

Figure 2. Membrane proteins channel exogenous fatty acids to phospholipid synthesis in Gram-positive bacteria. FakB, PlsY, and PlsC are monomeric proteins. PlsX is dimeric and the opposite protomer is shown with transparency. Membrane-binding domains are shown in blue and soluble domains are shown in orange (FakB), yellow (PlsX), or purple (PlsC). The structures shown are *Staphylococcus aureus* FakB1 (PDB ID:6MH9) [16], *Bacillus subtilis* PlsX (PDB ID: 6A1K) [36], *Aquifex aeolicus* PlsY (PDB ID: 5XJ9) [34], and *Thermotoga maritima* PlsC (PDB ID: 5KYM) [35]. Figure created using BioRender.com.

2.2. Acyl-ACP:Phosphate Transacylase (PlsX)

Immunofluorescent imaging of cell fractionation experiments using a PlsX antibody found PlsX associated with the membrane in intact *Bacillus subtilis* cells, but when *B. subtilis* cells were disrupted then PlsX was found in the soluble fraction [31]. Fluorescence microscopy of *B. subtilis* cells expressing green fluorescent protein (GFP) fused to PlsX showed PlsX accumulation at membrane foci during early log phase growth, followed by uniform distribution on the membrane during later stage growth [37]. These dynamic subcellular localization experiments implicated PlsX as a peripheral membrane protein that can reversibly associate with the membrane.

PlsX forms a soluble dimer, and the crystal structure shows that each protomer contains a core $\alpha/\beta/\alpha$ sandwich resembling a Rossmann fold, and an α-helical hairpin motif that extends away from the core domain (Figure 3A) [38]. The hairpins from each protomer, made up of helices α-9 and α-10 connected by a loop, combine to make an amphipathic helical bundle (Figure 3B) [39]. The loop at the tip of the hairpin contains hydrophobic residues needed for lipid binding. PlsX co-sediments with liposomes made from *B. subtilis* membranes, but loses the lipid binding property when residues in the hydrophobic loop are mutated to glutamates [40]. Fluorescence microscopy of *B. subtilis* cells expressing GFP fused to PlsX containing glutamate in the hairpin loop showed cytosolic accumulation of PlsX, confirming this region is essential for membrane association in vivo [40]. *B. subtilis* cells expressing PlsX containing glutamate in the hairpin have a growth defect, indicating that membrane association is necessary for function in vivo [40].

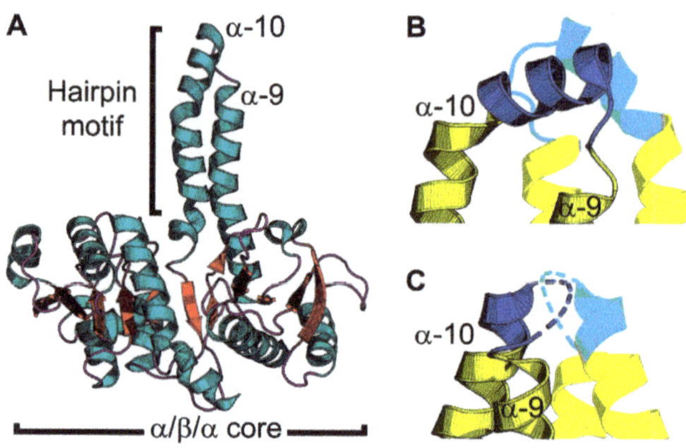

Figure 3. PlsX structure and conformational changes. (**A**) The PlsX protomer contains an $\alpha/\beta/\alpha$ sandwich core domain and protruding hairpin motif made by α-helices α-9 and α-10. PlsX is colored by secondary structure elements with teal α-helices and red β-strands and loops. (**B**,**C**) Zoomed-in views of the tip of the hairpin motif. PlsX is dimeric and the hairpin motif of the opposite protomer is shown with transparency. The membrane-binding segment is shown in blue and the soluble segment is yellow. (**B**) The helix conformation of the hairpin motif is proposed to be the membrane-bound conformation. (**C**) The loop conformation of the hairpin motif is proposed to be the cytosolic conformation. The structures in (**A**,**B**) are from *Bacillus subtilis* PlsX with the hairpin tip in the helix conformation (PDB ID: 6A1K) [36] and the structure in (**C**) is from *Enterococcus faecalis* PlsX with the hairpin tip in the loop conformation (PDB ID: 1U7N) [39].

Direct SPR measurements of PlsX–lipid binding confirmed that PlsX binds anionic phosphatidylglycerol with nanomolar affinity but does not bind zwitterionic phosphatidylcholine [41]. EPR analysis of PlsX interacting with spin-labeled lipids showed spectral perturbation from the lipid headgroup to the center of the bilayer [41]. SPR and EPR analysis of PlsX containing glutamate in the hairpin loop showed that the mutant PlsX still bound phosphatidylglycerol but only elicited minimal spectral perturbation at the lipid head group region, suggesting only superficial membrane association [41]. These data indicate that PlsX inserts into a membrane leaflet. PlsX was also crystallized in the presence of a product analog (palmitoyl phosphoramide) and although the analog was not resolved in the final structure, a new conformation of the hairpin loop was observed (Figure 3C) [36]. In this structure, the hairpin loop adopts an amphipathic α-helix conformation, and introduction of polar interfacial residues to disrupt the amphipathicity of this segment caused cytosolic accumulation of GFP-fused PlsX [36]. Mutations that compromised membrane association from this segment also correlated with growth defects [36], which agrees with a similar report [40]. These data suggest a model for PlsX membrane binding where the hairpin loop (Figure 3C) undergoes a conformational change at the membrane surface to become an amphipathic α-helix (Figure 3B) that inserts into the bilayer to transfer acyl phosphate to PlsY or extract acyl phosphate from the membrane. Additionally, the catalytic site is inferred to be located at the interface between the PlsX protomers, but the only structures available are ligand-free. A PlsX-acyl phosphate complex structure would validate this site, help clarify the catalytic mechanism, and determine the effect acyl phosphate binding has on the hairpin and membrane binding. It is still unclear if PlsX ever comes off the membrane in vivo and what steps in catalysis, if any, require membrane disengagement. Where is the acyl-ACP binding site and how does binding this substrate affect membrane binding? Docking calculations predict an acyl-ACP binding site on a positively charged surface patch on the opposite side of the hairpin tip [41]. This binding mode would enable PlsX to interact with acyl-ACP without disengaging from the membrane bilayer; however, direct binding experiments are needed to validate this prediction. How does acyl phosphate binding affect membrane binding? PlsX–membrane association is sensitive to lipid unsaturation [41] and may also be sensitive to changes in the membrane curvature elastic stress imposed by acyl phosphate in the bilayer. A combination of SPR and EPR could be helpful in clarifying the mechanism with this granularity.

2.3. Fatty Acid Kinase (FakAB)

The FakAB system in bacteria typically consists of one FakA and multiple FakBs to expand the spectrum of eFAs that can be utilized and establish FakB as the eFA selectivity filter. FakA forms a soluble dimer with a unique topology of a zinc finger-containing domain flanked by an amino terminal DhaL domain and carboxy terminal DegV domain [42,43]. The FakA domains can be independently expressed and purified, but there is no evidence that FakA is an interfacial enzyme or any of the domains contact the membrane bilayer. *S. aureus* encodes two FakB genes and biochemical assays show that FakB1 selectively binds the saturated fatty acids myristate and palmitate, and FakB2 selectivity binds the monounsaturated fatty acid oleate [33]. The Gram positive pathogen *Streptococcus pneumoniae* encodes three FakBs, and FakB3 can bind polyunsaturated fatty acid linoleate, linoleneate, and arachidonate in a biochemical assay [44].

The available FakB crystal structures all contain bound fatty acid since ligand-free FakB is insoluble [45]. It is unclear why ligand-free FakB is insoluble, and it could require lipid or detergent for stability. Cell compartment localization experiments have not been carried out with FakAB. The fatty acid length that FakB can accommodate is determined by an amino acid ruler at the end of the hydrophobic binding pocket, whereas fatty acid saturation selectivity arises from differences in the binding pocket shape. These differences do not impact the protein surface, so all FakBs can still interact with FakA.

The FakB structure is monomeric and consists of an amino terminal EDD fold domain and carboxy terminal six-stranded β-sheet flanked by α-helices that binds fatty acid at the interface between the domains (Figure 4A) [43,45]. The crystal structure of FakB1 shows that the fatty acid binding site is a slightly curved tunnel with an isoleucine ruler that cannot accommodate the kink of a fatty acid with a cis-double bond [45]. The crystal structure of FakB2 shows that phenylalanine and isoleucine side chains in the fatty acid binding site create a sharp turn in the tunnel at the cis-9 double bond position to accommodate the kink of monounsaturated fatty acids [46]. The crystal structure of FakB3 shows an expanded fatty acid binding site that can accommodate multiple kinks from the multiple cis-double bonds in polyunsaturated fatty acids [44]. In all cases, the fatty acid binding site is covered by a helix–loop "closed" cap made from helix α-8 surrounded by an electropositive surface (Figure 4B).

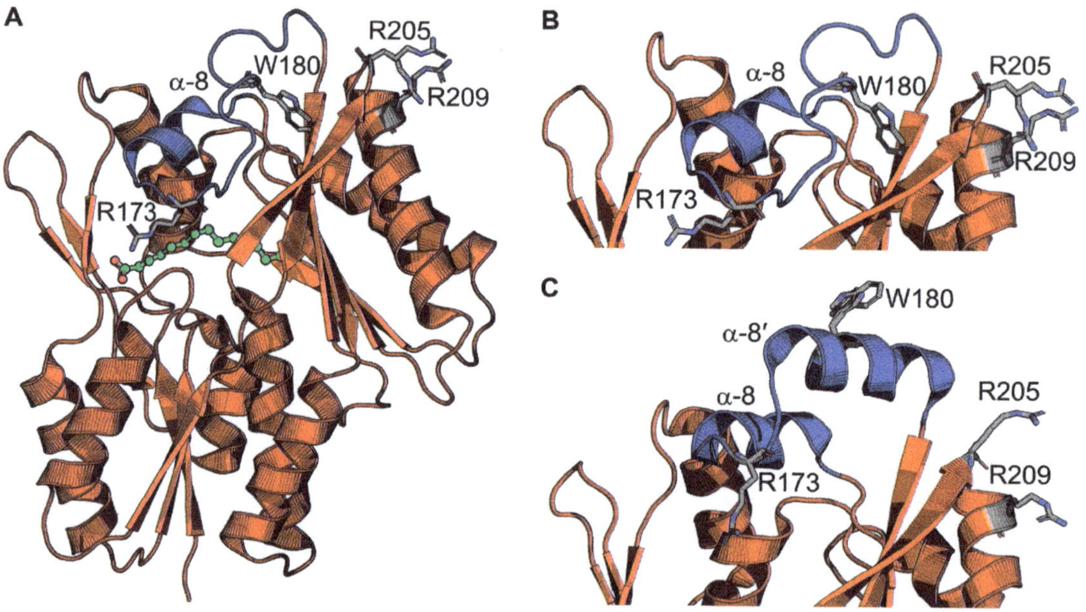

Figure 4. FakB structure and conformational changes. (**A**) The FakB structure envelopes the fatty acid (green). A cap (blue) closes the fatty acid binding site and is the membrane-binding segment. The soluble segment is orange. Cap amino acid R173 engages the fatty acid and seals the fatty acid binding site from bulk solvent. Amino acids W180, R205, and R209 are residues that bind or penetrate the membrane. (**B**,**C**) Zoomed-in views of the helix–loop "closed" cap conformation that is proposed to be the cytosolic conformation (**B**), and the helix–loop–helix "open" cap conformation that is proposed to be the membrane-bound conformation (**C**). The structures in (**A**,**B**) are from *Staphylococcus aureus* FakB1 with a "closed" cap (PDB ID: 6ALW) [45], and the structure in (**C**) is from *Staphylococcus aureus* FakB1 with an "open" cap (PDB ID: 6MH9) [16].

Liposome sedimentation experiments show FakB1 sediments with anionic phosphatidylglycerol but not zwitterionic phosphatidylcholine, and SPR direct measurements of FakB1–lipid binding measured micromolar affinity for phosphatidylglycerol and no binding to phosphatidylcholine [16]. FakB also has micromolar affinity for FakA [46]. The crystal structure of an FakB1 mutant (A121I) captured an "open" conformation of the cap where a portion of the loop forms a new amphipathic α-helix (helix α-8′) that rotates away from the protein and makes a helix–loop–helix cap that exposes the fatty acid binding site (Figure 4C) [16]. NMR analysis of FakB1 determined that the cap is dynamic in solution and MD simulations of FakB1 with an accelerated membrane model and full membrane/full lipid model of phosphatidylglycerol predicted that the new amphipathic α-helix of the "open" cap inserts below the phosphate plane of the bilayer [16]. These structural transitions are thought to create a diffusion channel for the hydrophobic fatty acid tail to access the hydrocarbon core and place the carboxyl group at the phosphate layer [16]. Mutagenesis of key amino acid side chains R205, R209, and W180 in the cap region that are predicted to insert into the membrane yield enzymes that do not bind phospholipid liposomes or catalyze the FakB1 reaction when substrate is presented in phosphatidylglycerol liposomes [16]. Cell localization experiments could be an approach to determine if FakB recruits FakA to the membrane for protein–protein interactions and catalysis at the cell membrane, or if FakB disengages from the membrane after fatty acid binding. The FakB–membrane affinity was measured with fatty acid-bound FakB and is log-fold weaker than PlsX. FakB disengagement from the membrane is likely needed to interact with FakA because site-directed mutagenesis experiments on the FakB cap implicate the cap in binding both the cell membrane and FakA. Coupled biochemical assays using FakAB, PlsX, and ACP suggest FakB can deliver fatty acid to PlsX, thus bypassing the need for a membrane [43], but the molecular mechanism of how that would work has not been determined. If FakB does disengage from the membrane after binding fatty acid in the "closed" cap conformation, then the acyl phosphate made by FakA could trigger the "open" cap conformation and enhance membrane affinity (Figure 5). NMR could be useful to compare the cap dynamics of fatty acid-bound FakB with acyl phosphate-bound FakB to better understand how the fatty acid binding site communicates with the cap. SPR and EPR could be used to study the impact of acyl phosphate on FakB–membrane association.

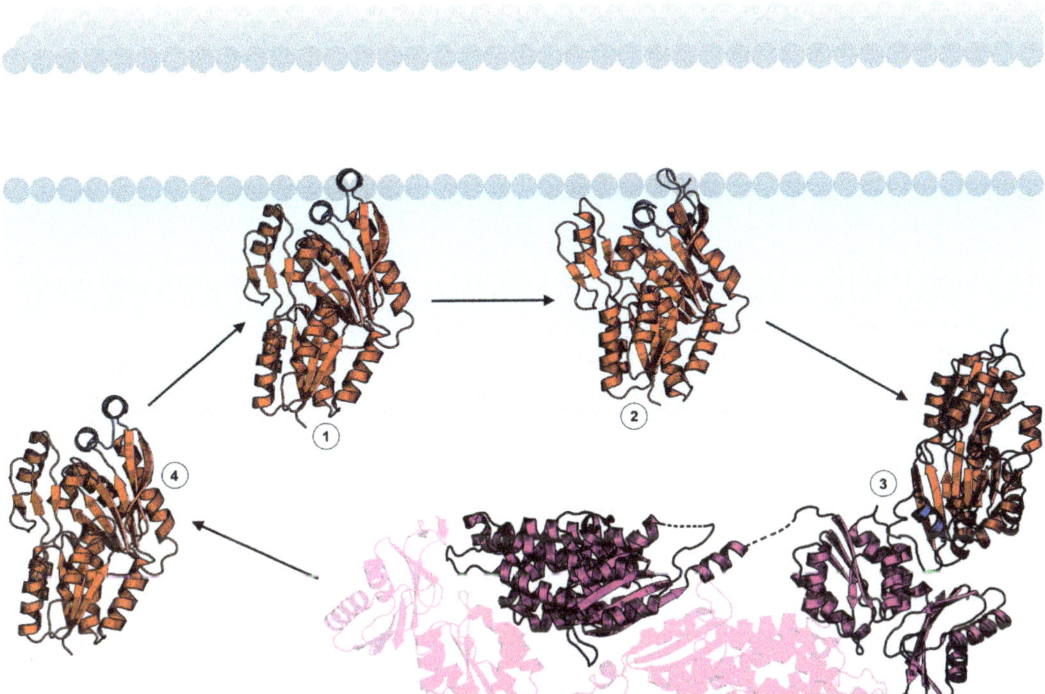

Figure 5. Hypothetical mechanism of FakB membrane binding. The FakB helix–loop–helix cap motif membrane binding domain is shown in blue and the soluble domain is orange. 1: FakB with an "open" cap scans the membrane for a fatty acid. 2: Upon binding fatty acid, the FakB cap transitions to the "closed" conformation. 3: FakB disengages from the membrane and binds FakA while in the "closed" cap conformation. FakA (magenta) is dimeric and the opposite protomer is shown with transparency. 4: After catalysis, FakB disengages from FakA in the "open" cap conformation to return to the membrane, deposit the acyl phosphate in the bilayer, and search for another fatty acid. The structures shown are *Staphylococcus aureus* FakB1 with an "open" cap (PDB ID: 6MH9) [16], *Staphylococcus aureus* FakB1 with a "closed" cap (PDB ID: 6ALW) [45], and *Streptococcus suis* FakAB2 complex (PDB ID: 7W7H) [43]. Figure created using BioRender.com.

3. Conclusions

Interfacial enzymes play a significant role in the bacterial strategy to eat environmental fatty acids, from their activation to making them compatible with the cell physiology program. Converting eFA nutrients to acyl-ACP enables their theoretical utilization in other biological processes, in addition to phospholipid synthesis such as polyketide synthesis and lipoprotein synthesis. The work discussed in this review presents current advances in understanding how key interfacial enzymes involved in eFA metabolism interact with the membrane. Studying PlsX and FakB adds to the molecular mechanistic understanding of how peripheral membrane proteins use conformational changes to precisely regulate their activation, localization, and integration into the membrane. More studies must be carried out to understand how these enzymes operate in their functional environment—the membrane. A major advance would be the structure of an interfacial enzyme embedded in the membrane. Complex visualization of an enzyme integrating with the membrane bilayer to extract or deposit its cargo would both enhance our biological understanding of this process and refine our computational capabilities to simulate these phenomena more accurately.

4. Discussion

Peripheral membrane proteins are readily purified from the cytosolic fraction of cells, which makes them suitable for detailed mechanistic studies; however, there is limited protein–membrane structural information because the amino acid sequences of their membrane binding segments are non-obvious and membrane binding studies are often omitted from their characterizations. Bioinformatic tools and workflows have been developed to predict the membrane-binding domains of modular proteins. These tools can be deployed to generate testable hypotheses about the structural elements that drive interfacial enzyme–membrane binding and add to this knowledgebase. There are also thousands of protein structures in the Protein Data Bank with the annotation "unknown function" and these tools could potentially identify peripheral membrane proteins among them. The protocol developed by Bhardwaj et al. [47] uses a support vector machine to classify proteins by net charge, distribution of cationic patches, and amino acid composition. This protocol predicts membrane binders based on the expectation that electrostatic complementarity between cationic proteins and anionic membranes is a major driver of binding. In a test set of 40 known membrane proteins and 230 non-binding proteins, the protocol was ~90% accurate in predicting membrane binding properties, and in a small sample of four structurally related C2 domains with unknown membrane binding properties, the protocol correctly predicted the single membrane binder that was confirmed by SPR. Another useful tool is the Drugging pRotein mEmbrAne Machine learning Method (DREAMM; https://dreamm.ni4os.eu/ (accessed on 1 April 2023)), which extracts physicochemical and biochemical information from a three-dimensional protein structure and predicts membrane-penetrating amino acids [48,49]. In a dataset of 54 known peripheral membrane proteins with known three-dimensional structures and experimentally known membrane-penetrating amino acids, DREAMM correctly predicted membrane-penetrating amino acids with a macro-averaged F_1 score = 0.92 and MCC = 0.84. In an independent test set with experimentally known protein–membrane regions, DREAMM demonstrated 91% precision in identifying membrane-penetrating amino acids. Bioinformatics tools like these will help advance understanding of the complexity of lipid–protein interactions at cell membranes by enabling the major step of identification of protein–membrane interaction sites of peripheral membrane proteins.

Funding: This work was supported by the National Institutes of Health, United States grant 1K99AI166116. The content is solely the responsibility of the author and does not necessarily represent the official views of the National Institutes of Health.

Data Availability Statement: Coordinates and structure factors for the structures discussed in this review are freely accessibility from the Protein Data Bank (www.rcsb.org (accessed on 1 March 2023)) using the PDB ID accession codes.

Conflicts of Interest: The author declares that they have no conflict of interest with the contents of this article. The funder had no role in the design of the study; in the collection, analyses, or interpretation of data; in the writing of the manuscript; or in the decision to publish the results.

References

1. Strahl, H.; Errington, J. Bacterial membranes: Structure, domains, and function. *Annu. Rev. Microbiol.* **2017**, *71*, 519–538. [CrossRef] [PubMed]
2. Lu, G.; Xu, Y.; Zhang, K.; Xiong, Y.; Li, H.; Cui, L.; Wang, X.; Lou, J.; Zhai, Y.; Sun, F.; et al. Crystal structure of *E. coli* apolipoprotein N-acyl transferase. *Nat. Commun.* **2017**, *8*, 15948. [CrossRef]
3. Gardiner, J.H.t.; Komazin, G.; Matsuo, M.; Cole, K.; Gotz, F.; Meredith, T.C. Lipoprotein N Acylation in *Staphylococcus aureus* Is Catalyzed by a Two-Component Acyl Transferase System. *mBio* **2020**, *11*, 01619–01620. [CrossRef] [PubMed]
4. Chen, X.; Alonzo, F., 3rd. Bacterial lipolysis of immune-activating ligands promotes evasion of innate defenses. *Proc. Natl. Acad. Sci. USA* **2019**, *116*, 3764–3773. [CrossRef] [PubMed]
5. Seelig, J. Thermodynamics of lipid-peptide interactions. *Biochim. Biophys. Acta* **2004**, *1666*, 40–50. [CrossRef] [PubMed]
6. Whited, A.M.; Johs, A. The interactions of peripheral membrane proteins with biological membranes. *Chem. Phys. Lipids* **2015**, *192*, 51–59. [CrossRef] [PubMed]
7. Malmberg, N.J.; Van Buskirk, D.R.; Falke, J.J. Membrane-docking loops of the cPLA2 C2 domain: Detailed structural analysis of the protein-membrane interface via site-directed spin-labeling. *Biochemistry* **2003**, *42*, 13227–13240. [CrossRef] [PubMed]
8. Gamsjaeger, R.; Johs, A.; Gries, A.; Gruber, H.J.; Romanin, C.; Prassl, R.; Hinterdorfer, P. Membrane binding of β2-glycoprotein I can be described by a two-state reaction model: An atomic force microscopy and surface plasmon resonance study. *Biochem. J.* **2005**, *389*, 665–673. [CrossRef]
9. Verma, M.L.; Azmi, W.; Kanwar, S.S. Microbial lipases: At the interface of aqueous and non-aqueous media. A review. *Acta Microbiol. Immunol. Hung.* **2008**, *55*, 265–294. [CrossRef]
10. Del Vecchio, K.; Stahelin, R.V. Using Surface Plasmon Resonance to Quantitatively Assess Lipid-Protein Interactions. *Methods Mol. Biol.* **2016**, *1376*, 141–153. [CrossRef]
11. Sahu, I.D.; Lorigan, G.A. Electron paramagnetic resonance as a tool for studying membrane proteins. *Biomolecules* **2020**, *10*, 763. [CrossRef] [PubMed]
12. Smirnova, T.I.; Smirnov, A.I. Peptide-membrane interactions by spin-labeling EPR. *Methods Enzymol.* **2015**, *564*, 219–258. [CrossRef] [PubMed]
13. Opella, S.J.; Marassi, F.M. Applications of NMR to membrane proteins. *Arch. Biochem. Biophys.* **2017**, *628*, 92–101. [CrossRef] [PubMed]
14. Oxenoid, K.; Chou, J.J. A functional NMR for membrane proteins: Dynamics, ligand binding, and allosteric modulation. *Protein Sci.* **2016**, *25*, 959–973. [CrossRef]
15. Pant, S.; Tajkhorshid, E. Microscopic characterization of GRP1 PH domain interaction with anionic membranes. *J. Comput. Chem.* **2020**, *41*, 489–499. [CrossRef]
16. Gullett, J.M.; Cuypers, M.G.; Grace, C.R.; Pant, S.; Subramanian, C.; Tajkhorshid, E.; Rock, C.O.; White, S.W. Identification of structural transitions in bacterial fatty acid binding proteins that permit ligand entry and exit at membranes. *J. Biol. Chem.* **2022**, *298*, 101676. [CrossRef]
17. Parsons, J.B.; Rock, C.O. Bacterial lipids: Metabolism and membrane homeostasis. *Prog. Lipid Res.* **2013**, *52*, 249–276. [CrossRef]
18. White, S.W.; Zheng, J.; Zhang, Y.-M.; Rock, C.O. The structural biology of type II fatty acid biosynthesis. *Annu. Rev. Biochem.* **2005**, *74*, 791–831. [CrossRef]
19. Radka, C.D.; Rock, C.O. Mining fatty acid biosynthesis for new antimicrobials. *Annu. Rev. Microbiol.* **2022**, *76*, 281–304. [CrossRef]
20. Wittke, F.; Vincent, C.; Chen, J.; Heller, B.; Kabler, H.; Overcash, J.S.; Leylavergne, F.; Dieppois, G. Afabicin, a first-in-class antistaphylococcal antibiotic, in the treatment of acute bacterial skin and skin structure infections: Clinical noninferiority to vancomycin/linezolid. *Antimicrob. Agents Chemother.* **2020**, *64*, e00250-20. [CrossRef]
21. Vuong, C.; Yeh, A.J.; Cheung, G.Y.; Otto, M. Investigational drugs to treat methicillin-resistant *Staphylococcus aureus*. *Expert Opin. Investig. Drugs* **2016**, *25*, 73–93. [CrossRef]
22. Quehenberger, O.; Dennis, E.A. The human plasma lipidome. *N. Engl. J. Med.* **2011**, *365*, 1812–1823. [CrossRef] [PubMed]
23. Radka, C.D.; Batte, J.L.; Frank, M.W.; Rosch, J.W.; Rock, C.O. Oleate hydratase (OhyA) is a virulence determinant in *Staphylococcus aureus*. *Microbiol. Spectr.* **2021**, *9*, e0154621. [CrossRef]
24. Brinster, S.; Lamberet, G.; Staels, B.; Trieu-Cuot, P.; Gruss, A.; Poyart, C. Type II fatty acid synthesis is not a suitable antibiotic target for Gram-positive pathogens. *Nature* **2009**, *458*, 83–86. [CrossRef] [PubMed]

25. Morvan, C.; Halpern, D.; Kenanian, G.; Pathania, A.; Anba-Mondoloni, J.; Lamberet, G.; Gruss, A.; Gloux, K. The *Staphylococcus aureus* FASII bypass escape route from FASII inhibitors. *Biochimie* **2017**, *141*, 40–46. [CrossRef]
26. Kenanian, G.; Morvan, C.; Weckel, A.; Pathania, A.; Anba-Mondoloni, J.; Halpern, D.; Gaillard, M.; Solgadi, A.; Dupont, L.; Henry, C.; et al. Permissive fatty acid incorporation promotes staphylococcal adaptation to FASII antibiotics in host environments. *Cell Rep.* **2019**, *29*, 3974–3982.e3974. [CrossRef]
27. Balemans, W.; Lounis, N.; Gilissen, R.; Guillemont, J.; Simmen, K.; Andries, K.; Koul, A. Essentiality of FASII pathway for *Staphylococcus aureus*. *Nature* **2010**, *463*, E3, discussion E4. [CrossRef] [PubMed]
28. Antimicrobial Resistance, C. Global burden of bacterial antimicrobial resistance in 2019: A systematic analysis. *Lancet* **2022**, *399*, 629–655. [CrossRef]
29. Lu, Y.-J.; Zhang, Y.-M.; Grimes, K.D.; Qi, J.; Lee, R.E.; Rock, C.O. Acyl-phosphates initiate membrane phospholipid synthesis in Gram-positive pathogens. *Mol. Cell* **2006**, *23*, 765–772. [CrossRef] [PubMed]
30. Yao, J.; Rock, C.O. Phosphatidic acid synthesis in bacteria. *Biochim. Biophys. Acta* **2013**, *1831*, 495–502. [CrossRef] [PubMed]
31. Paoletti, L.; Lu, Y.-J.; Schujman, G.E.; de Mendoza, D.; Rock, C.O. Coupling of fatty acid and phospholipid synthesis in *Bacillus subtilis*. *J. Bacteriol.* **2007**, *189*, 5816–5824. [CrossRef]
32. Zhang, Y.-M.; Rock, C.O. Acyltransferases in bacterial glycerophospholipid synthesis. *J. Lipid Res* **2008**, *49*, 1867–1874. [CrossRef] [PubMed]
33. Parsons, J.B.; Broussard, T.C.; Bose, J.L.; Rosch, J.W.; Jackson, P.; Subramanian, C.; Rock, C.O. Identification of a two-component fatty acid kinase responsible for host fatty acid incorporation by *Staphylococcus aureus*. *Proc. Natl. Acad. Sci. USA* **2014**, *111*, 10532–10537. [CrossRef]
34. Li, Z.; Tang, Y.; Wu, Y.; Zhao, S.; Bao, J.; Luo, Y.; Li, D. Structural insights into the committed step of bacterial phospholipid biosynthesis. *Nat. Commun.* **2017**, *8*, 1691. [CrossRef] [PubMed]
35. Robertson, R.M.; Yao, J.; Gajewski, S.; Kumar, G.; Martin, E.W.; Rock, C.O.; White, S.W. A two-helix motif positions the lysophosphatidic acid acyltransferase active site for catalysis within the membrane bilayer. *Nat. Struct. Mol. Biol.* **2017**, *24*, 666–671. [CrossRef] [PubMed]
36. Jiang, Y.; Dai, X.; Qin, M.; Guo, Z. Identification of an amphipathic peptide sensor of the *Bacillus subtilis* fluid membrane microdomains. *Commun. Biol.* **2019**, *2*, 316. [CrossRef]
37. Sastre, D.E.; Bisson-Filho, A.; de Mendoza, D.; Gueiros-Filho, F.J. Revisiting the cell biology of the acyl-ACP: Phosphate transacylase PlsX suggests that the phospholipid synthesis and cell division machineries are not coupled in *Bacillus subtilis*. *Mol. Microbiol.* **2016**, *100*, 621–634. [CrossRef]
38. Badger, J.; Sauder, J.M.; Adams, J.M.; Antonysamy, S.; Bain, K.; Bergseid, M.G.; Buchanan, S.G.; Buchanan, M.D.; Batiyenko, Y.; Christopher, J.A.; et al. Structural analysis of a set of proteins resulting from a bacterial genomics project. *Proteins* **2005**, *60*, 787–796. [CrossRef]
39. Kim, Y.; Li, H.; Binkowski, T.A.; Holzle, D.; Joachimiak, A. Crystal structure of fatty acid/phospholipid synthesis protein PlsX from *Enterococcus faecalis*. *J. Struct. Funct. Genomics* **2009**, *10*, 157–163. [CrossRef] [PubMed]
40. Sastre, D.E.; Pulschen, A.A.; Basso, L.G.M.; Benites Pariente, J.S.; Marques Netto, C.G.C.; Machinandiarena, F.; Albanesi, D.; Navarro, M.; de Mendoza, D.; Gueiros-Filho, F.J. The phosphatidic acid pathway enzyme PlsX plays both catalytic and channeling roles in bacterial phospholipid synthesis. *J. Biol. Chem.* **2020**, *295*, 2148–2159. [CrossRef] [PubMed]
41. Sastre, D.E.; Basso, L.G.M.; Trastoy, B.; Cifuente, J.O.; Contreras, X.; Gueiros-Filho, F.; de Mendoza, D.; Navarro, M.; Guerin, M.E. Membrane fluidity adjusts the insertion of the transacylase PlsX to regulate phospholipid biosynthesis in Gram-positive bacteria. *J. Biol. Chem.* **2020**, *295*, 2136–2147. [CrossRef]
42. Subramanian, C.; Cuypers, M.G.; Radka, C.D.; White, S.W.; Rock, C.O. Domain architecture and catalysis of the *Staphylococcus aureus* fatty acid kinase. *J. Biol. Chem.* **2022**, *298*, 101993. [CrossRef] [PubMed]
43. Shi, Y.; Zang, N.; Lou, N.; Xu, Y.; Sun, J.; Huang, M.; Zhang, H.; Lu, H.; Zhou, C.; Feng, Y. Structure and mechanism for streptococcal fatty acid kinase (Fak) system dedicated to host fatty acid scavenging. *Sci. Adv.* **2022**, *8*, eabq3944. [CrossRef] [PubMed]
44. Gullett, J.M.; Cuypers, M.G.; Frank, M.W.; White, S.W.; Rock, C.O. A fatty acid-binding protein of *Streptococcus pneumoniae* facilitates the acquisition of host polyunsaturated fatty acids. *J. Biol. Chem.* **2019**, *294*, 16416–16428. [CrossRef] [PubMed]
45. Cuypers, M.G.; Subramanian, C.; Gullett, J.M.; Frank, M.W.; White, S.W.; Rock, C.O. Acyl chain selectivity and physiological roles of *Staphylococcus aureus* fatty acid binding proteins. *J. Biol. Chem.* **2019**, *294*, 38–49. [CrossRef]
46. Broussard, T.C.; Miller, D.J.; Jackson, P.; Nourse, A.; White, S.W.; Rock, C.O. Biochemical roles for conserved residues in the bacterial fatty acid binding protein family. *J. Biol. Chem.* **2016**, *291*, 6292–6303. [CrossRef]
47. Bhardwaj, N.; Stahelin, R.V.; Langlois, R.E.; Cho, W.; Lu, H. Structural bioinformatics prediction of membrane-binding proteins. *J. Mol. Biol.* **2006**, *359*, 486–495. [CrossRef]

48. Chatzigoulas, A.; Cournia, Z. Predicting protein-membrane interfaces of peripheral membrane proteins using ensemble machine learning. *Brief. Bioinform.* **2022**, *23*, bbab518. [CrossRef]
49. Chatzigoulas, A.; Cournia, Z. DREAMM: A web-based server for drugging protein-membrane interfaces as a novel workflow for targeted drug design. *Bioinformatics* **2022**, *38*, 5449–5451. [CrossRef]

membranes

Article

Borate Transporters and SLC4 Bicarbonate Transporters Share Key Functional Properties

Jean L. Beltran, Lila G. McGrath, Sophia Caruso, Richara K. Bain, Claire E. Hendrix, Hana Kamran, Hartlee G. Johnston, Rebecca M. Collings, Menkara-Chinua N. Henry, Tsega-Ab L. Abera, Valeria A. Donoso, Erin C. Carriker and Bryan H. Thurtle-Schmidt *

Department of Biology, Davidson College, Davidson, NC 28035, USA
* Correspondence: brthurtleschmidt@davidson.edu

Abstract: Borate transporters are membrane transport proteins that regulate intracellular borate levels. In plants, borate is a micronutrient essential for growth but is toxic in excess, while in yeast, borate is unnecessary for growth and borate export confers tolerance. Borate transporters share structural homology with human bicarbonate transporters in the SLC4 family despite low sequence identity and differences in transported solutes. Here, we characterize the *S. cerevisiae* borate transporter Bor1p and examine whether key biochemical features of SLC4 transporters extend to borate transporters. We show that borate transporters and SLC4 transporters share multiple properties, including lipid-promoted dimerization, sensitivity to stilbene disulfonate-derived inhibitors, and a requirement for an acidic residue at the solute binding site. We also identify several amino acids critical for Bor1p function and show that disease-causing mutations in human SLC4A1 will eliminate in vivo function when their homologous mutations are introduced in Bor1p. Our data help elucidate mechanistic features of Bor1p and reveal significant functional properties shared between borate transporters and SLC4 transporters.

Keywords: membrane transporters; protein–lipid interactions; Bor1; SLC4A1; yeast

Citation: Beltran, J.L.; McGrath, L.G.; Caruso, S.; Bain, R.K.; Hendrix, C.E.; Kamran, H.; Johnston, H.G.; Collings, R.M.; Henry, M.-C.N., Abera, T.-A.L.; et al. Borate Transporters and SLC4 Bicarbonate Transporters Share Key Functional Properties. *Membranes* **2023**, *13*, 235. https://doi.org/10.3390/membranes13020235

Academic Editors: Yosuke Senju and Shiro Suetsugu

Received: 27 January 2023
Revised: 10 February 2023
Accepted: 13 February 2023
Published: 15 February 2023

1. Introduction

Bicarbonate transporters in the solute carrier (SLC) 4 family include electroneutral anion exchangers and sodium-coupled cotransporters. The archetypal SLC4 anion exchanger is SLC4A1, also known as Band 3 or Anion Exchanger 1 (AE1). SLC4A1 is the most abundant membrane protein in human red blood cells [1], and reversibly exchanges bicarbonate and chloride ions. Anion exchange can be blocked by stilbene disulfonate-derived inhibitors, and diseases such as renal tubular acidosis, hereditary spherocytosis, and hereditary stomatocytosis can be caused by mutations in the membrane transport domain of SLC4A1 [2]. Despite low sequence identity, SLC4 transporters share structural homology with borate efflux transporters [3,4], which were first discovered in plants [5,6]. Borate is an essential plant micronutrient that enters from the soil and participates in the formation of borate esters that contribute to plant cell wall stability [7,8]. However, excess borate levels are toxic to plants, and thus the regulation of borate by transporters is important for plant viability. Borate levels are regulated partly by BOR1, a borate exporter that promotes borate transport from roots to shoots and leaves by xylem loading [6]. BOR1 is active in *A. thaliana* under limiting borate conditions, but is degraded under high borate concentrations to avoid the accumulation of toxic borate levels in plant shoots [9]. In *S. cerevisiae*, borate is not required for growth, and the function of its borate transporter Bor1p (ScBor1p) is important for mediating resistance to toxic levels of borate [10]. Structures of borate transporters BOR1 from *A. thaliana* [3], Bor1p from *S. mikatae* [4], and BOR3 from *O. sativa* [11], along with structures of SLC4A1 [12–15], SLC4A4 [16], and SLC4A8 [17], all share the same homodimeric assembly, in which centralized Gate domains mediate dimerization and outer

Core domains contain the solute binding pocket between the transmembrane helices TM3 and TM10. Several of these structures occupy different states in the transport cycle, and structural comparisons show that the Gate domains remain relatively rigid throughout the transport cycle while the Core domains are mobile and slide to generate alternating access for bound solutes; these observations are most consistent with an elevator mechanism of transport [3,15,18]. The same architecture of homodimers, composed of Gate and Core domains, has also been seen in the SLC23 orthologs UraA [19,20] and UapA [21], as well as SLC26 proteins like prestin [22] and orthologs BicA [23] and SLC26Dg [24], showing that the SLC4, SLC23, and SLC26 families all share a conserved fold and that insights gained from studying one family may extend to a diverse set of membrane transport proteins.

Despite the structural similarities between borate transporters and SLC4 transporters, one striking difference is that they differ in solute; all but one of the ten human SLC4 transporters transport bicarbonate, while borate transporters instead transport borate. Thus, it remains an open question as to what other biophysical characteristics and functional similarities might extend from SLC4 transporters to borate transporters and vice versa. In particular, the role lipids play in mediating multimeric assembly, whether borate transporters display sensitivity to stilbene disulfonate-derived inhibitors, and structural and mechanistic similarities between amino acids at or near the solute binding site all remain outstanding questions. Because *S. cerevisiae* Bor1p can be overexpressed and purified and has a readily observable phenotype [25], we characterized Bor1p in more detail to learn if its similarities to human SLC4 transporters extend beyond sequence and structural homology.

2. Materials and Methods

2.1. Protein Overexpression and Purification

ScBor1p was overexpressed and purified as described previously [25]. Briefly, protein constructs bearing C-terminal deca-histidine tags in pRS423-derived plasmids were expressed under galactose-inducible control in *S. cerevisiae* strain BTSY1 (MATα his3::GAL1-GAL4 pep4 prb1-1122) [26]. Cells were lysed by bead beating, and membranes were harvested by differential centrifugation. Following protein solubilization in 1% (*w*/*v*) n-Dodecyl-beta-D-Maltopyranoside (DDM), protein was purified by performing nickel affinity chromatography followed by size exclusion chromatography. Protein was injected onto a Superdex 200 Increase 10/300 GL column (Cytiva) equilibrated in 20 mM Mes pH 6.5, 100 mM NaCl, 2% glycerol, and 0.03% DDM. To compare size exclusion chromatograms between wild-type and mutant ScBor1p, preparations of 1 L of cell culture were grown and harvested under identical conditions.

2.2. Proteoliposome Reconstitution and Crosslinking Assays

Purified ScBor1p was reconstituted into lipids by adapting a protocol using preformed vesicles permeabilized with Triton X-100 (EMD Millipore, Billerica, MA, USA) [27]. 1-palmitoyl-2-oleoyl-glycero-3-phosphocholine (PC) and 1-palmitoyl-2-oleoyl-sn-glycero-3-phospho-(1′-rac-glycerol) (PG) were each acquired suspended in chloroform (Avanti Polar Lipids, Alabaster, AL, USA). Lipid solutions were dried to remove chloroform first under nitrogen gas stream and then under vacuum. The dried lipid film was rehydrated at 10 mg/mL in 20 mM Mes (pH 6.5), 100 mM NaCl, and 2% glycerol by incubating at 37 °C for 30 min and resuspending through pipetting. To form unilamellar vesicles, the lipid suspension was subjected to 4 cycles of freeze–thaw, with sonication in a room-temperature water bath for 5 min between freeze–thaw cycles. The lipids were then extruded through a 100 nm filter with 9 passages using a Mini Extruder apparatus (Avanti Polar Lipids, Alabaster, AL, USA). To prepare proteoliposomes, the liposomes were destabilized by adding 0.6% (*v*/*v*) Triton X-100 and were incubated at 4 °C overnight under gentle nutation. Purified ScBor1p was added to the destabilized liposomes at a 1:20 protein:lipid (*w*:*w*) ratio and incubated for 1 h at 4 °C. To extract detergent, SM2 Bio-beads (Bio-Rad, Hercules, CA, USA) were added sequentially in 2 steps. First, 50 mg SM2 Bio-beads were added and incubated for 1 h at 4 °C, followed by the addition of another 50 mg of SM2 Bio-beads and another incubation for 1 h at 4 °C. The glutaraldehyde cross-linking assay was performed as described previously [25], with cross-linking performed for 5 min at room temperature with 0.15% glutaraldehyde (Electron Microscopy Sciences, Hatfield, PA, USA) in the presence or absence of a 5 min pre-treatment of 2% (*w*/*v*) sodium dodecyl sulfate (SDS) detergent (Sigma-Aldrich, St. Louis, MO, USA).

2.3. Genetic Plating Assay

The genetic plating assay was performed as described previously [26]. A yeast strain (BTSY2) with a knocked out BOR1 (MATα his3::GAL1-GAL4 pep4 prb1-1122 bor1::KanMX) was transformed with pRS423-derived plasmids that are under inducible control of the GAL1 promoter and possess a C-terminal deca-histidine tag. Plasmids differ only by the transporter they encode or the specified mutation they bear. Overnight cultures were grown in media consisting of yeast nitrogen base (YNB + nitrogen), complete supplement mixture lacking histidine, and were supplemented with adenine (CSM-His w/Ade40), with 2% raffinose for a sugar source. Cells were diluted to an OD600 of 0.5 and then serial fivefold dilutions were prepared and added to plates containing CSM-His w/Ade40, 2% raffinose, 0.05% galactose to induce expression, and either 0 mM, 10 mM, or 20 mM boric acid to challenge yeast growth. 10 µL of each dilution was plated, and plates were stored at 30 °C for 5 days until imaged. Images presented here are representative plates from at least three biologically independent replicates. For the SITS and H_2DIDS experiments, the protocol was identical. However, per L of media we used 1 g monosodium glutamate instead of 5 g ammonium sulfate for a nitrogen source. Additionally, inhibitors were prepared at 50 mM in dimethyl sulfoxide (DMSO), which permitted adding up to 300 µM inhibitors with only 0.6% DMSO present. Therefore, 0.6% DMSO was used in all other lesser inhibitor concentrations and control conditions.

2.4. Borate Quantification Assay

The spectrophotometric assay to quantify borate from cell lysates was adapted from a previously published protocol [28]. Yeast colonies were used to inoculate overnight cultures, which were then seeded to an OD600 of 0.25 the following morning and allowed to grow for 7 h until the OD600 was around 0.9. Protein expression was induced for 16 h by the addition of 2% galactose, after which cultures had 1 mM boric acid added for 90 min. Cells were pelleted and washed with water before being pelleted a second time, resuspended in 350 µL water, and lysed by incubating at 98 °C for 30 min. After a 5 min spin at 16,100× *g*, 300 µL of supernatant was prepared for curcumin addition per the previously reported protocol, with the data collected using a quartz cuvette that was washed twice with 91% isopropyl alcohol in between measurements. Standard curves were generated from standards containing 0, 0.625, 1.25, 2.5, and 5.0 mg/L of borate. To avoid the risk of borate entering the experiment via borosilicate glassware, no solutions in these experiments touched glass. 95% confidence intervals were generated for n = 7 biologically independent experiments.

2.5. Western Blot

Protein overexpression protocols were adapted from those described above, but with the following distinctions: 50 mL yeast cultures were grown using strain BTSY1, and after 16 h induction by 2% galactose the cells were lysed by passing once through an Emulsiflex-B15 homogenizer (Avestin, Ottawa, ON, Canada) at 22,500 psi. Membranes were collected by differential centrifugation and resuspended in 20 mM Tris pH 7.4, 1 mM EDTA, 1 mM PMSF, 10% glycerol, and 300 mM NaCl. For Western blot analysis, 10 µg total protein was loaded per lane, and proteins were transferred to a PVDF membrane in a wet tank. The anti-His_6, rabbit polyclonal primary antibody (Fisher, Waltham, MA, USA, Cat# PIPA1983B) was incubated at 4 °C overnight, and the HRP-conjugated goat anti-rabbit secondary antibody (Invitrogen, Waltham, MA, USA, Cat# 32460) was incubated at room temperature for one hour before exposure using chemiluminescent horseradish peroxidase substrate to image. Total protein loading controls were imaged using Bio-Rad stain-free gels.

3. Results and Discussion

3.1. A Dispensable N-Terminal Tail and a Conserved Functional Membrane Transport Fold

All ten transporters in the human SLC4 family are composed of an N-terminal cytoplasmic domain linked to a C-terminal membrane transport domain. A sequence alignment of ScBor1p with human SLC4A1 shows just over 24% sequence identity between their transporter domains (Figure S1). Fungal borate transporters lack the cytoplasmic domain of SLC4 transporters, and ScBor1p instead has a ~50 amino acid N-terminal tail that differentiates it from plant borate transporters, as well as human SLC4 orthologs (Figure 1A). The structural model of ScBor1p from the AlphaFold server predicts this N-terminal region to be composed of a long alpha helix running parallel to the membrane and preceding helix H1 (Figure 1B) [29], but any functional significance of this region is unknown. To test whether this N-terminal region is important for function, we generated N-terminally truncated constructs and tested them for phenotypes in a genetic plating assay. In this experiment and those to follow, plates containing no boric acid are controls that are expected to show equivalent growth for all samples. Only cells that express a functional borate transporter can grow on the plates with 20 mM boric acid, while plates containing 10 mM boric acid can identify more subtle phenotypic changes. We used the nonfunctional yeast aquaporin AQY1 as a negative control [30]. Our results show that the N-terminal region of ScBor1p is not essential through the first 53 residues but that truncating the first 61 amino acids results in a total loss of ability to grow on plates containing 20 mM boric acid (Figure 1C). A previous study in the membrane domain of human SLC4A1 showed that deleting amino acids 381–385 results in lost transport activity [31]. Residues 381–385 in human SLC4A1 and residues 54–61 in ScBor1p overlap by one amino acid in a multiple sequence alignment (Figure S1), and both regions encapsulate the beginning of helix H1 (Figure 1B). Our data demonstrate that the first 53 amino acids of ScBor1p are dispensable for in vivo function, and that borate transporters and SLC4 transporters share a conserved functional core in which the beginning of helix H1 is important for function.

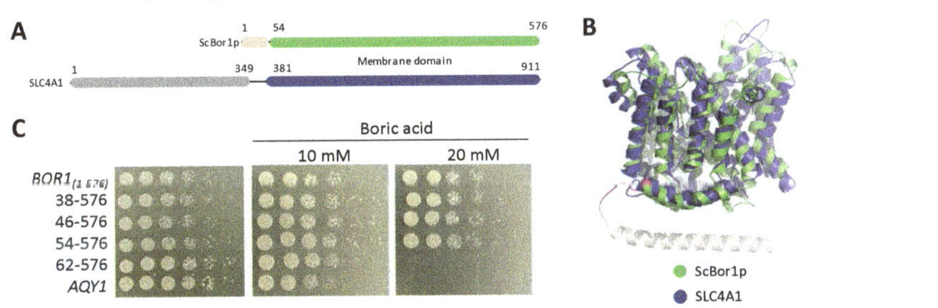

Figure 1. An N-terminal region is dispensable for ScBor1p function. (**A**) Schematic showing the domain arrangements of ScBor1p and human SLC4A1. The membrane domains are in green and blue, respectively. SLC4A1 cytosolic ankyrin-binding domain is in gray. (**B**) Superposition of a cryoEM structure of the membrane domain (blue) of human SLC4A1 (PDB ID: 8CT3) [14] and the AlphaFold model of ScBor1p (green) rendered in PyMOL with RMSD = 2.993 Å. The dispensable region of the N-terminal tail is tan; the region that results in no function when deleted is pink and marks the beginning of helix H1. (**C**) Plasmids encoding the specified ScBor1p truncated construct or negative control *AQY1* were transformed into *bor1* deletion cells and the ability of each to rescue growth was tested by plating fivefold serial dilutions onto plates containing CSM-His selective media and the indicated boric acid concentrations. Plates were incubated at 30 °C and imaged after 5 days.

3.2. Lipids Promote ScBor1p Dimerization

ScBor1p could be readily purified as described previously [25]. When solubilized and purified in DDM, our group and others have shown that ScBor1p is almost entirely monomeric [25,32]. However, previous studies have determined experimental structures of *A. thaliana* Bor1 and the yeast *S. mikatae* Bor1p (SmBor1p) in dimeric assemblies [3,4], and SmBor1p has 89% sequence identity to ScBor1p (Figure S1). To better understand the physical basis of ScBor1p monomer-dimer equilibrium, we prepared proteoliposomes and performed a glutaraldehyde covalent cross-linking assay that we have previously used to assess dimerization in detergent [25]. We tested phosphatidylcholine (PC) because it is the most abundant glycerophospholipid in *S. cerevisiae* [33], and we used phosphatidylglycerol (PG) as a control because a previous study showed that it promoted ScBor1p dimerization, despite its typical absence in yeast membranes [32]. The PC and PG preparations each contain identical mixed acyl fatty acid chains of 16:0 and 18:1, so the lipids differ only in their head group identity. Our results show that PC partially shifts the sample towards dimerization, while PG results in a more modest shift than PC (Figure 2). Controls in which SDS was added prior to the addition of glutaraldehyde demonstrate that the dimerization we observe is minimal in denaturing conditions.

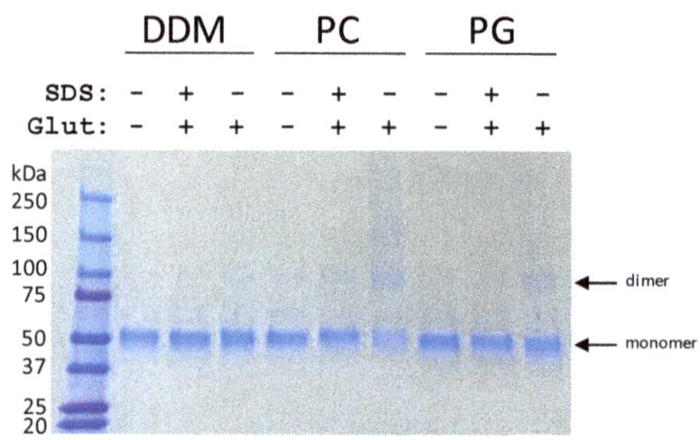

Figure 2. Cross-linking shows lipid-promoted dimerization of ScBor1p. It is indicated above each lane whether the sample is in DDM or reconstituted with PC or PG lipid. "SDS" indicates a 5-min pre-treatment of 2% sodium dodecyl sulfate before the addition of 0.15% glutaraldehyde for 5 min.

A previous study on ScBor1p showed that PS and PE each shift ScBor1p to just under 50% dimer, while PC did not support dimerization [32]. There are several notable differences between the two approaches. In the prior study, detergent-solubilized protein had lipid added to it and then was subjected to mass spectrometry. In our study, protein was reconstituted into proteoliposomes, detergent was removed, and dimerization was assessed by a glutaraldehyde cross-linking assay [20,25]. The glutaraldehyde cross-linking assay depends on the presence of primary amino groups from lysines in proximity to one another. Phosphatidylserine (PS) and phosphatidylethanolamine (PE) each possess a primary amino group and are therefore incompatible with a glutaraldehyde-based crosslinking assay. There are several possible interpretations of our observed shift from PC compared to the absence of dimer observed from PC in the former study. One interpretation is that proteoliposomes are closer to physiological conditions than those performed in the other study, which could explain the greater dimerization seen for PC in our experiment. Likewise, perhaps ScBor1p assembly into PS or PE proteoliposomes would shift the equilibria further towards dimer, but we cannot assess that due to PS and PE's incompatibility with glutaraldehyde. A second possibility is that no one lipid drives the equilibrium towards complete dimerization, but that a more complex lipid mixture, like what occurs in vivo, could. Further studies will be required to understand the relationships between lipid binding and multimerization. Our data and the prior study are in agreement that multiple lipids can promote dimerization, though not to more than half of the sample in tested conditions. Interestingly, experimental structures of human SLC4A1 show lipids bind in the crevice in between Gate domains at the dimerization interface [13,14]. Additionally, the structurally similar SLC23 homolog UapA has been shown to display lipid-mediated dimerization [34], and the human SLC26 protein prestin has been shown to have cholesterol bridge contacts between its dimer interface [22]. Likewise, the fumarate transporter SLC26Dg purifies as a monomer in DDM but is observed to show dimers in lipids [24,35], just as we observe for ScBor1p. Lipid-mediated multimeric assembly between Gate domains may therefore be broadly conserved among the SLC4, SLC23, and SLC26 families.

3.3. Sensitivity to Derivatives of Stilbene Disulfonate

One hallmark feature of human SLC4 transporters is the inhibition of their anion transport by stilbene disulfonate-derived inhibitors [36,37]. If ScBor1p is a strong model for studying SLC4 transporters, we predict that borate transporters would be inhibited by stilbene disulfonate-derived inhibitors such as 4-acetamido-4′-isothiocyanatostilbene-2,2′-disulfonic acid (SITS). To determine whether borate transporters can be inhibited by SITS, we tested the sensitivity of ScBor1p and the *A. thaliana* Bor4 transporter (AtBOR4) to SITS through genetic plating assays. We chose AtBOR4 both because the sensitivity of plant borate transporters to SITS is unknown and because a prior study showed that among all seven *A. thaliana* borate transporters, BOR4 shows the strongest growth phenotype against boric acid in *bor1* deletion cells [26]. Our results show that ScBor1p and AtBOR4 are each inhibited by SITS (Figure 3). Importantly, negative controls show that the highest SITS concentration tested has no deleterious effect on growth in plates lacking boric acid, demonstrating that SITS toxicity is specific to the cellular context in which borate transport is necessary for survival. A plating assay with 4,4′-diisothiocyanatodihydrostilbene-2,2′-disulfonic acid (H_2DIDS) shows borate transporter sensitivity, though H_2DIDS also shows some toxicity in non-borate testing conditions. Therefore, the effect of H_2DIDS on ScBor1p is less clear (Figure S2). A previous study showed that ScBor1p could bind to a resin conjugated to SITS [38]; here we show in vivo evidence demonstrating that both ScBor1p and AtBOR4 are inhibited by SITS. Borate transporter similarities with SLC4 transporters thus include sensitivity to the same small molecule inhibitors.

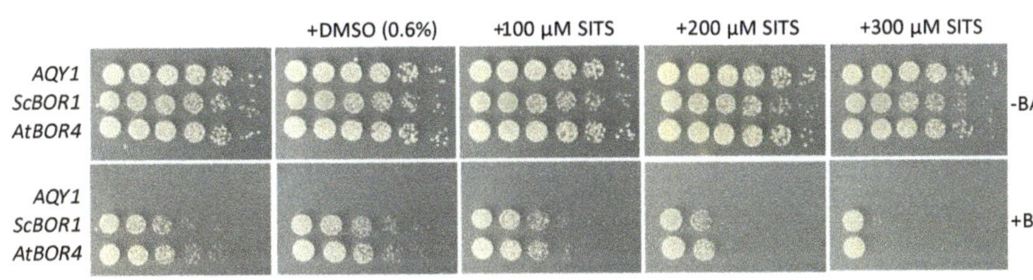

Figure 3. Borate transporters are sensitive to SITS. Plasmids encoding the specified gene were tested against increasing concentrations of SITS in 0.6% dimethyl sulfoxide (DMSO) in the presence or absence of 20 mM boric acid and plated on CSM-His selective media. Plates were incubated at 30 °C and imaged after 5 days.

3.4. Identifying Functional Amino Acids at the Solute Binding Site

Previous studies in SLC4A1 show that the solute binding site contains a glutamate, E681 in human numbering, that is essential for function [39–41]. An acidic residue in this position is perfectly conserved in the SLC4 family (Figure 4A), and previous studies have suggested that the difference between the presence of a glutamate or aspartate in this position could influence whether the transporter is an anion exchanger or a sodium co-transporter, respectively [16]. In borate transporters, this position is an invariant aspartate. Previous work has shown that alanine substitutions at position D347 in ScBor1p and the homologous D311 in AtBor1 each abolish function [3,11]. To test how more conservative changes to D347 might impact function, we assayed the effect of D347E and D347N in genetic plating assays. Interestingly, we show that the D347E mutation results in a hypomorphic phenotype, while a mutation as conservative as D347N results in no growth (Figure 4B). Western blot analysis shows that the least conservative mutation,

D347A, shows robust expression like the wild-type protein, and therefore a decrease in expression cannot explain the phenotype (Figure 4C). Because we previously reported that AtBOR1 supports no growth in yeast plated on media containing 20 mM boric acid but can nevertheless display borate efflux activity through assaying borate quantities in cell lysates [26], we tested whether these amino acid changes to D347 directly affected borate efflux activity by using a spectrophotometric assay for quantifying borate [28]. Our results show that the D347E mutation results in decreased but significant borate transport relative to negative controls, while the D347N mutation abolishes borate transport to levels indistinguishable from negative controls (Figure 4D). Taken together, the multiple sequence alignment, genetic data, and borate quantification data suggest a requirement for an acidic residue at this location in borate transporters and SLC4 transporters alike. That the most conservative mutation tested, D347N, results in a total loss of function suggests that D347 may be involved as a proton donor/acceptor.

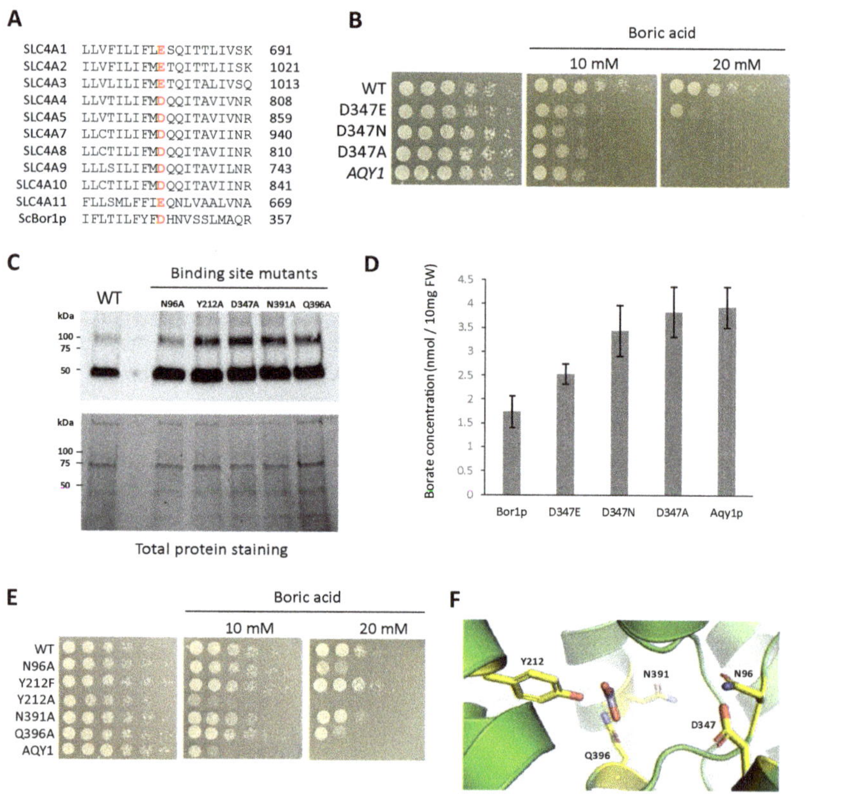

Figure 4. Identification of amino acids important for Bor1p function. (**A**) Multiple sequence alignment of a region of TM8 for all 10 human SLC4 transporters and ScBor1p. (**B**,**E**) Plasmids encoding the specified ScBor1p mutant were tested against the indicated boric acid concentrations. Plates were incubated at 30 °C and imaged after 5 days. (**C**) Western immunoblotting analysis of His-tagged ScBor1p. (**Top**) Samples were analyzed with an anti-His6-tag antibody. (**Bottom**) Total protein stain-free gel imaging of 12µg protein per lane served as loading controls. (**D**) Quantification of borate efflux activity in yeast cells expressing each indicated protein. Intracellular borate contents are reported as nmol per 10 mg fresh weight (FW) of yeast cells. Error bars represent 95% confidence intervals for n = 7 biologically independent experiments. (**F**) View of the solute-binding site of the ScBor1p AlphaFold model with tested amino acids in yellow. Bicarbonate is displayed from a superposed human SLC4A1 structure (RMSD = 3.149 Å).

Having examined D347, the amino acid most established to be critical for transport, we next wanted to identify other proximal amino acids that might be important for the solute binding and transport mechanism of ScBor1p. We performed alanine scanning mutagenesis on amino acids that are conserved in borate transporters (Figure S1) and adjacent to the cavity predicted to bind to borate and any co-transported ions. We selected hydrophilic amino acids with the assumption that they were most likely to interact directly with a polar solute. As controls we included N391A and Q396A, which were previously shown to result in reduced rescue on plates with boric acid [3]. Here we identify N96A as displaying a hypomorphic phenotype. Additionally, the Y212A substitution eliminates growth entirely, while the more conservative Y212F mutation has no effect (Figure 4E). Western blot analysis of membrane fractions shows that these mutants have robust expression and therefore their phenotypes cannot be explained by reduced protein expression (Figure 4C). Our data here identify a constellation of amino acids important for ScBor1p function. Might any of these residues interact with solute? There is no structure of a borate transporter bound to its solute, so we performed a superposition of a structure of bicarbonate-bound human SCL4A1 [42] with the AlphaFold ScBor1p model (RMSD = 3.149 Å) (Figure 4F) [29]. As suspected, the bicarbonate places in the cavity formed where TM3 and TM10 meet. Among amino acids identified in this study, Q396 and Y212 are closest to where bicarbonate superposes. In human SLC4A1, R730 is homologous with Q396 and is observed to interact directly with bicarbonate [42]. Interestingly, the carboxylate group of D347 is located 8.2 Å away from the superposed bicarbonate carbon. There is room to accommodate a sodium ion in the space between, but a previous study suggests that sodium is not coupled with borate transport in ScBor1p [43]. That same study suggested that protons are the ion whose favorable transport is coupled with the pumping of borate against its gradient. Our observation of no function for the D347N suggests it is possible that D347 is involved in proton binding.

3.5. Disease-Causing Mutations from SLC4A1 Also Eliminate ScBor1p Function

Mutations in human SLC4A1 are known to lead to several genetic disorders, including hereditary spherocytosis and hereditary stomatocytosis [2]. Prior work has shown that two disease-causing mutations in human SLC4A1, S762R and G796R, result in loss of borate transport when the homologous mutations, S466R and A500R, are introduced in AtBOR1 [11]. To test whether these same homologous disease-causing mutations impact ScBor1p and whether additional disease-causing mutations have deleterious effects on ScBor1p function, we tested the above two mutations, as well as an additional three amino acid substitutions that are linked to disease in humans and are conserved in ScBor1p (Figures 5A and S1). The aforementioned S762R and G796R are linked to hereditary stomatocytosis [44,45], and their homologous mutations in ScBor1p are T422R and G458R, respectively. Additionally, we identified three mutations in SLC4A1 that are linked to hereditary spherocytosis—G455R, D705Y, and R760Q [46–48]—which have homologous mutations in ScBor1p—G135R, D371Y, and R420Q, respectively (Figures 5A and S1). We tested all five mutations in a genetic plating assay and show that all five mutants fail to rescue growth on plates containing 20 mM boric acid (Figure 5B). To see what effect these mutations have on protein expression, we subjected each mutant ScBor1p to identical protein expression, solubilization, and purification protocols [25], and then compared their size-exclusion chromatograms. The data show a severe decrease in protein for all mutants (Figure 5C). Interestingly, the R420Q mutant promotes growth on the 10 mM plates but does not express better than the other four mutant proteins. A previous study of the SLC4A1 mutant R760Q (homologous with R420Q) showed an absence of R760Q mutant protein detected in the red blood cell membranes of a patient bearing the R760Q mutation in one of their alleles [48]. These data suggest that the five mutations have impacts on the folding, stability, or trafficking of ScBor1p that could lead to lost in vivo function, though it is possible the mutant proteins express at meaningful levels in the cell but cannot be readily solubilized and purified in DDM detergent. Nonetheless, mutations equivalent to disease-causing mutations in SLC4A1 have significant deleterious effects on the in vivo function of ScBor1p.

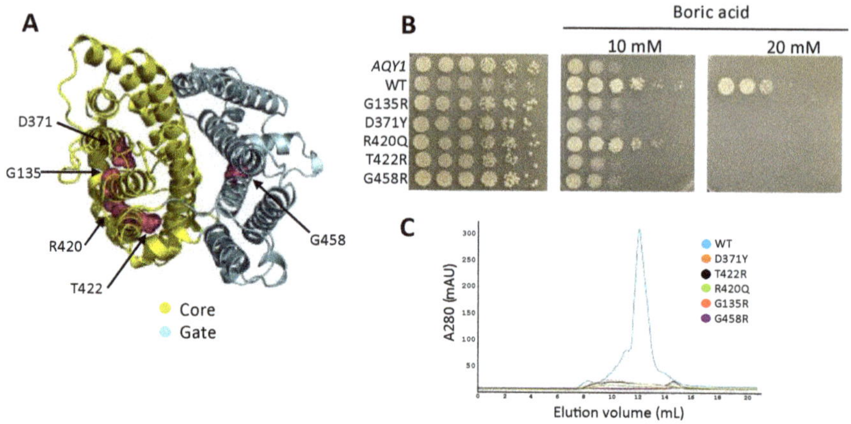

Figure 5. Impacts of disease-causing mutations in human SLC4A1 transfer to ScBor1p. (**A**) AlphaFold model of ScBor1p with homologous locations of disease-causing mutations in SLC4A1 in pink. (**B**) Plasmids encoding the specified ScBor1p mutant were tested against the indicated boric acid concentrations. Plates were incubated at 30 °C and imaged after 5 days. (**C**) Superposed chromatograms from size-exclusion chromatography performed for wild-type and mutant ScBor1p.

4. Conclusions

When borate transporters were first discovered, the sequence similarity between them and human SLC4 transporters was noted immediately [6]. More than a decade later, experimentally determined structures began to emerge for SLC4A1 [12–15], SLC4A4 [16], SLC4A8 [17], AtBOR1 [3], OsBOR3 [11], and SmBor1p [4]. Although an obvious difference exists between the solutes of these transporters—borate transporters transport borate while all but one of the ten SLC4 transporters transport bicarbonate—the structural similarities are striking, as all of them share the same homodimeric assembly composed of Gate and Core domains. Our work here shows that the similarities extend beyond their sequence and structure to include lipid-promoted dimerization, sensitivity to stilbene disulfonate-derived inhibitors, a requirement for an acidic amino acid at the solute binding site, and the conservation of deleterious impact from disease-causing mutations.

Studies of the one SLC4 transporter that does not transport bicarbonate, SLC4A11, reveal additional connections between the SLC4 family and borate transport. When SLC4A11 was first characterized it was initially proposed to transport borate [49], but this claim proved controversial and evidence has since emerged that human SLC4A11 cannot transport borate but rather is likely to transport H^+/OH^- [26,50–52]. A recent report, however, shows that seawater fish use their SLC4A11 ortholog to excrete boric acid in the kidneys, suggesting that, during the history of vertebrate evolution, either mammalian SLC4A11 lost boric acid transport activity or saltwater fish acquired boric acid transport activity [53]. Further studies are likely to establish more evolutionary, structural, and mechanistic connections between borate transporters and the SLC4 family.

Supplementary Materials: The following supporting information can be downloaded at: https://www.mdpi.com/article/10.3390/membranes13020235/s1, Figure S1: Alignment of borate transporters with human SLC4A1; Figure S2: Sensitivity to H_2DIDS.

Author Contributions: B.H.T.-S. conceived, administered, and supervised the project. J.L.B., L.G.M., S.C., R.K.B., C.E.H., H.K., H.G.J., R.M.C., M.-C.N.H., T.-A.L.A., V.A.D. and E.C.C. designed and performed experiments and interpreted data. B.H.T.-S. wrote the manuscript with input from all authors. All authors have read and agreed to the published version of the manuscript.

Funding: Research reported in this study was supported by the National Institute of General Medical Sciences of the National Institutes of Health under Award Number R15GM132786.

Institutional Review Board Statement: Not applicable.

Data Availability Statement: The data that support the findings of this study are available on request from the corresponding author, B.H.T.-S.

Conflicts of Interest: The authors declare no conflict of interest.

References

1. Fairbanks, G.; Steck, T.L.; Wallach, D.F. Electrophoretic Analysis of the Major Polypeptides of the Human Erythrocyte Membrane. *Biochemistry* **1971**, *10*, 2606–2617. [CrossRef] [PubMed]
2. Reithmeier, R.A.F.; Casey, J.R.; Kalli, A.C.; Sansom, M.S.P.; Alguel, Y.; Iwata, S. Band 3, the Human Red Cell Chloride/Bicarbonate Anion Exchanger (AE1, SLC4A1), in a Structural Context. *Biochim. Biophys. Acta* **2016**, *1858*, 1507–1532. [CrossRef] [PubMed]
3. Thurtle-Schmidt, B.H.; Stroud, R.M. Structure of Bor1 Supports an Elevator Transport Mechanism for SLC4 Anion Exchangers. *Proc. Natl. Acad. Sci. USA* **2016**, *113*, 10542–10546. [CrossRef] [PubMed]
4. Coudray, N.; Seyler, S.L.; Lasala, R.; Zhang, Z.; Clark, K.M.; Dumont, M.E.; Rohou, A.; Beckstein, O.; Stokes, D.L. Structure of the SLC4 Transporter Bor1p in an Inward-Facing Conformation. *Protein Sci.* **2017**, *26*, 130–145. [CrossRef]
5. Noguchi, K.; Yasumori, M.; Imai, T.; Naito, S.; Matsunaga, T.; Oda, H.; Hayashi, H.; Chino, M.; Fujiwara, T. Bor1-1, an Arabidopsis Thaliana Mutant That Requires a High Level of Boron. *Plant Physiol.* **1997**, *115*, 901–906. [CrossRef]
6. Takano, J.; Noguchi, K.; Yasumori, M.; Kobayashi, M.; Gajdos, Z.; Miwa, K.; Hayashi, H.; Yoneyama, T.; Fujiwara, T. Arabidopsis Boron Transporter for Xylem Loading. *Nature* **2002**, *420*, 337–340. [CrossRef]
7. Kobayashi, M.; Matoh, T.; Azuma, J. Two Chains of Rhamnogalacturonan II Are Cross-Linked by Borate-Diol Ester Bonds in Higher Plant Cell Walls. *Plant Physiol.* **1996**, *110*, 1017–1020. [CrossRef]
8. O'Neill, M.A.; Eberhard, S.; Albersheim, P.; Darvill, A.G. Requirement of Borate Cross-Linking of Cell Wall Rhamnogalacturonan II for Arabidopsis Growth. *Science* **2001**, *294*, 846–849. [CrossRef]
9. Takano, J.; Miwa, K.; Yuan, L.; von Wirén, N.; Fujiwara, T. Endocytosis and Degradation of BOR1, a Boron Transporter of Arabidopsis Thaliana, Regulated by Boron Availability. *Proc. Natl. Acad. Sci. USA* **2005**, *102*, 12276–12281. [CrossRef]
10. Takano, J.; Kobayashi, M.; Noda, Y.; Fujiwara, T. *Saccharomyces Cerevisiae* Bor1p Is a Boron Exporter and a Key Determinant of Boron Tolerance. *FEMS Microbiol. Lett.* **2007**, *267*, 230–235. [CrossRef]
11. Saouros, S.; Mohan, T.C.; Cecchetti, C.; Lehmann, S.; Barrit, J.D.; Scull, N.J.; Simpson, P.; Alguel, Y.; Cameron, A.D.; Jones, A.M.E.; et al. Structural and Functional Insights into the Mechanism of Action of Plant Borate Transporters. *Sci. Rep.* **2021**, *11*, 12328. [CrossRef]
12. Arakawa, T.; Kobayashi-Yurugi, T.; Alguel, Y.; Iwanari, H.; Hatae, H.; Iwata, M.; Abe, Y.; Hino, T.; Ikeda-Suno, C.; Kuma, H.; et al. Crystal Structure of the Anion Exchanger Domain of Human Erythrocyte Band 3. *Science* **2015**, *350*, 680–684. [CrossRef]
13. Xia, X.; Liu, S.; Zhou, Z.H. Structure, Dynamics and Assembly of the Ankyrin Complex on Human Red Blood Cell Membrane. *Nat. Struct. Mol. Biol.* **2022**, *29*, 698–705. [CrossRef]
14. Vallese, F.; Kim, K.; Yen, L.Y.; Johnston, J.D.; Noble, A.J.; Calì, T.; Clarke, O.B. Architecture of the Human Erythrocyte Ankyrin-1 Complex. *Nat. Struct. Mol. Biol.* **2022**, *29*, 706–718. [CrossRef]
15. Zhekova, H.R.; Jiang, J.; Wang, W.; Tsirulnikov, K.; Kayık, G.; Khan, H.M.; Azimov, R.; Abuladze, N.; Kao, L.; Newman, D.; et al. CryoEM Structures of Anion Exchanger 1 Capture Multiple States of Inward- and Outward-Facing Conformations. *Commun. Biol.* **2022**, *5*, 1372. [CrossRef]
16. Huynh, K.W.; Jiang, J.; Abuladze, N.; Tsirulnikov, K.; Kao, L.; Shao, X.; Newman, D.; Azimov, R.; Pushkin, A.; Zhou, Z.H.; et al. CryoEM Structure of the Human SLC4A4 Sodium-Coupled Acid-Base Transporter NBCe1. *Nat. Commun.* **2018**, *9*, 900. [CrossRef]
17. Wang, W.; Tsirulnikov, K.; Zhekova, H.R.; Kayık, G.; Khan, H.M.; Azimov, R.; Abuladze, N.; Kao, L.; Newman, D.; Noskov, S.Y.; et al. Cryo-EM Structure of the Sodium-Driven Chloride/Bicarbonate Exchanger NDCBE. *Nat. Commun.* **2021**, *12*, 5690. [CrossRef]
18. Ficici, E.; Faraldo-Gómez, J.D.; Jennings, M.L.; Forrest, L.R. Asymmetry of Inverted-Topology Repeats in the AE1 Anion Exchanger Suggests an Elevator-like Mechanism. *J. Gen. Physiol.* **2017**, *149*, 1149–1164. [CrossRef]
19. Lu, F.; Li, S.; Jiang, Y.; Jiang, J.; Fan, H.; Lu, G.; Deng, D.; Dang, S.; Zhang, X.; Wang, J.; et al. Structure and Mechanism of the Uracil Transporter UraA. *Nature* **2011**, *472*, 243–246. [CrossRef]
20. Yu, X.; Yang, G.; Yan, C.; Baylon, J.L.; Jiang, J.; Fan, H.; Lu, G.; Hasegawa, K.; Okumura, H.; Wang, T.; et al. Dimeric Structure of the Uracil:Proton Symporter UraA Provides Mechanistic Insights into the SLC4/23/26 Transporters. *Cell Res.* **2017**, *27*, 1020–1033. [CrossRef]
21. Alguel, Y.; Amillis, S.; Leung, J.; Lambrinidis, G.; Capaldi, S.; Scull, N.J.; Craven, G.; Iwata, S.; Armstrong, A.; Mikros, E.; et al. Structure of Eukaryotic Purine/H(+) Symporter UapA Suggests a Role for Homodimerization in Transport Activity. *Nat. Commun.* **2016**, *7*, 11336. [CrossRef] [PubMed]
22. Ge, J.; Elferich, J.; Dehghani-Ghahnaviyeh, S.; Zhao, Z.; Meadows, M.; von Gersdorff, H.; Tajkhorshid, E.; Gouaux, E. Molecular Mechanism of Prestin Electromotive Signal Amplification. *Cell* **2021**, *184*, 4669–4679.e13. [CrossRef] [PubMed]
23. Wang, C.; Sun, B.; Zhang, X.; Huang, X.; Zhang, M.; Guo, H.; Chen, X.; Huang, F.; Chen, T.; Mi, H.; et al. Structural Mechanism of the Active Bicarbonate Transporter from Cyanobacteria. *Nat. Plants* **2019**, *5*, 1184–1193. [CrossRef] [PubMed]

24. Geertsma, E.R.; Chang, Y.-N.; Shaik, F.R.; Neldner, Y.; Pardon, E.; Steyaert, J.; Dutzler, R. Structure of a Prokaryotic Fumarate Transporter Reveals the Architecture of the SLC26 Family. *Nat. Struct. Mol. Biol.* **2015**, *22*, 803–808. [CrossRef]
25. Flores, S.; Feld, A.P.; Thurtle-Schmidt, B.H. Expression, Solubilization, and Purification of Eukaryotic Borate Transporters. *J. Vis. Exp.* **2019**, *145*, e59166. [CrossRef]
26. Beltran, J.L.; Bain, R.K.; Stiefel, M.J.; McDonnell, A.S.; Carr, N.N.; Thurtle-Schmidt, B.H. Human SLC4A11 Does Not Complement BOR1 or Support Borate Transport in *Saccharomyces Cerevisiae*. *MicroPubl. Biol.* **2022**. [CrossRef]
27. Leano, J.B.; Batarni, S.; Eriksen, J.; Juge, N.; Pak, J.E.; Kimura-Someya, T.; Robles-Colmenares, Y.; Moriyama, Y.; Stroud, R.M.; Edwards, R.H. Structures Suggest a Mechanism for Energy Coupling by a Family of Organic Anion Transporters. *PLoS Biol.* **2019**, *17*, e3000260. [CrossRef]
28. Mohan, T.C.; Jones, A.M.E. Determination of Boron Content Using a Simple and Rapid Miniaturized Curcumin Assay. *Bio-Protoc.* **2018**, *8*, e2703. [CrossRef]
29. Varadi, M.; Anyango, S.; Deshpande, M.; Nair, S.; Natassia, C.; Yordanova, G.; Yuan, D.; Stroe, O.; Wood, G.; Laydon, A.; et al. AlphaFold Protein Structure Database: Massively Expanding the Structural Coverage of Protein-Sequence Space with High-Accuracy Models. *Nucleic Acids Res.* **2022**, *50*, D439–D444. [CrossRef]
30. Bonhivers, M.; Carbrey, J.M.; Gould, S.J.; Agre, P. Aquaporins in Saccharomyces. Genetic and Functional Distinctions between Laboratory and Wild-Type Strains. *J. Biol. Chem.* **1998**, *273*, 27565–27572. [CrossRef]
31. Kanki, T.; Young, M.T.; Sakaguchi, M.; Hamasaki, N.; Tanner, M.J.A. The N-Terminal Region of the Transmembrane Domain of Human Erythrocyte Band 3. Residues Critical for Membrane Insertion and Transport Activity. *J. Biol. Chem.* **2003**, *278*, 5564–5573. [CrossRef]
32. Pyle, E.; Guo, C.; Hofmann, T.; Schmidt, C.; Ribiero, O.; Politis, A.; Byrne, B. Protein-Lipid Interactions Stabilize the Oligomeric State of BOR1p from *Saccharomyces Cerevisiae*. *Anal. Chem.* **2019**, *91*, 13071–13079. [CrossRef]
33. Prasad, R. Lipids in the Structure and Function of Yeast Membrane. *Adv. Lipid Res.* **1985**, *21*, 187–242. [CrossRef]
34. Pyle, E.; Kalli, A.C.; Amillis, S.; Hall, Z.; Lau, A.M.; Hanyaloglu, A.C.; Diallinas, G.; Byrne, B.; Politis, A. Structural Lipids Enable the Formation of Functional Oligomers of the Eukaryotic Purine Symporter UapA. *Cell Chem. Biol.* **2018**, *25*, 840–848.e4. [CrossRef]
35. Chang, Y.-N.; Jaumann, E.A.; Reichel, K.; Hartmann, J.; Oliver, D.; Hummer, G.; Joseph, B.; Geertsma, E.R. Structural Basis for Functional Interactions in Dimers of SLC26 Transporters. *Nat. Commun.* **2019**, *10*, 2032. [CrossRef]
36. Cabantchik, Z.I.; Rothstein, A. The Nature of the Membrane Sites Controlling Anion Permeability of Human Red Blood Cells as Determined by Studies with Disulfonic Stilbene Derivatives. *J. Membr. Biol.* **1972**, *10*, 311–330. [CrossRef]
37. Cabantchik, Z.I.; Rothstein, A. Membrane Proteins Related to Anion Permeability of Human Red Blood Cells. I. Localization of Disulfonic Stilbene Binding Sites in Proteins Involved in Permeation. *J. Membr. Biol.* **1974**, *15*, 207–226. [CrossRef]
38. Zhao, R.; Reithmeier, R.A. Expression and Characterization of the Anion Transporter Homologue YNL275w in *Saccharomyces Cerevisiae*. *Am. J. Physiol. Cell Physiol.* **2001**, *281*, C33–C45. [CrossRef]
39. Jennings, M.L.; Smith, J.S. Anion-Proton Cotransport through the Human Red Blood Cell Band 3 Protein. Role of Glutamate 681. *J. Biol. Chem.* **1992**, *267*, 13964–13971. [CrossRef]
40. Chernova, M.N.; Jiang, L.; Crest, M.; Hand, M.; Vandorpe, D.H.; Strange, K.; Alper, S.L. Electrogenic Sulfate/Chloride Exchange in Xenopus Oocytes Mediated by Murine AE1 E699Q. *J. Gen. Physiol.* **1997**, *109*, 345–360. [CrossRef]
41. Tang, X.B.; Fujinaga, J.; Kopito, R.; Casey, J.R. Topology of the Region Surrounding Glu681 of Human AE1 Protein, the Erythrocyte Anion Exchanger. *J. Biol. Chem.* **1998**, *273*, 22545–22553. [CrossRef] [PubMed]
42. Capper, M.J.; Yang, S.; Stone, A.C.; Vatansever, S.; Zilberg, G.; Mathiharan, Y.K.; Habib, R.; Hutchinson, K.; Schlessinger, A.; Mezei, M.; et al. Substrate Binding and Inhibition of the Anion Exchanger 1 Transporter. *bioRxiv* **2022**. [CrossRef]
43. Jennings, M.L.; Howren, T.R.; Cui, J.; Winters, M.; Hannigan, R. Transport and Regulatory Characteristics of the Yeast Bicarbonate Transporter Homolog Bor1p. *Am. J. Physiol. Cell Physiol.* **2007**, *293*, C468–C476. [CrossRef] [PubMed]
44. Guizouarn, H.; Borgese, F.; Gabillat, N.; Harrison, P.; Goede, J.S.; McMahon, C.; Stewart, G.W.; Bruce, L.J. South-East Asian Ovalocytosis and the Cryohydrocytosis Form of Hereditary Stomatocytosis Show Virtually Indistinguishable Cation Permeability Defects. *Br. J. Haematol.* **2011**, *152*, 655–664. [CrossRef] [PubMed]
45. Iolascon, A.; De Falco, L.; Borgese, F.; Esposito, M.R.; Avvisati, R.A.; Izzo, P.; Piscopo, C.; Guizouarn, H.; Biondani, A.; Pantaleo, A.; et al. A Novel Erythroid Anion Exchange Variant (Gly796Arg) of Hereditary Stomatocytosis Associated with Dyserythropoiesis. *Haematologica* **2009**, *94*, 1049–1059. [CrossRef]
46. Yawata, Y.; Kanzaki, A.; Yawata, A.; Doerfler, W.; Ozcan, R.; Eber, S.W. Characteristic Features of the Genotype and Phenotype of Hereditary Spherocytosis in the Japanese Population. *Int. J. Hematol.* **2000**, *71*, 118–135.
47. Bruce, L.J.; Robinson, H.C.; Guizouarn, H.; Borgese, F.; Harrison, P.; King, M.-J.; Goede, J.S.; Coles, S.E.; Gore, D.M.; Lutz, H.U.; et al. Monovalent Cation Leaks in Human Red Cells Caused by Single Amino-Acid Substitutions in the Transport Domain of the Band 3 Chloride-Bicarbonate Exchanger, AE1. *Nat. Genet.* **2005**, *37*, 1258–1263. [CrossRef]
48. Jarolim, P.; Rubin, H.L.; Brabec, V.; Chrobak, L.; Zolotarev, A.S.; Alper, S.L.; Brugnara, C.; Wichterle, H.; Palek, J. Mutations of Conserved Arginines in the Membrane Domain of Erythroid Band 3 Lead to a Decrease in Membrane-Associated Band 3 and to the Phenotype of Hereditary Spherocytosis. *Blood* **1995**, *85*, 634–640. [CrossRef]
49. Park, M.; Li, Q.; Shcheynikov, N.; Zeng, W.; Muallem, S. NaBC1 Is a Ubiquitous Electrogenic Na+-Coupled Borate Transporter Essential for Cellular Boron Homeostasis and Cell Growth and Proliferation. *Mol. Cell* **2004**, *16*, 331–341. [CrossRef]
50. Ogando, D.G.; Jalimarada, S.S.; Zhang, W.; Vithana, E.N.; Bonanno, J.A. SLC4A11 Is an EIPA-Sensitive Na(+) Permeable PHi Regulator. *Am. J. Physiol. Cell Physiol.* **2013**, *305*, C716–C727. [CrossRef]

51. Kao, L.; Azimov, R.; Abuladze, N.; Newman, D.; Kurtz, I. Human SLC4A11-C Functions as a DIDS-Stimulatable $H^+(OH^-)$ Permeation Pathway: Partial Correction of R109H Mutant Transport. *Am. J. Physiol. Cell Physiol.* **2015**, *308*, C176–C188. [CrossRef]
52. Loganathan, S.K.; Schneider, H.-P.; Morgan, P.E.; Deitmer, J.W.; Casey, J.R. Functional Assessment of SLC4A11, an Integral Membrane Protein Mutated in Corneal Dystrophies. *Am. J. Physiol. Cell Physiol.* **2016**, *311*, C735–C748. [CrossRef]
53. Kato, A.; Kimura, Y.; Kurita, Y.; Chang, M.-H.; Kasai, K.; Fujiwara, T.; Hirata, T.; Doi, H.; Hirose, S.; Romero, M.F. Seawater Fish Use an Electrogenic Boric Acid Transporter, Slc4a11A, for Boric Acid Excretion by the Kidney. *J. Biol. Chem.* **2022**, *299*, 102740. [CrossRef]

membranes

Review

Entamoeba histolytica: Membrane and Non-Membrane Protein Structure, Function, Immune Response Interaction, and Vaccine Development

Nurhana Jasni [1], Syazwan Saidin [2], Wong Weng Kin [3], Norsyahida Arifin [1] and Nurulhasanah Othman [1,*]

1 Institute for Research in Molecular Medicine (INFORMM), Universiti Sains Malaysia, Gelugor 11800, Malaysia
2 Department of Biology, Faculty of Science and Mathematics, Universiti Pendidikan Sultan Idris, Tanjung Malim 35900, Malaysia
3 School of Health Sciences, Universiti Sains Malaysia, Kubang Kerian 16150, Malaysia
* Correspondence: nurulhasanah@usm.my

Abstract: *Entamoeba histolytica* is a protozoan parasite that is the causative agent of amoebiasis. This parasite has caused widespread infection in India, Africa, Mexico, and Central and South America, and results in 100,000 deaths yearly. An immune response is a body's mechanism for eradicating and fighting against substances it sees as harmful or foreign. *E. histolytica* biological membranes are considered foreign and immunogenic to the human body, thereby initiating the body's immune responses. Understanding immune response and antigen interaction are essential for vaccine development. Thus, this review aims to identify and understand the protein structure, function, and interaction of the biological membrane with the immune response, which could contribute to vaccine development. Furthermore, the current trend of vaccine development studies to combat amoebiasis is also reviewed.

Keywords: biological membrane; structure; function; immune response; *Entamoeba histolytica*; vaccine

Citation: Jasni, N.; Saidin, S.; Kin, W.W.; Arifin, N.; Othman, N. *Entamoeba histolytica*: Membrane and Non-Membrane Protein Structure, Function, Immune Response Interaction, and Vaccine Development. *Membranes* **2022**, *12*, 1079. https://doi.org/10.3390/membranes12111079

Academic Editors: Yosuke Senju and Shiro Suetsugu

Received: 23 September 2022
Accepted: 26 October 2022
Published: 31 October 2022

1. Introduction

Amoebiasis is a parasitic infection caused by *Entamoeba histolytica* and constitutes an alarming public health concern in developing countries [1,2]. This parasite has recorded the highest rates of infection in Central and South America, Africa, and India [2,3]. Children, immigrants, travelers returning from endemic countries, patients with HIV, and homosexual men are among the groups with the most significant risk of infection [4–6]. The most common infection route is from ingesting contaminated food and water. Around 90% of the infected individuals are asymptomatic, whereas the other 10% show a wide range of clinical manifestations, such as colitis, dysentery, and extraintestinal amoebiasis [7–9]. The factors that turn the parasite virulent in asymptomatic individuals are still uncertain; however, the gut microbiota is considered one of the factors triggering the virulence [10]. Metronidazole has been the drug of choice against amoebiasis for decades; however, the toxic effects of metronidazole and recent failures in treating several intestinal protozoan parasites have led to a search for other drugs and mechanisms to combat this parasite [11,12].

The innate and adaptive immune systems work together to eliminate this parasite naturally. The innate immune system is the body's first defense mechanism against germs and is non-specific. The acid in the stomach and a thick layer of mucin mucus are examples of innate immune responses that amoebas will first encounter during the invasion [13,14]. Intrusive parasite control includes the secretion of cytokines or chemokines, the recruitment of immune cells, such as neutrophils and macrophages, and inflammasome activation [15]. An adaptive immune response is highly specific to the antigen that induced them, and the response is 'remembered' by the immune system, resulting in long-lasting protection [16,17]. Adaptive immune responses toward *E. histolytica* invasion include the generation of specific IgG antibodies [13,14,18]. *E. histolytica* biological membrane and non-membrane proteins are the molecules responsible for triggering immune responses. These membrane proteins are galactose/N-acetylgalactosamine (Gal/GalNAc) lectin, cysteine proteinase, thioredoxin, lipopeptidophosphoglycan, and protein disulfide isomerase. Meanwhile, the non-membrane proteins include *E. histolytica* ubiquitin, calreticulin, *E. histolytica* migration inhibitory factor (EhMIF), enolase, actins, alcohol dehydrogenase, and extracellular vesicle proteins.

As more is learned about the pathogenesis of *E. histolytica* infection and the host's immune reaction, the potential for developing an effective vaccine holds promise [3,18]. Modern subunit vaccines rely on combining a highly purified antigen with an adjuvant to increase their efficacy [4]. A successful vaccine must have the ability to induce long-term protective immunity and should involve a simultaneous response of antibodies and T cells [19,20]. Vaccination using native and recombinant Gal/GalNAc lectin was found to be successful in protecting animals against intestinal and amoebic liver abscesses [18]. However, no clinical trial validates its human efficiency [18]. To the best of our knowledge, there have been no licensed vaccines for the prevention of amoebiasis until now. Currently, work is in progress to develop a vaccine, and recent experimental studies seem promising.

Thus, this review outlines the protein structure, function, and interaction of immune response with the *E. histolytica* biological membrane and vaccine studies. The study was conducted by reviewing research publications from Google Scholar, PubMed, and Scopus. The keywords used were *E. histolytica* protein and antibody, structure, function, antigenic proteins, immune responses, and vaccine studies.

2. Biological Membrane Proteins and Antigens

Membrane proteins contain a signal peptide and transmembrane domain [21]. The proteins in the membrane can be divided into integral, peripheral, and lipid-anchored proteins [22,23]. The integral membrane protein is permanently anchored in the phospholipid bilayer. Meanwhile, peripheral membrane proteins are temporarily attached to a lipid bilayer or other integral proteins. Some *E. histolytica* biological membrane proteins are antigenic and were reported to induce a host immune response. These membrane proteins and components are Gal/GalNAc lectin, cysteine proteinase, EhLPPG, thioredoxin, and protein disulfide isomerase.

2.1. Gal/GalNAc Lectin

Researchers widely study galactose/N-acetylgalactosamine (Gal/GalNAc) lectin as a membrane protein. It plays several roles in the pathogenicity of intestinal amoebiasis, including adherence to the host cell, cytotoxicity, complement resistance, induction of encystation, and generation of the cyst wall [24,25]. This protein contains a signal peptide and transmembrane domain; thus, it can be categorized as a membrane protein. The molecular weight of this protein is 260 kDa, and all *E. histolytica* strains express this protein on their surface [26]. Structurally, this protein is a heterotrimer composed of heavy (Hgl), intermediate (Igl), and light (Lgl) subunits [27–29]. The Hgl and Lgl are covalently attached to each other and interact non-covalently with the Igl. The Igl, which is the intermediate subunit, is constitutively localized at the lipid rafts; meanwhile, the heavy subunits (Hgl) and light subunits (Lgl) are transiently associated with rafts [30]. The Hgl has lectin activity; meanwhile, Lgl does not [28]. The intermediate subunits Igl have been shown to have hemolytic and cytotoxic activities that reside in the C-terminus of the protein [31].

The Gal/GalNAc lectin is antigenically conserved and can be characterized as immunogenic [26]. It is also considered to be the starting point of the invasion of *E. histolytica*, via attachment to the host mucin and colonic epithelial cells. The mucin forms high-affinity binding with the Gal/GalNAc lectin, allows the parasite to colonize the gut, and concurrently acts as a physical barrier [13,15].

In macrophages, *E. histolytica* Gal/GalNAc lectin up-regulates the expression of the mRNA of different cytokines [15]. Furthermore, it induces the mRNA expression of pattern recognition receptors, such as TLR-2 and TLR-4, modulated by nuclear factor NF-κB and the MAPK pathway [15]. The binding of Gal/GalNAc lectin to the TLR2 leads to NF-κB activation and the release of pro-inflammatory cytokines [32]. The lectin activates the CD4 cells, natural killer T cells, and CD8 T cells, enhancing protective cellular immunity [32].

The CD4 T cells secrete IL-13, IFN-γ, IL-4, and IL-5. Meanwhile, the CD8 T cells secrete the IL-17 cytokines. This IL-17 will induce the infiltration of neutrophils and enhance the secretion of antimicrobial peptides, mucin, and IgA into the colonic lumen [32].

Gal/GalNAc lectin and its components are widely used in vaccine development. Vaccines are biological products that elicit an immune response to a specific antigen, derived from an infectious disease-causing pathogen. Abhyankar et al. (2017) [4] evaluated the nanoliposome adjuvant system in triggering a strong mucosal immune response to the Gal/GalNAc lectin Lec A antigen of *E. histolytica* by immunizing mice with synthetic TLR agonists—alum, emulsion, or liposome formulations. Their results showed that the formulation of liposomes containing TLR4 and TLR7/8 agonists could generate intestinal IgA, plasma IgG2a/IgG1, IFN-γ, and IL-17A. This finding suggested that the nanoliposome delivery system, combined with TLR agonists, was promising for vaccine development against enteric pathogens.

In a subsequent study, Abhyankar et al. (2018) [33] formulated a stable PEGylated liposomal adjuvant containing synthetic GLA (TLR4) and 3M-052 (TLR7/8) ligands, which they administered intranasally to mice along with the Lec A antigen. According to their findings, the liposome containing both GLA and 3M-052 could elicit the secretion of Lec A-specific fecal IgA and Th1 immune responses. In addition, they conducted studies that modified the liposomal formulation at the PEG length. The optimized liposome increased the murine models' fecal IgA, serum IgG2a, systemic IFN-γ, and IL-17A levels. Furthermore, the adjuvant's improved formulation could reduce parasite antigens in the colon by more than 80%, which helped to protect against disease. The dose and excipient composition of each vaccine formulation component was optimized using the statistical design of experiment (DOE) and desirability approach, according to recent studies by Abhyankar et al. (2021) [1]. The optimization was conducted to maximize the desired characteristics while maintaining the physicochemical stability of the vaccine. This method has allowed for identifying the GLA and 3M-052 compositions, which exhibit high durability and protective immunity in the mouse model.

In a study conducted by Martinez-Hernandez et al. (2017; 2021) [34,35], the newly developed recombinant chimeric vaccine (PEΔIII-LC3-KDEL3) produced from the LC3 fragment of *E. histolytica*, domains I and II of *Pseudomonas aeruginosa*, and the carboxy-terminal signal KDEL3 in *Pichia pastoris* has potential as an immunogen and showed the inhibition of the adherence of trophozoites to the HepG2 cell monolayer in a hamster model. Furthermore, the vaccine protects against liver tissue damage and uncontrolled inflammation. Meneses-Ruiz et al. (2011; 2015) [36,37] demonstrated oral and intramuscular immunization using the *Autographa californica* baculovirus; expressing the *E. histolytica* Gal/GalNAc lectin LC3 fragment (AcNPV-LC3) could help to protect against ALA development in hamsters.

Next, the 150-kDa intermediate subunit (Igl) of the Gal/GalNAc lectin is gaining more attention in numerous studies and has been shown to have hemolytic activity [38]. Min et al. (2016) [39] studied the effect of immunization with recombinant Igl on amoebic liver abscess (ALA) formation. Their findings showed that recombinant Igl, particularly the C-terminal fragment, represents a promising vaccine against amoebiasis. It produced a significant humoral immune response with high antibodies, inhibiting 80% of the trophozoites' adherence to mammalian cells and inducing the complement-mediated lysis of 80% of the trophozoites. Thus, the result shows that this subunit should be considered in vaccine development. The predicted protein structure of Gal/GalNAc lectin Igl1 can be found in Figure 1.

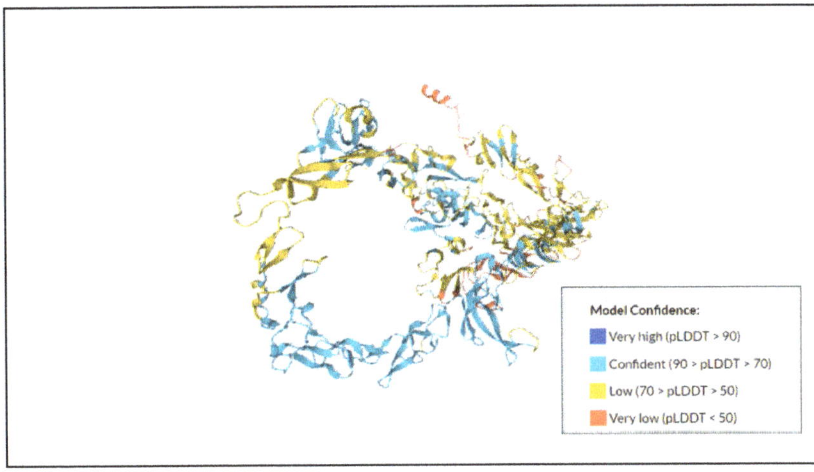

Figure 1. Predicted protein structure of Gal/GalNAc lectin Igl1 [40,41] (https://www.uniprot.org/uniprotkb/Q964D2/entry, accessed on 25 October 2022).

2.2. Cysteine Proteinase

Cysteine proteinases are virulence factors in *E. histolytica;* some of them were predicted to have transmembrane and glycosylphosphatidylinositol anchor attachment domains. Thus, they can be classified as membrane proteins [42,43]. This cysteine proteinase has several types; for example, EhCP1 and EhCP4 were found to colonize the large cytoplasmic vesicles that differ from the sites of other proteinases [43,44]. Fifty genes encode cysteine proteinases; *ehcp-a4* is the gene that is most up-regulated during the invasion and colonization of *E. histolytica* in the mouse cecal model [43]. Furthermore, the protein plays a role in tissue invasion by disrupting the colonic epithelial barrier, disrupting host defenses, and digesting extracellular matrix components such as immunoglobulins, complement, and cytokines [43–45]. Furthermore, cysteine proteinases also form a complex heterodimer protein (EhCPADH) by a combination of *E. histolytica* cysteine protease (EhCP112) and adhesin (EhADH), which is involved in the cytopathic mechanism and virulence of the parasite [46,47]. The EhCP112 alone can be inhibited by thiol inhibitors, such as E-64, TLCK, and chymostatin [47].

Meanwhile, a recombinant EhCP4 can be inhibited by a vinyl sulfone inhibitor, WRR605 [44]. The inhibition of EhCP4 reduced the parasite burden and inflammation in the mouse cecal model [44]. Cysteine proteinase 5 (EhCP5) was reported as a virulence factor in live *E. histolytica* that elicits a fast release of mucin [48].

The pro-mature cysteine proteinase 5 (PCP5) that was usually secreted or found on the surface of the amoeba stimulated NF-κB-mediated pro-inflammatory responses by the binding of the RGD motif to the $\alpha V\beta 3$ integrin on Caco-2 colonic cells [49]. The predicted protein structure of cysteine proteinase 5 and its model confidence can be found in Figure 2.

2.3. *Entamoeba Histolytica* Lipopeptidophosphoglycan (EhLPPG)

E. histolytica lipopeptidophosphoglycan (EhLPPG) is one of the virulence factors of *E. histolytica;* this macromolecule can promote tissue invasion by causing inflammatory damage to the host cells. It can be found on the *E. histolytica* membrane [51]. This molecule consists of 85% carbohydrate, 8% peptide, 2.5% lipid, and 1% phosphate [52] Moody et al. (1997) [53] and Wong-Baeza (2010) [52] reported the under-expression of LPPG and LPG in the non-virulent strain of *E. histolytica,* compared to the virulent strain. No LPG, and a modified version of LPPG, were found in the low-virulence Rahman strain and the non-pathogenic *Entamoeba dispar,* respectively. Thus, it can be inferred that the protein expression of LPPG and LPG influenced the virulence of the *Entamoeba* species.

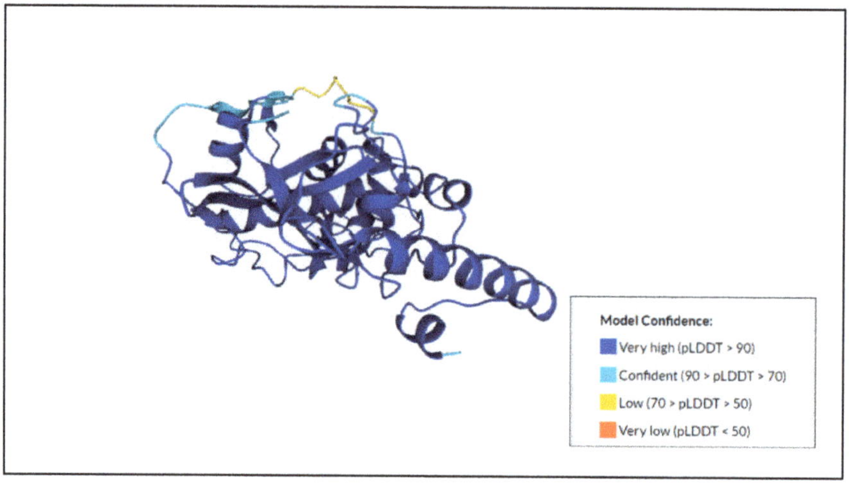

Figure 2. Predicted protein structure of cysteine proteinase 5 [41,49,50] (https://www.uniprot.org/uniprotkb/Q06FF8/entry, accessed on 25 October 2022).

In the *E. histolytica* immune responses, the interaction of EhLPPG with TLR2 and TLR4 resulted in the activation of NF-κB and the release of IL-10, IL-12p40, TNF-α, and IL-8 from human monocytes [54]. Furthermore, the purified phosphoinositol moiety of EhLPPG is capable of inducing IFN-γ production [51]. In particular, a diacylated PI, (1-O-[(28:0)-lyso-glycero-3-phosphatidyl-]2-O-(C16:0)-Ins) is a primary component of EhLPPG that is responsible for the activation of natural killer T cells [51].

In addition, the activation of dendritic cells by LPPG increased the expression of co-stimulatory molecules CD80, CD86, and CD40. Besides this, it produces pro-inflammatory cytokines, such as TNF-α, IL-8, and IL-12 [55]. This molecule can also induce antibody production and the formation of in vitro neutrophil extracellular traps (NET) [52,56].

2.4. Protein Disulfide Isomerase

Protein disulfide isomerase (PDI) is one of the antigenic membrane proteins found in *E. histolytica* [57]. The predicted protein disulfide isomerase structure and its model confidence can be found in Figure 3. This protein is classified as a membrane protein as it contains a transmembrane and signal peptide, as indicated by the Uniprot database and in studies reported by Azmi and Othman in 2020 [21]. It was also up-regulated in the membrane fractions [21]. Structurally, PDIs share a common feature associated with at least one active thioredoxin-like domain [58]. The predicted size of the protein is 41.73 kDa, with a length of 368 aa. The functions of this protein are to catalyze the oxidation, reduction, and isomerization of disulfide bonds in polypeptide substrates [59,60]. Furthermore, it also exhibits chaperone-like activity and can be inhibited by bacitracin [60]. The recombinant type of EhPDI shows the specific properties of PDI enzymes, such as performing oxidase and reductase activities [59]. This protein also participates in the cellular mechanism related to protein homeostasis [61]. For example, during stressful conditions, this enzyme is involved in holding, refolding, and degrading unfolded or misfolded proteins [58]. It also helps to protect heat-labile enzymes against thermal inactivation [58].

Mares et al. (2015) [58] used the PDI model of 38 kDa polypeptide with two active thioredoxin-like domains to study protein folding and disulfide bond formation in *E. histolytica*. Their studies revealed a significant difference in the functional role of each thioredoxin-like domain. Thus, their study indirectly indicated that the amount or type of thioredoxin domain might influence the PDI functions.

Next, this protein was classified as antigenic by Kumarasamy et al. (2020), as this protein was recognized by sera from patients with an amoebic liver abscess [57]. However, no other study of immune response interaction with this *E. histolytica* protein could be found. Only a study of protein disulfide isomerases from other organisms was reported. One example was the immunization of BALB/c mice, with *Leishmania donovani* protein disulfide isomerase (LdPDI) as a DNA vaccine that elicited protective immunity against visceral leishmaniasis through the production of two pro-inflammatory cytokines, CD8 and CD4 (Th1 and Th17) [62]. Despite the lack of study on this protein, it merits further investigation as a drug target for anti-amoebic therapy and as a vaccine candidate.

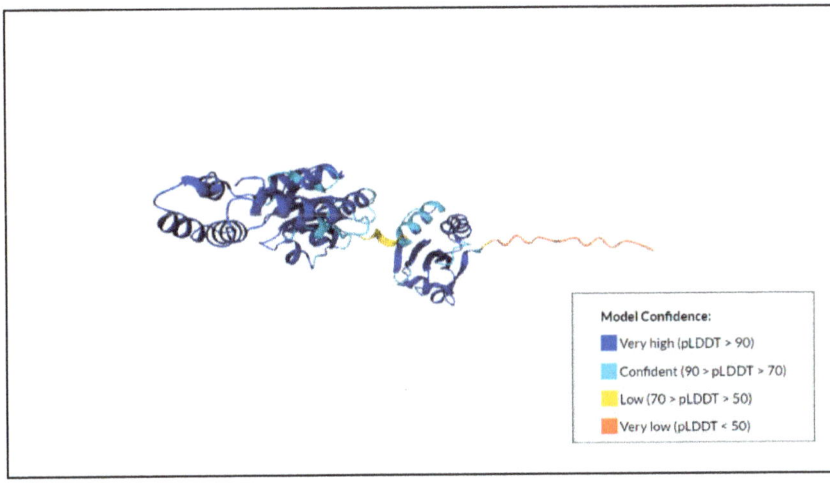

Figure 3. Predicted protein structure of protein disulfide isomerase [41,62,63] (https://www.uniprot.org/uniprotkb/A0A5K1UZD0/entry, accessed on 25 October 2022).

2.5. Thioredoxin

Thioredoxin is an antigenic protein that has been found in the *E. histolytica* membrane (Figure 4) [57]. Structurally, it contains a Toll/IL-1R/Resistance (TIR)-like domain and is predicted to have a signal peptide [64]. This protein is associated with thioredoxin reductase, forming the thioredoxin system. The two work together and play essential roles in *E. histolytica*. Thioredoxin is also one of the proteins that form an adduct with metronidazole metabolites and other proteins: thioredoxin reductase, superoxide dismutase, purine nucleoside phosphorylase, and a previously unknown protein [65]. Thioredoxin is crucial as it has various functions: acting as an antioxidative defense, modulating the intracellular and extracellular signaling pathways, cell growth, regulating DNA synthesis and transcription factor, modulating the immune response, apoptosis, and involvement in post-translational redox modifications on the target protein [57,66,67]. The silence of thioredoxin transcripts decreased the phagocytosis of erythrocytes and *Escherichia coli* [64]. The roles it fulfills make it suitable for use as a drug target or vaccine candidate, although no human immune response study was reported. Further study is suggested to elucidate its immune response interactions and potential as a vaccine candidate.

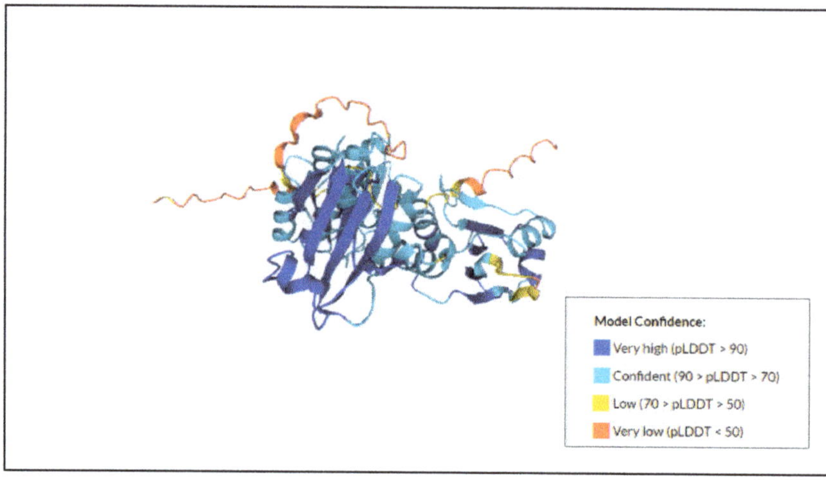

Figure 4. Predicted protein structure of thioredoxin (putative) [41,65,68] https://www.uniprot.org/uniprotkb/S0AYD1/entry, accessed on 25 October 2022.

3. Non-Membrane Protein

Non-membrane proteins have been found in other parts of *E. histolytica*, such as the cytoplasm and nucleus. It does not contain a signal peptide or transmembrane domain and thus excluded as a membrane protein. However, recent findings indicated that the *E. histolytica* cytosolic proteins were present in the membrane fraction and a few of them were antigenic [21,57]. Non-membrane proteins also contribute to initiating host immune responses. These proteins are ubiquitin, calreticulin, *E. histolytica* migration inhibitory factor (EhMIF), actin, alcohol dehydrogenase, enolase, and extracellular vesicles.

3.1. Entamoeba Histolytica Ubiquitin (Ehub)

Ubiquitin is one of the proteins found in *E. histolytica* and can trigger immune responses to produce antibodies. It is also immunogenic and antigenic [69]. According to the UniProt database, this protein can be found in the cytoplasm and nucleus of the parasite. The ubiquitin structure consists of protein moieties and glycosylated structures [69]. The sequence of *E. histolytica* ubiquitin amino acids, as deduced from a cDNA nucleotide sequence, was found to deviate by six positions from the consensus of all other known ubiquitins, the sources of which range from *Trypanosoma cruzi* to *Homo sapiens* [70]. Furthermore, the ubiquitin is also usually associated with the proteasome forming the ubiquitin-proteasome system, which is important for numerous cellular processes and for maintaining parasite homeostasis [69,71]. This ubiquitin-proteasome pathway presents a viable therapeutic target [72].

Infection with *E. histolytica* in humans induces strong IgG antibodies toward ubiquitin [69]. Human antibodies can recognize ubiquitin's protein moieties and glycosylated structure [70]. Notably, the ubiquitin's glycan moieties are immunodominant and induce IgG. Furthermore, antibodies to the ubiquitin Ehub are induced exclusively in patients with invasive amoebiasis and mainly to glycoprotein [73]. It indicates that glycan is immunodominant and could be a potential target for diagnostic tests, drugs, and vaccines to combat parasitic diseases [69,73].

3.2. Calreticulin

Calreticulin (CRT) and its model confidence, as shown in Figure 5, is one of the immunogenic proteins in *E. histolytica* that is localized in the endoplasmic reticulum [74,75]. It can also be found in the membrane fraction [74,75]. It is essential for regulating critical cellular functions, such as migration, apoptotic cell phagocytosis, and cytotoxic T lymphocyte- or natural killer T cell-mediated lysis [76]. It is also crucial in properly folding and transporting the protein through the endoplasmic reticulum [76]. CRT also interacts with human C1q and inhibits the activation of the classical complement pathway [77]. Both can be observed in the CRT of *E. histolytica* and *E. dispar* [77]. Furthermore, a higher level of CRT expression in the in-situ lesions of amoebic liver abscess (ALA) in the hamster model was found in *E. histolytica* compared to *E. dispar*. Additionally, CRT plays an important role in the early stages of the host-parasite relationship, in which the parasite needs to adapt to a new environment [74].

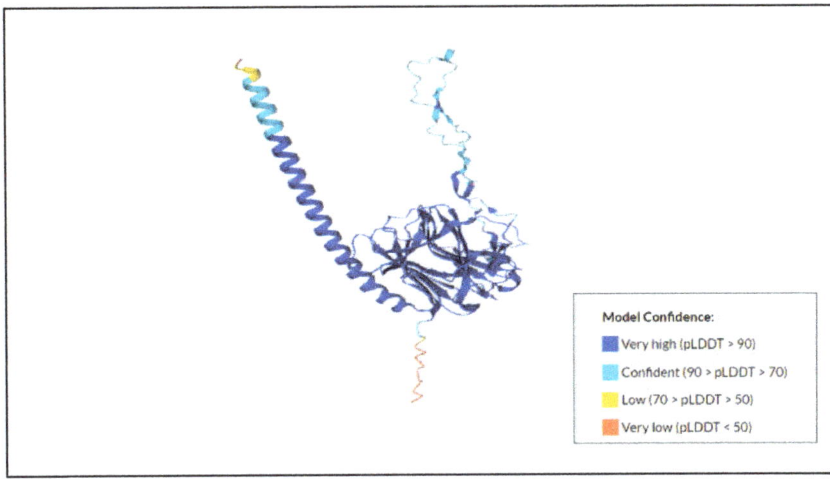

Figure 5. Predicted protein structure of calreticulin [41,75,78] (https://www.uniprot.org/uniprotkb/C4M296/entry, accessed on 25 October 2022).

EhCRT is also an excellent immunogen for activating specific peripheral blood mononuclear cells (PBMC) [75]. PBMCs are a variety of immune cells, such as lymphocytes, monocytes, and dendritic cells, that works together to protect humans from harmful pathogens. EhCRT helps determine the immune response by inducing the differential expression of Ils, depending on the outcome of the disease. For example, patients with an amoebic liver abscess (ALA) during its acute phase (AP-ALA) show a Th2 cytokine profile; meanwhile, patients with amoebic liver abscess (ALA) during the resolution phase (R-ALA) show a Th1 cytokine profile [75]. In patients with AP-ALA, a higher level of the interleukins IL-6 and IL-10, granulocyte colony-stimulating factor (GCSF), and transforming growth factor β1 (TGFβ1)] were also observed. In contrast, higher levels of IFN-γ were detected in patients with R-ALA [75].

3.3. *Entamoeba Histolytica Migration Inhibitory Factor (EhMIF)*

The *E. histolytica* migration inhibitory factor (EhMIF) is a 12-kDa protein localized to the cytoplasm of trophozoites [79]. Figure 6 depict the predicted protein structure of the migration inhibitory factor and its model confidence. It is a homolog of the human cytokine MIF [80]. This cytokine is one of the virulence factors linked to severe disease and is secreted by several medically important protozoan parasites, such as *Plasmodium, Entamoeba*, *Toxoplasma*, and *Leishmania* [81]. Structurally, EhMIF lacks signal peptides [80]. Antibodies toward EhMIF were found in the sera of children living in an area where the *E. histolytica* infection was endemic [79].

EhMIF interacts with the MIF receptor CD74 and binds to macrophages [79]. EhMIF induces IL-6 production and can also enhance TNF-α secretion. There are two ways of improving the secretion of TNF-α by inhibiting the glucocorticoid-mediated suppression of TNF-α secretion and by amplifying TNF-α production via lipopolysaccharide (LPS)-stimulated macrophages [79].

Ngobeni et al. (2017) [80] reported that mice infected with parasites overexpressing EhMIF increased chemokine expression, the neutrophil influx, and mucosal damage. In contrast, blocking the EhMIF gene decreased chemokine expression and reduced neutrophil infiltration. Furthermore, a combination of antiparasite MIF-blocking antibodies and metronidazole significantly reduced intestinal inflammation and tissue damage in mice [81].

In addition, Ghosh et al. (2018) [82] identified the EhJAB 1 protein as a potential binding partner and an EhMIF positive-negative regulator. The binding of EhJAB1 to EhMIF blocked the production of IL-8 by the human epithelial cells.

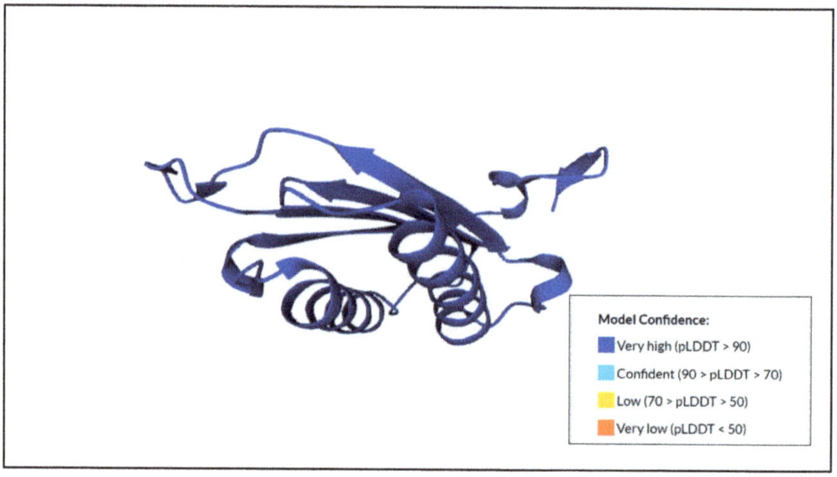

Figure 6. Predicted protein structure of migration inhibitory factor (EhMIF) [41,80,83] (https://www.uniprot.org/uniprotkb/A0A5K1URL6/entry, accessed on 25 October 2022).

3.4. Extracellular Vesicles

The extracellular vesicle (EV) is a heterogeneous, membrane-limited structure secreted by prokaryotic and eukaryotic cells to transport lipids, proteins, and nucleic acids to the external environment [84]. They play essential roles in cellular communication, information transfer through cargo, and modulating the host's immune system [85]. Furthermore, virulence factors and effector molecules can be transferred using EV to the host. In *E. histolytica*, the secretion of EV influences encystation efficiency [84,86]. Moreover, Nievas et al. (2020) [86] suggested using EVs in vaccine studies as it has more stable conformational conditions, has the ability to circulate in bodily fluids, and uses the body's natural mechanism for transporting antigens between cells. However, thus far, no study of immune response interaction with this protein has been reported.

3.5. Enolase

E. histolytica enolase (EhENO) is one of the proteins found in the nucleus and cytoplasm of *E. histolytica* [87]. The single copy of the *E. histolytica* enolase gene (*Ehenl-1*), is located on the 1.6 Mb chromosome [88]. This protein is also predicted to have a TATA box function, as the sequence TATAAG, at −31, interacts with nuclear proteins [88]. The crystal structure EhENO manifests as an asymmetric dimer with one active site in the open conformation and the other active site in the closed conformation [89]. Both of the active sites also contain a 2-PGA molecule [89]. It catalyzes the conversion of 2-phosphoglycerate (2-PG) to phosphoenolpyruvate (PEP), which is a part of the pathway by which *E. histolytica* obtains its energy [87]. It also acts as an inhibitor of DNA methyltransferase 2 (Dnmt2) [87].

In studies by Kumarasamy et al. (2020) [57] this protein was found in the membrane fraction and was reported to be antigenic since it could be recognized by the sera of patients with amoebic liver abscesses. However, no study of human immune response interaction with this *E. histolytica* protein has been reported. There are only studies of enolase from different organisms to be found. In a study by Xue et al. (2021) [90] mice and piglets were immunized using recombinant *Mycoplasma suis* alpha-enolase (rMseno). In mice, immunizations with rMseno caused increased levels of IFN-γ and IL-4 cytokines and increased the T lymphocyte proliferation index [90]. Meanwhile, in piglets, the results showed an increased level of specific IgG1, IgG2a, CD4, and CD8 cells [90]. A study by Téllez-Martínez et al. (2019) [91] demonstrated that BALB/c mice, immunized with enolase-based vaccine and Montanide PetGel A (PGA) as an adjuvant, demonstrated a strong specific Th1 response and protective immunity against *Sporothrix schenckii*. In addition, the oral delivery of *Bacillus subtilis* spores expressing the enolase of *Clonorchis sinensis* resulted in increased specific IgM levels in sera, intestine mucus, and skin mucus in grass carps (*Ctenopharyngodon idellus*) [92]. From all the studies conducted, it can be concluded that enolase has the potential to be used as a vaccine candidate and therapeutic target, as previous studies from other organisms have demonstrated promising results. The protein structure of enolase 1, the N-terminal domain, and C-terminal Tim Barrel domain can be found in Figure 7.

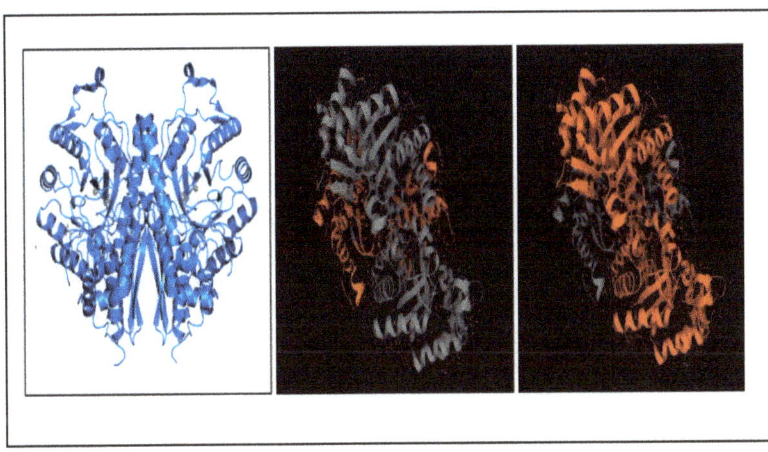

Figure 7. Figure showing two copies of enolase 1, N-terminal domain and C-terminal domain, respectively [89,91–95] (https://www.ebi.ac.uk/pdbe/entry/pdb/3qtp/analysis, accessed on 25 October 2022).

3.6. Actin

As shown in Figure 8, actin is one of the most conserved, abundant, and ubiquitous proteins in all eukaryotes [96]. *E. histolytica* has a single actin protein, the structure of which differs significantly from human homologs. *E. histolytica* actin was reported to be antigenic and has been predicted as a suitable vaccine candidate [57,97]. It is localized in the cytoplasm and membrane of *E. histolytica* [57]. Structurally, *E. histolytica* actin contains glycine residue (Gly2) at position 2 [96]. This Gly2 is not found in most other eukaryotic actins. Still, it is crucial for cell morphology and division, host invasion in an in vitro model of human amoebic infection, and parasite-host cell adhesion [96]. Actin is an essential protein in many cellular processes, including directing the process of phagocytosis [98].

Several drug treatments have indirectly targeted actin, such as *Adenophyllum aurantium* extract, linearolactone, and kaempferol, which affects the structuration of the actin cytoskeleton [99,100]. However, until now, no immune response study has been reported regarding its interaction with *E. histolytica* actin.

A vaccine study from Li et al. (2011) reported using a DNA vaccine encoding *Toxoplasma gondii* actin against BALB/c mice [101]. Their study reported a high titer of specific antibodies and increased CD4 and CD8 T percentages, which showed that the actin could generate specific humoral and cellular immune responses [101]. Thus, we postulate that the actin from *E. histolytica* could have the same potential results as other parasites, inducing an immune response.

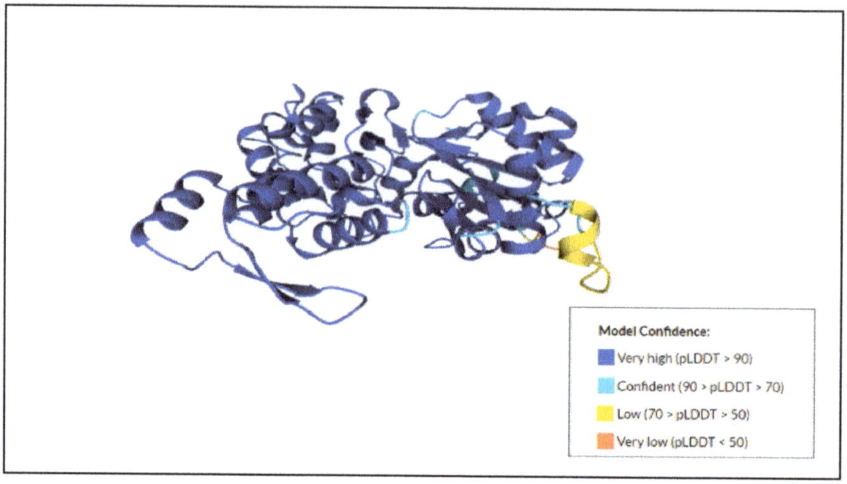

Figure 8. Predicted protein structure of actin [41,99,102] (https://www.uniprot.org/uniprotkb/P11426/entry, accessed on 25 October 2022).

3.7. Alcohol Dehydrogenase (ADH)

Alcohol dehydrogenase (ADH) is one of the enzymes in the cytoplasm of *E. histolytica*. It has various isoforms, but the most reported ones are alcohol dehydrogenase 2 (ADH2) and alcohol dehydrogenase 3 (Figure 9). *E. histolytica* alcohol dehydrogenase 3 (EhADH3) was reported to be antigenic in a study by Kumarasamy et al. (2020) [57], as it could be recognized by the sera of patients with amoebic liver abscesses. Some EhADH3 have been found to localize at the surface of *E. histolytica* [103]. This enzyme was also expressed at higher levels in non-pathogenic than in pathogenic amoebae [104]. The overexpression or silence of *ehadh3bb* using transfectant was found to have no or little effect on the parasite growth, size, erythrophagocytosis, motility, and hemolytic or cysteine peptidase activity [104].

Another alcohol dehydrogenase, *E. histolytica* alcohol dehydrogenase 2, called EhADH2, is a bifunctional enzyme that is essential for growth and survival [105–107]. It is called bifunctional because it also has aldehyde dehydrogenase (ALDH) and alcohol dehydrogenase (ADH) activities [105]. The cofactor of this enzyme, EhADH2, is Fe^{2+} [108]. Furthermore, EhADH2 is similar to other protist and bacterial bifunctional enzymes, as suggested by the phylogenetic tools [106]. This enzyme plays a role in catalyzing the conversion of acetyl CoA to acetaldehyde and the final reduction of acetaldehyde to ethanol [106,107,109]. It was also suggested by Adeoti et al. (2021) [97] as a vaccine candidate, based on bioinformatic analyses using GEPTOP CELLO NCBI. Other studies also presented it as a target for anti-amoebic inhibitors and the substrate and cofactor (NADH, Fe) binding sites could act as the inhibition targets [106]. Several studies reported that the enzymes could be inhibited by cyclopropyl (CPC) and cyclobutyl (CBC) carbinols, laboratory-tested pyrazoline derivatives, and S-nitroso-glutathione [106,110,111].

In addition, only an immune response study related to alcohol dehydrogenase was found that was not associated with *E. histolytica* but instead with *Drosophila*. Their experimental results showed that alcohol dehydrogenase played no role in the *Drosophila* immune response [112]. Thus, this protein has an equal chance of success or is not being used as a vaccine candidate.

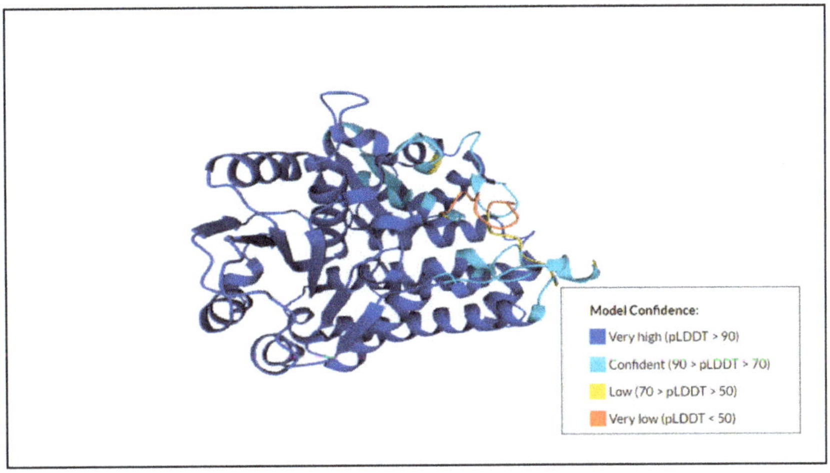

Figure 9. Predicted protein structure of alcohol dehydrogenase 3 [41,107,113] (https://www.uniprot.org/uniprotkb/Q24857/entry, accessed on 25 October 2022).

4. Docking of Membrane Protein and Vaccine Development

The vaccine development of *E. histolytica* in studies is limited to Gal/GalNAc lectin and its components. To our knowledge, no other membrane or non-membrane proteins have been utilized as vaccine candidates. Table 1 summarizes the immune response interaction of biological membrane and non-membrane *E. histolytica* proteins and the various vaccination studies.

Table 1. Summary of the immune response interaction of biological membrane and non-membrane *E. histolytica* and their vaccination studies.

Protein	Types of Molecule	Antibody Interaction Study	Signal Peptide/ Transmembrane Domain	Used in Vaccine Studies	References
Gal/GalNAc lectin	Membrane protein	IgA	Signal peptide	Yes	[13,15,32]
Cysteine proteinases	Membrane protein	Antigenic	Putative, transmembrane	No	[48,49]
Lipopeptidophosphoglycan (LPPG)	Macromolecule	The antibody involved was not specified	Signal peptide	No	[51,52,54–56]
Protein disulfide isomerase (PDI)	Membrane protein	Antigenic	Signal peptide	No	[57]
Thioredoxin	Membrane protein	Antigenic	Signal peptide	No	[57]
E. histolytica Ubiquitin (Ehub)	Cytosolic protein	IgG	-	No	[69]
E. histolytica migration inhibitory factor (EhMIF)	Cytosolic protein	The antibody involved was not specified	-	No	[79–81]
Enolase	Cytosolic protein	Antigenic	-	No	[57]
Actin	Cytosolic protein	Antigenic	-	No	[57]
Alcohol dehydrogenase (ADH)	Cytosolic protein	Antigenic	-	No	[57]
Calreticulin (CRT)	Cytosolic protein	Antigenic	-	No	[75–77]
Extracellular vesicles (EVs)	Macromolecule	Modulates the immune system	-	No	[86]

Docking studies elucidate how two or more molecular structures fit together to form a stable complex. It has become an important drug discovery tool and has been used for over three decades [114,115]. It is one of the most commonly used virtual screening methods, especially when the target protein's three-dimensional (3D) structure is available [114]. Aside from that, it is one of the efforts made to shorten the research timeline and reduce costs by reducing wet-laboratory experiments.

Generally, the docking of membrane proteins for vaccination involves five key stages: (1) the selection and antigenic evaluation of proteins, (2) the prediction of B-cell epitopes and T-cell epitopes, (3) the non-allergenicity and non-toxicity prediction of selected T-cell epitopes, (4) structural modeling and molecular docking, and finally, (5) an evaluation of the docking results [97,116]. The selection of proteins was conducted by prioritizing these criteria in terms of the reported antigenicity, virulence, and proteins related to the adhesion mechanisms [97]. Different databases were used, such as PATRIC3.5.16, T TMHMM v2.0, and VaxiJen V2.0. The PATRIC3.5.16 database was used to investigate the role of virulence proteins, while TMHMM v2.0 was used to predict transmembrane (TM) helixes, and VaxiJen V2.0 served to predict antigenic properties [97,116].

The second stage is the prediction of T-cell and B-cell epitopes. Identification of the MHC-I binding epitopes can be obtained using the NetMHC 4.0 server or the Propred-1 server; meanwhile, for MHC-II binding epitopes, the servers that can be used are NetMHCII 2.3 server or the Propred tool [97,116]. A server named the immune-epitope database (IEDB), which is an analysis resource, can be utilized for B-cell epitope prediction [116]. B-cell epitope prediction identifies potential antigens interacting with the B lymphocytes and initiating an immune response. The identification and selection of the B-cell epitope were conducted by referring to antigenicity, linear epitope prediction, hydrophilicity, surface accessibility, and flexibility [116].

Next, the allergenicity and toxicity of T-cell epitopes can be predicted using the AllerCatPro server and the toxpred tool, respectively. The 3D structure modeling of protein interest can be obtained from Uniprot/PDB/ProsA. The T-cell epitopes of MHC class-I peptides can be modeled using the RPBS MOBYL portal from the PEPFOLD server [116]. At this point, protein and peptide docking were then carried out. Several types of docking software can be used, including SANJEEVINI, GOLD, AUTODOCK, GemDock, Hex Protein Docking, and GRAMM. The scoring function computes scores based on the best-fitting ligand. The docking results were analyzed; the best-predicted epitopes from the study can be further tested for therapeutic potency for vaccine development.

5. Conclusions

To summarize, this review discusses those studies regarding the *E. histolytica* biological membrane and non-membrane protein structure, function, interaction with immune response, and recent vaccine studies. From all the membrane and non-membrane proteins, Gal/GalNAc lectin, ubiquitin, lipopeptidophosphoglycan, migration inhibitory factor, enolase, actin, ADH, and CRT interacted with the antibody. However, the extracellular vesicle was found to modulate the immune system. Furthermore, *E. histolytica* vaccine studies reported using Gal/GalNAc lectin and its components in vaccine development, such as the Gal/GalNAc lectin Lec A, LC3 fragment, and intermediate subunit Igl. Vaccine development utilizing those elements seems to have the potential to be used in combating *E. histolytica*; therefore, it should be evaluated in the clinical trial. Other membrane and non-membrane proteins can be further researched as vaccine candidates because they showed immune response interactions with the host, induced antibodies, and modulated the immune response.

Author Contributions: Conceptualization, N.J. and N.O.; methodology, N.J.; formal analysis, N.J. and N.O.; investigation, N.J.; resources, N.O., S.S., N.A. and W.W.K.; data curation, N.J.; writing—original draft preparation, N.J.; writing—review and editing, N.O., S.S., N.A. and W.W.K.; supervision, N.O.; project administration, N.O.; funding acquisition, N.O., S.S. and N.A. All authors have read and agreed to the published version of the manuscript.

Funding: This study was funded by the Malaysian Ministry of Higher Education Fundamental research grant (FRGS) Ref code no. FRGS/1/2020/STG01/USM/02/11.

Institutional Review Board Statement: Not applicable.

Data Availability Statement: Not applicable.

Acknowledgments: We would like to thank Universiti Sains Malaysia for providing facility support.

Conflicts of Interest: The authors declare no conflict of interest.

References

1. Abhyankar, M.M.; Orr, M.T.; Kinsey, R.; Sivananthan, S.; Nafziger, A.J.; Oakland, D.N.; Young, M.K.; Farr, L.; Uddin, M.J.; Leslie, J.L.; et al. Optimizing a multi-component intranasal *Entamoeba histolytica* vaccine formulation using a design of experiments strategy. *Front. Immunol.* **2021**, *12*, 683157. [CrossRef] [PubMed]
2. WHO/PAHO/UNESCO Report. A consultation with experts on amoebiasis. Mexico City, Mexico 28–29 January, 1997. *Epidemiol. Bull.* **1997**, *18*, 13–14.
3. Kantor, M.; Abrantes, A.; Estevez, A.; Schiller, A.; Torrent, J.; Gascon, J.; Hernandez, R.; Ochner, C. *Entamoeba histolytica*: Updates in clinical manifestation, pathogenesis, and vaccine development. *Can. J. Gastroenterol. Hepatol.* **2018**, *2018*, 4601420. [CrossRef] [PubMed]
4. Abhyankar, M.M.; Noor, Z.; Tomai, M.A.; Elvecrog, J.; Fox, C.B.; Petri, W.A. Nanoformulation of synergistic TLR ligands to enhance vaccination against *Entamoeba histolytica*. *Vaccine* **2017**, *35*, 916–922. [CrossRef]
5. Haque, R.; Mondal, D.; Duggal, P.; Kabir, M.; Roy, S.; Farr, B.M.; Sack, R.B.; Petri, W.A., Jr. *Entamoeba histolytica* infection in children and protection from subsequent amebiasis. *Infect. Immun.* **2006**, *74*, 904–909. [CrossRef]
6. Swaminathan, A.; Torresi, J.; Schlagenhauf, P.; Thursky, K.; Wilder-Smith, A.; Connor, B.A.; Schwartz, E.; Vonsonnenberg, F.; Keystone, J.; O'Brien, D.P. A global study of pathogens and host risk factors associated with infectious gastrointestinal disease in returned international travellers. *J. Infect.* **2009**, *59*, 19–27. [CrossRef]
7. Fotedar, R.; Stark, D.; Beebe, N.; Marriott, D.; Ellis, J.; Harkness, J. Laboratory diagnostic techniques for *Entamoeba* species. *Clin. Microbiol. Rev.* **2007**, *20*, 511–532. [CrossRef]
8. Noordin, R.; Yunus, M.H.; Saidin, S.; Mohamed, Z.; Fuentes Corripio, I.; Rubio, J.M.; Golkar, M.; Hisam, S.; Lee, R.; Mahmud, R. Multi-laboratory evaluation of a lateral flow rapid test for detection of amebic liver abscess. *Am. J. Trop. Med.* **2020**, *103*, 2233–2238. [CrossRef]
9. Saidin, S.; Othman, N.; Noordin, R. Update on laboratory diagnosis of amoebiasis. *Eur. J. Clin. Microbiol. Infect. Dis.* **2019**, *38*, 15–38. [CrossRef]
10. Padilla-Vaca, F.; Ankri, S.; Bracha, R.; Koole, L.A.; Mirelman, D. Down regulation of *Entamoeba histolytica* virulence by monoxenic cultivation with *Escherichia coli* O55 is related to a decrease in expression of the light (35-kilodalton) subunit of the Gal/GalNAc lectin. *Infect. Immun.* **1999**, *67*, 2096–2102. [CrossRef]
11. Chacín-Bonilla, L. Current pharmacotherapy of amebiasis, advances in new drugs, and design of a vaccine. *SCIELO* **2012**, *53*, 301–314.
12. Shrivastav, M.T.; Malik, Z.; Somlata. Revisiting drug development against the neglected tropical disease, amebiasis. *Front. Cell. Infect. Microbiol.* **2020**, *10*, 628257. [CrossRef] [PubMed]
13. Moonah, S.N.; Jiang, N.M.; Petri, W.A., Jr. Host immune response to intestinal amebiasis. *PLoS Pathog.* **2013**, *9*, e1003489. [CrossRef] [PubMed]
14. Uribe-Querol, E.; Rosales, C. Immune response to the enteric parasite *Entamoeba histolytica*. *Physiol. J.* **2020**, *35*, 244–260. [CrossRef]
15. Begum, S.; Quach, J.; Chadee, K. Immune evasion mechanisms of *Entamoeba histolytica*: Progression to disease. *Front. Microbiol.* **2015**, *6*, 1394. [CrossRef]
16. Alberts, B.; Johnson, A.; Lewis, J.; Raff, M.; Roberts, K.; Walter, P. Innate Immunity. In *Molecular Biology of the Cell*, 4th ed.; Garland Science: New York, NY, USA, 2002.
17. Snyder, P.W. Immunology for the toxicologic pathologist. *Toxicol. Pathol.* **2012**, *40*, 143–147. [CrossRef]
18. Quach, J.; St-Pierre, J.; Chadee, K. The future for vaccine development against *Entamoeba histolytica*. *Hum. Vaccines Immunother.* **2014**, *10*, 1514–1521. [CrossRef]
19. Amanna, I.J.; Slifka, M.K. Successful vaccines. *Curr. Top. Microbiol. Immunol.* **2020**, *428*, 1. [CrossRef]
20. Bayat, M.; Asemani, Y.; Najafi, S. Essential considerations during vaccine design against COVID-19 and review of pioneering vaccine candidate platforms. *Int. Immunopharmacol.* **2021**, *97*, 107679. [CrossRef]
21. Azmi, N.; Othman, N. *Entamoeba histolytica*: Proteomics bioinformatics reveal predictive functions and protein–protein interactions of differentially abundant membrane and cytosolic proteins. *Membranes* **2021**, *11*, 376. [CrossRef]
22. Chou, K.-C.; Shen, H.-B. MemType-2L: A web server for predicting membrane proteins and their types by incorporating evolution information through Pse-PSSM. *Biochem. Biophys. Res. Commun.* **2007**, *360*, 339–345. [CrossRef]
23. Guo, L.; Wang, S.; Li, M.; Cao, Z. Accurate classification of membrane protein types based on sequence and evolutionary information using deep learning. *BMC Bioinform.* **2019**, *20*, 700. [CrossRef] [PubMed]
24. Frederick, J.R.; Petri, W.A., Jr. Roles for the galactose-/N-acetylgalactosamine-binding lectin of *Entamoeba* in parasite virulence and differentiation. *J. Glycobiol.* **2005**, *15*, 53R–59R. [CrossRef] [PubMed]
25. Mann, B.J. Structure and function of the *Entamoeba histolytica* Gal/GalNAc lectin. *Int. Rev. Cytol.* **2002**, *216*, 59–80. [CrossRef] [PubMed]

26. Dodson, J.M.; Lenkowski, P.W.; Eubanks, A.C.; Jackson, T.F.G.H.; Napodano, J.; Lyerly, D.M.; Lockhart, L.A.; Mann, B.J.; Petri, W.A. Infection and immunity mediated by the carbohydrate recognition domain of the *Entamoeba histolytica* Gal/GalNAc lectin. *J. Infect. Dis.* **1999**, *179*, 460–466. [CrossRef] [PubMed]
27. Cheng, X.-J.; Hughes, M.A.; Huston, C.D.; Loftus, B.; Gilchrist, C.A.; Lockhart, L.A.; Ghosh, S.; Miller-Sims, V.; Mann, B.J.; Petri, W.A.; et al. Intermediate subunit of the Gal/GalNAc lectin of *Entamoeba histolytica* is a member of a gene family containing multiple CXXC sequence motifs. *Infect. Immun.* **2001**, *69*, 5892–5898. [CrossRef]
28. Kato, K.; Yahata, K.; Gopal Dhoubhadel, B.; Fujii, Y.; Tachibana, H. Novel hemagglutinating, hemolytic and cytotoxic activities of the intermediate subunit of *Entamoeba histolytica* lectin. *Sci. Rep.* **2015**, *5*, 13901. [CrossRef]
29. Pérez-Hernández, J.; Retana-González, C.; Ramos-Martínez, E.; Cruz-Colín, J.; Saralegui-Amaro, A.; Baltazar-Rosario, G.; Gutiérrez-Ruíz, C.; Aristi-Urista, G.; López-Vancell, R. *Entamoeba histolytica* trophozoites interact with the c-Met receptor at the surfàce of liver origin cells through the Gal/GalNAc amoebic lectin. *Life* **2021**, *11*, 923. [CrossRef]
30. Welter, B.H.; Walters, H.A.; Temesvari, L.A. Reduced expression of a rhomboid protease, EhROM1, correlates with changes in the submembrane distribution and size of the Gal/GalNAc lectin subunits in the human protozoan parasite, *Entamoeba histolytica*. *PLoS ONE* **2020**, *15*, e0219870. [CrossRef]
31. Kato, K.; Tachibana, H. Identification of multiple domains of *Entamoeba histolytica* intermediate subunit lectin-1 with hemolytic and cytotoxic activities. *Int. J. Mol. Sci.* **2022**, *23*, 7700. [CrossRef]
32. Nakada-Tsukui, K.; Nozaki, T. Immune response of amebiasis and immune evasion by *Entamoeba histolytica*. *Front. Immunol.* **2016**, *7*, 175. [CrossRef] [PubMed]
33. Abhyankar, M.M.; Orr, M.T.; Lin, S.; Suraju, M.O.; Simpson, A.; Blust, M.; Pham, T.; Guderian, J.A.; Tomai, M.A.; Elvecrog, J.; et al. Adjuvant composition and delivery route shape immune response quality and protective efficacy of a recombinant vaccine for *Entamoeba histolytica*. *NPJ Vaccines* **2018**, *3*, 22. [CrossRef] [PubMed]
34. Martínez-Hernández, S.L.; Cervantes-García, D.; Muñoz-Ortega, M.; Aldaba-Muruato, L.R.; Loera-Muro, V.M.; Ascacio-Martínez, J.A.; de Jesús Loera-Arias, M.; de Oca-Luna, R.M.; Ventura-Juárez, J. An anti-amoebic vaccine: Generation of the recombinant antigen LC3 from *Entamoeba histolytica* linked to mutated exotoxin A (PEΔIII) via the *Pichia pastoris* system. *Biotechnol. Lett.* **2017**, *39*, 1149–1157. [CrossRef] [PubMed]
35. Martínez-Hernández, S.L.; Becerra-González, V.M.; Muñoz-Ortega, M.H.; Loera-Muro, V.M.; Ávila-Blanco, M.E.; Medina-Rosales, M.N.; Ventura-Juárez, J. Evaluation of the PEΔIII-LC3-KDEL3 chimeric protein of *Entamoeba histolytica*-lectin as a vaccine candidate against amebic liver abscess. *J. Immunol. Res.* **2021**, *2021*, 6697900. [CrossRef]
36. Meneses-Ruiz, D.M.; Laclette, J.P.; Aguilar-Díaz, H.; Hernández-Ruiz, J.; Luz-Madrigal, A.; Sampieri, A.; Vaca, L.; Carrero, J.C. Mucosal delivery of ACNPV baculovirus driving expression of the Gal-lectin LC3 fragment confers protection against amoebic liver abscess in hamster. *Int. J. Biol. Sci.* **2011**, *7*, 1345–1356. [CrossRef] [PubMed]
37. Meneses-Ruiz, D.M.; Aguilar-Diaz, H.; Bobes, R.J.; Sampieri, A.; Vaca, L.; Laclette, J.P.; Carrero, J.C. Protection against amoebic liver abscess in hamster by intramuscular immunization with an *Autographa californica* baculovirus driving the expression of the Gal-lectin LC3 fragment. *Biomed Res. Int.* **2015**, *2015*, 760598. [CrossRef]
38. Kato, K.; Makiuchi, T.; Cheng, X.; Tachibana, H. Comparison of hemolytic activity of the intermediate subunit of *Entamoeba histolytica* and *Entamoeba dispar* lectins. *PLoS ONE* **2017**, *12*, e0181864. [CrossRef]
39. Min, X.; Feng, M.; Guan, Y.; Man, S.; Fu, Y.; Cheng, X.; Tachibana, H. Evaluation of the c-terminal fragment of *Entamoeba histolytica* Gal/GalNAc lectin intermediate subunit as a vaccine candidate against amebic liver abscess. *PLoS Negl. Trop. Dis.* **2016**, *10*, e0004419. [CrossRef]
40. UNIPROT. Gal/GalNAc Lectin Igl1. Available online: https://www.uniprot.org/uniprotkb/Q964D2/entry (accessed on 19 September 2022).
41. Jumper, J.; Evans, R.; Pritzel, A.; Green, T.; Figurnov, M.; Ronneberger, O.; Tunyasuvunakool, K.; Bates, R.; Žídek, A.; Potapenko, A.; et al. Highly accurate protein structure prediction with AlphaFold. *Nature* **2021**, *596*, 583–589. [CrossRef]
42. Bruchhaus, I.; Loftus, B.J.; Hall, N.; Tannich, E. The intestinal protozoan parasite *Entamoeba histolytica* contains 20 cysteine protease genes, of which only a small subset is expressed during in vitro cultivation. *Eukaryot. Cell* **2003**, *2*, 501–509. [CrossRef]
43. He, C.; Nora, G.P.; Schneider, E.L.; Kerr, I.D.; Hansell, E.; Hirata, K.; Gonzalez, D.; Sajid, M.; Boyd, S.E.; Hruz, P.; et al. A novel *Entamoeba histolytica* cysteine proteinase, EhCP4, is key for invasive amebiasis and a therapeutic target. *J. Biol. Chem.* **2010**, *285*, 18516–18527. [CrossRef] [PubMed]
44. Meléndez-López, S.G.; Herdman, S.; Hirata, K.; Choi, M.-H.; Choe, Y.; Craik, C.; Caffrey, C.R.; Hansell, E.; Chávez-Munguía, B.; Chen, Y.T.; et al. Use of recombinant *Entamoeba histolytica* cysteine proteinase 1 to identify a potent inhibitor of amebic invasion in a human colonic model. *Eukaryot. Cell* **2007**, *6*, 1130–1136. [CrossRef] [PubMed]
45. Singh, D.; Naik, S.R.; Naik, S. Role of cysteine proteinase of *Entamoeba histolytica* in target cell death. *J. Parasitol.* **2004**, *129*, 127–135. [CrossRef]
46. Ocádiz-Ruiz, R.; Fonseca, W.; Linford, A.S.; Yoshino, T.P.; Orozco, E.; Rodríguez, M.A. The knockdown of each component of the cysteine proteinase-adhesin complex of *Entamoeba histolytica* (EhCPADH) affects the expression of the other complex element as well as the in vitro and in vivo virulence. *J. Parasitol.* **2016**, *143*, 50–59. [CrossRef] [PubMed]
47. Quintas-Granados, L.I.; Orozco, E.; Brieba, L.G.; Arroyo, R.; Ortega-López, J. Purification, refolding and autoactivation of the recombinant cysteine proteinase EhCP112 from *Entamoeba histolytica*. *Protein Expr. Purif.* **2009**, *63*, 26–32. [CrossRef]

48. Cornick, S.; Moreau, F.; Chadee, K. *Entamoeba histolytica* cysteine proteinase 5 evokes mucin exocytosis from colonic goblet cells via $\alpha v\beta 3$ integrin. *PLoS Pathog.* **2016**, *12*, e1005579. [CrossRef] [PubMed]
49. Hou, Y.; Mortimer, L.; Chadee, K. *Entamoeba histolytica* cysteine proteinase 5 binds integrin on colonic cells and stimulates NFkappaB-mediated pro-inflammatory responses. *J. Biol. Chem.* **2010**, *285*, 35497–35504. [CrossRef]
50. UNIPROT. Cysteine Proteinase 5. Available online: https://www.uniprot.org/uniprotkb/Q06FF8/entry (accessed on 19 September 2022).
51. Lotter, H.; González-Roldán, N.; Lindner, B.; Winau, F.; Isibasi, A.; Moreno-Lafont, M.; Ulmer, A.J.; Holst, O.; Tannich, E.; Jacobs, T. Natural killer T cells activated by a lipopeptidophosphoglycan from *Entamoeba histolytica* are critically important to control amebic liver abscess. *PLoS Pathog.* **2009**, *5*, e1000434. [CrossRef]
52. Wong-Baeza, I.; Alcántara-Hernández, M.; Mancilla-Herrera, I.; Ramírez-Saldívar, I.; Arriaga-Pizano, L.; Ferat-Osorio, E.; López-Macías, C.; Isibasi, A. The role of lipopeptidophosphoglycan in the immune response to *Entamoeba histolytica. J. Biomed. Biotechnol.* **2010**, *2010*, 254521. [CrossRef]
53. Moody, S.; Becker, S.; Nuchamowitz, Y.; Mirelman, D. Virulent and avirulent *Entamoeba histolytica* and *Entamoeba dispar* differ in their cell surface phosphorylated glycolipids. *J. Parasitol.* **1997**, *114*, 95–104. [CrossRef]
54. Maldonado-Bernal, C.; Kirschning, C.J.; Rosenstein, Y.; Rocha, L.M.; Rios-Sarabia, N.; Espinosa-Cantellano, M.; Becker, I.; Estrada, I.; Salazar-González, R.M.; López-Macías, C.; et al. The innate immune response to *Entamoeba histolytica* lipopeptidophosphoglycan is mediated by toll-like receptors 2 and 4. *Parasite Immunol.* **2005**, *27*, 127–137. [CrossRef] [PubMed]
55. Vivanco-Cid, H.; Alpuche-Aranda, C.; Wong-Baeza, I.; Rocha-RamÍRez, L.M.; Rios-Sarabia, N.; Estrada-Garcia, I.; Villasis-Keever, M.A.; Lopez-Macias, C.; Isibasi, A. Lipopopeptidephosphoglycan from *Entamoeba histolytica* activates human macrophages and dendritic cells and reaches their late endosomes. *Parasite Immunol.* **2007**, *29*, 467–474. [CrossRef] [PubMed]
56. Ávila, E.E.; Salaiza, N.; Pulido, J.; Rodríguez, M.C.; Díaz-Godínez, C.; Laclette, J.P.; Becker, I.; Carrero, J.C. *Entamoeba histolytica* trophozoites and lipopeptidophosphoglycan trigger human neutrophil extracellular traps. *PLoS ONE* **2016**, *11*, e0158979. [CrossRef] [PubMed]
57. Kumarasamy, G.; Abdus Sani, A.A.; Olivos-García, A.; Noordin, R.; Othman, N. Antigenic membrane proteins of virulent variant of *Entamoeba histolytica* HM-1:IMSS. *Pathog. Glob. Health* **2020**, *114*, 333–342. [CrossRef] [PubMed]
58. Mares, R.E.; Minchaca, A.Z.; Villagrana, S.; Melendez-Lopez, S.G.; Ramos, M.A. Analysis of the isomerase and chaperone-like activities of an amebic PDI (EhPDI). *Biomed Res. Int.* **2015**, *2015*, 286972. [CrossRef] [PubMed]
59. Mares, R.E.; Magaña, P.D.; Meléndez-López, S.G.; Licea, A.F.; Cornejo-Bravo, J.M.; Ramos, M.A. Oxidative folding and reductive activities of EhPDI, a protein disulfide isomerase from *Entamoeba histolytica. Parasitol. Int.* **2009**, *58*, 311–313. [CrossRef]
60. Ramos, M.A.; Mares, R.E.; Magaña, P.D.; Rivas, I.D.; Meléndez-López, S.G. *Entamoeba histolytica*: Biochemical characterization of a protein disulfide isomerase. *Exp. Parasitol.* **2011**, *128*, 76–81. [CrossRef]
61. Mares, R.E.; Ramos, M.A. An amebic protein disulfide isomerase (PDI) complements the yeast PDI1 mutation but is unable to support cell viability under ER or thermal stress. *FEBS Open Bio* **2018**, *8*, 49–55. [CrossRef]
62. Amit, A.; Vijayamahantesh; Dikhit, M.R.; Singh, A.K.; Kumar, V.; Suman, S.S.; Singh, A.; Kumar, A.; Thakur, A.K.; Das, V.R.; et al. Immunization with *Leishmania donovani* protein disulfide isomerase DNA construct induces Th1 and Th17 dependent immune response and protection against experimental *Visceral leishmaniasis* in BALB/c mice. *Mol. Immunol.* **2017**, *82*, 104–113. [CrossRef]
63. UNIPROT. Protein Disulfide Isomerase. Available online: https://www.uniprot.org/uniprotkb/A0A5K1UZD0/entry (accessed on 19 September 2022).
64. Mancilla-Herrera, I.; Méndez-Tenorio, A.; Wong-Baeza, I.; Jiménez-Uribe, A.P.; Alcántara-Hernández, M.; Ocadiz-Ruiz, R.; Moreno-Eutimio, M.A.; Arriaga-Pizano, L.A.; López-Macías, C.; González-y-Merchand, J.; et al. A Toll/IL-1R/resistance domain-containing thioredoxin regulates phagocytosis in *Entamoeba histolytica. Parasites Vectors* **2012**, *5*, 224. [CrossRef]
65. Leitsch, D. Drug susceptibility testing in microaerophilic parasites: Cysteine strongly affects the effectivities of metronidazole and auranofin, a novel and promising antimicrobial. *Int. J. Parasitol. Drugs Drug Resist.* **2017**, *7*, 321–327. [CrossRef] [PubMed]
66. Andrade, R.M.; Reed, S.L. New drug target in protozoan parasites: The role of thioredoxin reductase. *Front. Microbiol.* **2015**, *6*, 975. [CrossRef] [PubMed]
67. Holmgren, A.; Johansson, C.; Berndt, C.; Lönn, M.E.; Hudemann, C.; Lillig, C.H. Thiol redox control via thioredoxin and glutaredoxin systems. *Biochem. Soc. Trans.* **2005**, *33*, 1375–1377. [CrossRef]
68. UNIPROT. Thioredoxin, Putative. Available online: https://www.uniprot.org/uniprotkb/S0AYD1/entry (accessed on 19 September 2022).
69. Flores, M.S.; Obregón-Cardenas, A.; Rangel, R.; Tamez, E.; Flores, A.; Trejo-Avila, L.; Quintero, I.; Arévalo, K.; Maldonado, M.G.; Gandarilla, F.L.; et al. Glycan moieties in *Entamoeba histolytica* ubiquitin are immunodominant. *Parasite Immunol.* **2021**, *43*, e12812. [CrossRef]
70. Wöstmann, C.; Tannich, E.; Bakker-Grunwald, T. Ubiquitin of *Entamoeba histolytica* deviates in six amino acid residues from the consensus of all other known ubiquitins. *FEBS Lett.* **1992**, *308*, 54–58. [CrossRef]
71. Obregón, A.; Flores, M.S.; Rangel, R.; Arévalo, K.; Maldonado, G.; Quintero, I.; Galán, L. Characterization of N-glycosylations in *Entamoeba histolytica* ubiquitin. *Exp. Parasitol.* **2019**, *196*, 38–47. [CrossRef] [PubMed]
72. Bosch, D.E.; Siderovski, D.P. Structural determinants of ubiquitin conjugation in *Entamoeba histolytica. Exp. Mol. Med.* **2013**, *288*, 2290. [CrossRef]

73. Flores, M.S.; Tamez, E.; Rangel, R.; Monjardin, J.; Bosques, F.; Obregon, A.; Trejo-Avila, L.; Quintero, I.; Gandarilla, F.; Arevalo, K.; et al. Ubiquitin of *Entamoeba histolytica* induces antibody response in patients with invasive amoebiasis. *Parasite Immunol.* **2022**, *44*, e12919. [CrossRef]
74. González, E.; de Leon Mdel, C.; Meza, I.; Ocadiz-Delgado, R.; Gariglio, P.; Silva-Olivares, A.; Galindo-Gómez, S.; Shibayama, M.; Morán, P.; Valadez, A.; et al. *Entamoeba histolytica* calreticulin: An endoplasmic reticulum protein expressed by trophozoites into experimentally induced amoebic liver abscesses. *Parasitol. Res.* **2011**, *108*, 439–449. [CrossRef]
75. González-Rivas, E.; Ximenez, C.; Nieves-Ramirez, M.E.; Moran Silva, P.; Partida-Rodríguez, O.; Hernandez, E.H.; Rojas Velázquez, L.; Serrano Vázquez, A.; Magaña Nuñez, U. *Entamoeba histolytica* calreticulin induces the expression of cytokines in peripheral blood mononuclear cells isolated from patients with amebic liver abscess. *Front. Cell. Infect. Microbiol.* **2018**, *8*, 358. [CrossRef]
76. Girard-Misguich, F.; Sachse, M.; Santi-Rocca, J.; Guillén, N. The endoplasmic reticulum chaperone calreticulin is recruited to the uropod during capping of surface receptors in *Entamoeba histolytica*. *Mol. Biochem. Parasitol.* **2008**, *157*, 236–240. [CrossRef] [PubMed]
77. Ximénez, C.; González, E.; Nieves, M.E.; Silva-Olivares, A.; Shibayama, M.; Galindo-Gómez, S.; Escobar-Herrera, J.; García de León, M.d.C.; Morán, P.; Valadez, A.; et al. *Entamoeba histolytica* and *E. dispar* calreticulin: Inhibition of classical complement pathway and differences in the level of expression in amoebic liver abscess. *Biomed Res. Int.* **2014**, *2014*, 127453. [CrossRef] [PubMed]
78. UNIPROT. Calreticulin. Available online: https://www.uniprot.org/uniprotkb/C4M296/entry (accessed on 19 September 2022).
79. Moonah, S.N.; Abhyankar, M.M.; Haque, R.; Petri, W.A. The macrophage migration inhibitory factor homolog of *Entamoeba histolytica* binds to and immunomodulates host macrophages. *Infect. Immun.* **2014**, *82*, 3523–3530. [CrossRef] [PubMed]
80. Ngobeni, R.; Abhyankar, M.M.; Jiang, N.M.; Farr, L.A.; Samie, A.; Haque, R.; Moonah, S.N. *Entamoeba histolytica*-encoded homolog of macrophage migration inhibitory factor contributes to mucosal inflammation during amoebic colitis. *J. Infect. Dis.* **2017**, *215*, 1294–1302. [CrossRef] [PubMed]
81. Ghosh, S.; Padalia, J.; Ngobeni, R.; Abendroth, J.; Farr, L.; Shirley, D.A.; Edwards, T.; Moonah, S. Targeting parasite-produced macrophage migration inhibitory factor as an antivirulence strategy with antibiotic-antibody combination to reduce tissue damage. *J. Infect. Dis.* **2020**, *221*, 1185–1193. [CrossRef]
82. Ghosh, S.; Leaton, L.A.; Farr, L.; Barfield, A.; Moonah, S. Interaction between parasite-encoded JAB1/CSN5 and macrophage migration inhibitory factor proteins attenuates its proinflammatory function. *Sci. Rep.* **2018**, *8*, 10241. [CrossRef]
83. UNIPROT. Migration Inhibitory Factor Protein. Available online: https://www.uniprot.org/uniprotkb/A0A5K1URL6/entry (accessed on 19 September 2022).
84. Sharma, M.; Morgado, P.; Zhang, H.; Ehrenkaufer, G.; Manna, D.; Singh, U. Characterization of extracellular vesicles from *Entamoeba histolytica* identifies roles in intercellular communication that regulates parasite growth and development. *Infect. Immun.* **2020**, *88*, e00349-20. [CrossRef]
85. Sabatke, B.; Gavinho, B.; Coceres, V.; de Miguel, N.; Ramirez, M.I. Unveiling the role of EVs in anaerobic parasitic protozoa. *Mol. Immunol.* **2021**, *133*, 34–43. [CrossRef]
86. Nievas, Y.R.; Lizarraga, A.; Salas, N.; Cóceres, V.M.; de Miguel, N. Extracellular vesicles released by anaerobic protozoan parasites: Current situation. *Cell. Microbiol.* **2020**, *22*, e13257. [CrossRef]
87. Tovy, A.; Siman Tov, R.; Gaentzsch, R.; Helm, M.; Ankri, S. A new nuclear function of the *Entamoeba histolytica* glycolytic enzyme enolase: The metabolic regulation of cytosine-5 methyltransferase 2 (Dnmt2) activity. *PLoS Pathog.* **2010**, *6*, e1000775. [CrossRef]
88. Hidalgo, M.E.; Sánchez, R.; Pérez, D.G.; Rodríguez, M.A.; García, J.; Orozco, E. Molecular characterization of the *Entamoeba histolytica* enolase gene and modelling of the predicted protein. *FEMS Microbiol. Lett.* **1997**, *148*, 123–129. [CrossRef] [PubMed]
89. Schulz, E.C.; Tietzel, M.; Tovy, A.; Ankri, S.; Ficner, R. Structure analysis of *Entamoeba histolytica* enolase. *Acta Crystallogr. D* **2011**, *67*, 619–627. [CrossRef] [PubMed]
90. Xue, S.; Seo, K.; Yang, M.; Cui, C.; Yang, M.; Xiang, S.; Yan, Z.; Wu, S.; Han, J.; Yu, X.; et al. *Mycoplasma suis* alpha-enolase subunit vaccine induces an immune response in experimental animals. *Vaccines* **2021**, *9*, 1506. [CrossRef] [PubMed]
91. Téllez-Martínez, D.; Leandro Portuondo, D.; Loesch, M.L.; Batista-Duharte, A.; Zeppone Carlos, I. A recombinant enolase-montanide™ petgel a vaccine promotes a protective Th1 immune response against a highly virulent *Sporothrix schenckii* by toluene exposure. *Pharmaceutics* **2019**, *11*, 144. [CrossRef]
92. Jiang, H.; Chen, T.; Sun, H.; Tang, Z.; Yu, J.; Lin, Z.; Ren, P.; Zhou, X.; Huang, Y.; Li, X.; et al. Immune response induced by oral delivery of *Bacillus subtilis* spores expressing enolase of *Clonorchis sinensis* in grass carps (*Ctenopharyngodon idellus*). *Fish Shellfish Immunol.* **2017**, *60*, 318–325. [CrossRef]
93. DataBank, P. Enolase 1. Available online: https://www.ebi.ac.uk/pdbe/entry/pdb/3qtp/analysis (accessed on 25 October 2022).
94. DataBank, P. Pfam Enolase, N-Terminal Domain. Available online: https://www.ebi.ac.uk/pdbe/entry/view3D/3qtp/?view=entry_index&viewer=litemol (accessed on 25 October 2022).
95. DataBank, P. Pfam Enolase, C-Terminal Tim Barrel Domain. Available online: https://www.ebi.ac.uk/pdbe/entry/view3D/3qtp/?view=entry_index&viewer=litemol (accessed on 25 October 2022).
96. Hernández-Cuevas, N.A.; Jhingan, G.D.; Petropolis, D.; Vargas, M.; Guillen, N. Acetylation is the most abundant actin modification in *Entamoeba histolytica* and modifications of actin's amino-terminal domain change cytoskeleton activities. *Cell. Microbiol.* **2019**, *21*, e12983. [CrossRef]

97. Adeoti, O.M.; Aderinto, A.O.; Adesina, D.A.; Adesanya, V.O.; Olaoye, O.J. A bioinformatics approach to designing a multi-variant vaccine against *Entamoeba histolytica*. *Biomed. Res. J.* **2021**, *5*, 360–370.
98. Rath, P.P.; Gourinath, S. The actin cytoskeleton orchestra in *Entamoeba histolytica*. *Proteins Struct. Funct. Genet.* **2020**, *88*, 1361–1375. [CrossRef]
99. Velázquez-Domínguez, J.A.; Hernández-Ramírez, V.I.; Calzada, F.; Varela-Rodríguez, L.; Pichardo-Hernández, D.L.; Bautista, E.; Herrera-Martínez, M.; Castellanos-Mijangos, R.D.; Matus-Meza, A.S.; Chávez-Munguía, B.; et al. Linearolactone and kaempferol disrupt the actin cytoskeleton in *Entamoeba histolytica:* Inhibition of amoebic liver abscess development. *J. Nat. Prod.* **2020**, *83*, 3671–3680. [CrossRef]
100. Herrera-Martínez, M.; Hernández-Ramírez, V.I.; Hernández-Carlos, B.; Chávez-Munguía, B.; Calderón-Oropeza, M.A.; Talamás-Rohana, P. Antiamoebic activity of *Adenophyllum aurantium* (L.) *Strother* and its effect on the actin cytoskeleton of *Entamoeba histolytica*. *Front. Pharmacol.* **2016**, *7*, 169. [CrossRef]
101. Li, J.; Huang, X.; Zhang, G.; Gong, P.; Zhang, X.; Wu, L. Immune response and protective efficacy against homologous challenge in BALB/c mice vaccinated with DNA vaccine encoding *Toxoplasma gondii* actin depolymerizing factor gene. *Vet. Parasitol.* **2011**, *179*, 1–6. [CrossRef] [PubMed]
102. UNIPROT. Actin. Available online: https://www.uniprot.org/uniprotkb/P11426/entry (accessed on 19 September 2022).
103. Davis, P.H.; Chen, M.; Zhang, X.; Clark, C.G.; Townsend, R.R.; Stanley, S.L., Jr. Proteomic comparison of *Entamoeba histolytica* and *Entamoeba dispar* and the role of *E. histolytica* alcohol dehydrogenase 3 in virulence. *PLoS Negl. Trop. Dis.* **2009**, 3, e415. [CrossRef]
104. König, C.; Meyer, M.; Lender, C.; Nehls, S.; Wallaschkowski, T.; Holm, T.; Matthies, T.; Lercher, D.; Matthiesen, J.; Fehling, H.; et al. An alcohol dehydrogenase 3 (ADH3) from *Entamoeba histolytica* is involved in the detoxification of toxic aldehydes. *Microorganisms* **2020**, *8*, 1608. [CrossRef] [PubMed]
105. Lowerre, K.; Hemme, C.; Espinosa, A. Prediction of the *Entamoeba histolytica* alcohol dehydrogenase 2 (EhADH2) protein structure using bioinformatics tools. *FASEB J.* **2017**, *31*, 603.23. [CrossRef]
106. Gabrielle, M.; Leito, J.; Espinosa, A. Structural and biochemical analyses of alcohol dehydrogenase E enzymes from *Entamoeba invadens* IP-1, *E. invadens* VK-1:NS and *E. dispar*. *FASEB J.* **2017**, *31*, 606.16. [CrossRef]
107. Hackey, M.; Rossi, L.; Espinosa, A. Inhibitory effects of substituted pyrazoline derivatives on *Entamoeba histolytica* alcohol and acetaldehyde dehydrogenase (EhADH2) activities. *FASEB J.* **2017**, *31*, 921.4. [CrossRef]
108. Espinosa, A.; Perdrizet, G.; Paz-y-Miño, C.G.; Lanfranchi, R.; Phay, M. Effects of iron depletion on *Entamoeba histolytica* alcohol dehydrogenase 2 (EhADH2) and trophozoite growth: Implications for antiamoebic therapy. *J. Antimicrob. Chemother.* **2009**, *63*, 675–678. [CrossRef]
109. Leitao, J.; Oduaran, E.; Rossi, L.; Espinosa, A. Isolation and purification of *Entamoeba histolytica* alcohol dehydrogenase 2 (EhADH2) enzymatic activities and inhibition by pyrazoline derivatives. *FASEB J.* **2015**, *29*, 722.1. [CrossRef]
110. Espinosa, A.; Clark, D.; Stanley, S.L. *Entamoeba histolytica* alcohol dehydrogenase 2 (EhADH2) as a target for anti-amoebic agents. *J. Antimicrob. Chemother.* **2004**, *54*, 56–59. [CrossRef]
111. Siman-Tov, R.; Ankri, S. Nitric oxide inhibits cysteine proteinases and alcohol dehydrogenase 2 of *Entamoeba histolytica*. *Parasitol. Res.* **2003**, *89*, 146–149. [CrossRef]
112. Schlenke, T.A.; McKean, K.A. A role for alcohol dehydrogenase in the *Drosophila* immune response? *Insect Mol. Biol.* **2005**, *14*, 175–178. [CrossRef] [PubMed]
113. UNIPROT. Alcohol Dehydrogenase 3. Available online: https://www.uniprot.org/uniprotkb/Q24857/entry (accessed on 19 September 2022).
114. Wang, G.; Zhu, W. Molecular docking for drug discovery and development: A widely used approach but far from perfect. *Future Med. Chem.* **2016**, *8*, 1707–1710. [CrossRef] [PubMed]
115. Meng, X.Y.; Zhang, H.X.; Mezei, M.; Cui, M. Molecular docking: A powerful approach for structure-based drug discovery. *Curr. Comput. Aided Drug Des.* **2011**, *7*, 146–157. [CrossRef] [PubMed]
116. Kumar, S. Drug and vaccine design against novel coronavirus (2019-nCoV) spike protein through computational approach. *Preprints* **2020**. [CrossRef]

MDPI
St. Alban-Anlage 66
4052 Basel
Switzerland
www.mdpi.com

Membranes Editorial Office
E-mail: membranes@mdpi.com
www.mdpi.com/journal/membranes

www.ingramcontent.com/pod-product-compliance
Lightning Source LLC
Chambersburg PA
CBHW061454210926
51850CB00008B/1716
9783725800285